T0305530

ILLUMINATION ENGINEERING

IEEE Press
445 Hoes Lane
Piscataway, NJ 08854

IEEE Press Editorial Board 2012
John Anderson, *Editor in Chief*

Ramesh Abhari
George W. Arnold
Flavio Canavero
Dmitry Goldgof
Bernhard M. Haemmerli
David Jacobson
Mary Lanzerotti
Om P. Malik
Saeid Nahavandi
Tariq Samad
George Zobrist

Kenneth Moore, *Director of IEEE Book and Information Services (BIS)*

ILLUMINATION ENGINEERING

Design with Nonimaging Optics

R. JOHN KOSHEL
Photon Engineering, LLC, and
College of Optical Sciences, The University of Arizona
Tucson, Arizona

IEEE Press

A JOHN WILEY & SONS, INC., PUBLICATION

Copyright © 2013 by the Institute of Electrical and Electronics Engineers. All rights reserved.

Published by John Wiley & Sons, Inc., Hoboken, New Jersey.
Published simultaneously in Canada.

No part of this publication may be reproduced, stored in a retrieval system, or transmitted in any form or by any means, electronic, mechanical, photocopying, recording, scanning, or otherwise, except as permitted under Section 107 or 108 of the 1976 United States Copyright Act, without either the prior written permission of the Publisher, or authorization through payment of the appropriate per-copy fee to the Copyright Clearance Center, Inc., 222 Rosewood Drive, Danvers, MA 01923, (978) 750-8400, fax (978) 750-4470, or on the web at www.copyright.com. Requests to the Publisher for permission should be addressed to the Permissions Department, John Wiley & Sons, Inc., 111 River Street, Hoboken, NJ 07030, (201) 748-6011, fax (201) 748-6008, or online at http://www.wiley.com/go/permissions.

Limit of Liability/Disclaimer of Warranty: While the publisher and author have used their best efforts in preparing this book, they make no representations or warranties with respect to the accuracy or completeness of the contents of this book and specifically disclaim any implied warranties of merchantability or fitness for a particular purpose. No warranty may be created or extended by sales representatives or written sales materials. The advice and strategies contained herein may not be suitable for your situation. You should consult with a professional where appropriate. Neither the publisher nor author shall be liable for any loss of profit or any other commercial damages, including but not limited to special, incidental, consequential, or other damages.

For general information on our other products and services or for technical support, please contact our Customer Care Department within the United States at (800) 762-2974, outside the United States at (317) 572-3993 or fax (317) 572-4002.

Wiley also publishes its books in a variety of electronic formats. Some content that appears in print may not be available in electronic formats. For more information about Wiley products, visit our web site at www.wiley.com.

Library of Congress Cataloging-in-Publication Data:

Koshel, R. John, author.
Illumination engineering : design with nonimaging optics / John Koshel.
pages cm
ISBN 978-0-470-91140-2 (hardback)
1. Optical engineering. 2. Lighting. I. Title.
TA1520.K67 2013
621.36–dc23

2012020167

Printed in the United States of America.

10 9 8 7 6 5 4 3 2 1

To my family
Dee, Gina, Abe, Lucy, Tanya, Fred, Frankie, and Trudy

CONTENTS

PREFACE

This book was started some time ago. At first I thought to write an introductory book in the field of illumination engineering, but the book has since evolved to have more breadth. The original title was "Advanced Nonimaging/Illumination Optics," but that title just did not convey the intent of the publication. I felt it important to have both "nonimaging" and "illumination" in the final title. Illumination, as discussed in the first chapter, indicates a light distribution from a source as used or detected by an observer. "Nonimaging" denotes that the imaging constraint is not required, but rather, as will be discussed in detail, pertains to the efficient transfer of radiation from a source to a target. The difference is subtle, but in the simplest sense, illumination demands an observer, while nonimaging optics strives to obtain a desired distribution and/or efficiency. Thus, I struggled for a better title, but just prior to submission, I came upon "Illumination Engineering: Design with Nonimaging Optics." Of course, not all illumination systems use nonimaging design principles (e.g., a number of projection systems use Köhler illumination, which is based on imaging principles). Additionally, not all illumination systems must have an "observer," but they have a target (e.g., solar power generation uses a photovoltaic cell as the target). Therefore, the terms "illumination" and "nonimaging" are a bit broad in their presentation herein. Such a broad interpretation is appropriate because, as will be seen, most nonimaging principles, such as the edge-ray, have imaging as the base of their design methods. Also, for illumination, consider that the source is "illuminating" the target rather than providing "illumination" for an observer. Thus, the major focus of this book is the use of nonimaging optics in illumination systems.

While the field of illumination is old, only recently have individuals started researching the utility of nonimaging optics to provide a desired distribution of radiation with high transfer efficiency. Increasingly, our society faces environmental and energy use issues, so optimally designed illumination/nonimaging optics are especially attractive. Additionally, solar power generation uses a number of nonimaging optics design methods to provide the power that we can use to drive our illumination systems. In essence, we have the potential "to have our cake and eat it too": with nonimaging/illumination optics, we can create and use the electrical power we use in our daily lives. Nonimaging optics will only gain in importance as the field and technology advance. One might say the illumination field is comparable to the lens design field of the early 20th century, so there is large room for improvement.

There is a wide breadth of topics that could be considered for this book, from design methods to sources to applications to fabrication. Ten years ago most of these topics were barely within the literature, with only one book actually addressing the topic of nonimaging optics. There was a bit more literature in the illumination field.

However, the illumination books primarily dealt with the application of lighting, which limited their scope to design methods and suggestions rather than theoretically developed design principles. Fortunately, this dearth of literature is decreasing, especially with the burgeoning growth of solid-state lighting and solar power generation. This book introduces a number of topics that have not been pursued much in the literature while also expanding on the fundamental limits. In the first chapter, I present a discussion of the units, design method, design types, and a short history. In the second chapter, I focus solely on the topic of étendue through its conservation, and its extension to the skew invariant. Étendue is the limit of what is possible with optics; therefore, it is imperative that the reader understand the term and what it implies. There is a wealth of theory presented in that chapter, including a number of proofs, while I also use an example to build the reader's understating of étendue. The proofs are geared to readers who come from varied physics backgrounds, from radiometry to thermodynamics to ray tracing. Chapter 3, by Pablo Benítez, Juan Carlos Miñano, and José Blen, continues the treatment of étendue by looking at its squeezing into a desired phase space. Juan Carlos Miñano, Pablo Benítez, Aleksandra Cvetkovic, and Rubén Mohedano continue by using methods to develop freeform optics that provide high efficiency into a desired distribution. Next, two application areas are presented: solar concentrators by Julio Chaves and Maikel Hernández (Chapter 5) and lightpipes (Chapter 6) by William Cassarly and Thomas Davenport. Chapter 5 investigates nonimaging optics that provide high efficiency and high uniformity at the solar cell with less demanding tolerances, which are necessary in light of tracking limitations. Chapter 6 highlights lightpipes and lightguides that are used frequently in our lives without us even knowing it. Lightpipes and lightguides appear in our car dashboards, laptop displays, indicator lights on a wide range of electronics, and so forth. Finally, this book ends with a lengthy discussion of sampling requirements (i.e., spatial distribution pixelization), ray tracing needs (i.e., ray and distribution sampling), optimization methods, and tolerancing. Also note that some of the material is repeated from one chapter to another, in particular, the concept of étendue is virtually in all chapters and the topic of freeform optics is prevalent in both Chapters 4 and 5. I did not want to limit the presentations of the chapter writers, so in some sense, each chapter is self-contained. However, as alluded to above, étendue is the driving force herein. By no means do I think this book presents the complete story. There are numerous areas that are not addressed herein: source modeling, projectors, color, fabrication, and measurement. Therefore, it is expected that future editions or new volumes will be released to meet the demands. There are other sources of literature that can be sought to address some of these needs, but I expect future editions/volumes of this book will expand greatly upon the available literature. I welcome suggestions from the readers on what should be added in future editions and/or volumes.

A number of individuals have helped with the writing of this book. First and foremost have been my employers as I was writing this book. I started while at Lambda Research Corporation, but finished while I was at Photon Engineering, LLC. Ed Freniere and Rich Pfisterer, at those companies, respectively, encouraged the writing of the book. This book would not have been possible without access to the optical analysis software from each of these two firms. I also received encouragement and feedback from individuals at the College of Optical Sciences at the University of Arizona. Dean Jim Wyant and Professor José Sasián encouraged me while providing feedback on the material. Additionally, I need to thank the individuals at Wiley/IEEE Press and Toppan Best-set Premedia Limited who had to persevere through the slow process of my writing: Taisuke Soda, Mary Hatcher, Christine Punzo, and Stephanie Sakson. All strongly enticed me to finish on time, someday, before the Sun burned out—so thank you for enduring my continued tardiness. My biggest sources were my students at Optical Sciences, where I teach a dual undergraduate-graduate course on illumination engineering. The students helped by being the first to see some of the material—finding errors and typos while challenging me to convey my points better. I have had over 50 students since this became a for-credit course, and around 70 more when it was a seminar, no credit course. My students learned that 95% of the time when I asked a question, the answer was étendue. However, although I told my students I was still learning how to apply the concepts of étendue, or in other words, I was still learning étendue—they likely did not believe me. I believe étendue is a fickle entity, that to fully grasp it is a life-long task. I may never fully understand all of its nuances, but I feel this book assists me and I trust the readers on this journey. Future volumes/editions will expand upon the topic of étendue, especially how it is applied to applications. Finally, I need to thank my family for the many days and nights I was not able to do anything. I dedicate this effort to all of you.

R. John Koshel

COVER

The cover shows two designs entered in the first Illumination Design Problem of the 2006 International Optical Design Conference.* The goal was to transform the emitted light from a square emitter into a cross pattern with the highest efficiency possible. Bill Cassarly (see Chapter 6) developed a method based initially on imaging principles, while Julio Chaves developed a solely nonimaging approach that uses rotationally asymmetric transformers: that is, an asymmetric lightpipe array for the outer regions and a bulk lightpipe in the central region. In 2010, the second IODC Illumination Design Problem was presented, and in 2014 will be the third competition. I encourage the readers to consult the literature to learn more about these design challenges.

* P. Benítez, 2006 IODC illumination design problem, *SPIE Proc. of the Intl. Opt. Des. Conf. 2006* **6342**, 634201V (2006). Society of Photo-Optical Instrumentation Engineers, Bellingham, WA.

CONTRIBUTORS

Pablo Benítez, Universidad Politécnica de Madrid, Cedint, Madrid, Spain, and LPI, Altadena, California

José Blen, LPI Europe, SL, Madrid, Spain

William Cassarly, Synopsys, Inc., Wooster, Ohio

Julio Chaves, LPI Europe, SL, Madrid, Spain

Aleksandra Cvetkovic, LPI Europe, SL, Madrid, Spain

Maikel Hernández, LPI Europe, SL, Madrid, Spain

R. John Koshel, Photon Engineering, LLC, Tucson, Arizona, and College of Optical Sciences, the University of Arizona, Tucson, Arizona

Juan C. Miñano, Universidad Politécnica de Madrid, Cedint, Madrid, Spain, and LPI, Altadena, California

Rubén Mohedano, LPI Europe, SL, Madrid, Spain

GLOSSARY

PARAMETERS

Term	Description	Units	First Use
A_{proj}	Projected area	m^2	Chapter 1
C	Concentration		Chapter 2
CAP	Concentration acceptance product		Chapter 4
C_g	Geometrical concentration		Chapter 4
Cx, Cy	Chromaticity coordinates		Chapter 6
E	Irradiance or illuminance		Chapter 1
E_e	e-Subscript: irradiance	W/m^2	
E_v	v-Subscript: illuminance	$lm/m^2 = lx$	
F	Configuration factor		Chapter 2
$f/\#$	F-number		Chapter 1
H	Lagrange invariant	m	Chapter 2
I	Intensity		Chapter 1
I_e	e-Subscript: radiant intensity	W/sr	
I_v	v-Subscript: luminous intensity	lm/sr	
$\mathbf{J}$	Jacobian matrix		Chapter 2
$\vec{k}$	Ray direction vector	1/m	Chapter 2
$K(\lambda)$	Luminous efficacy	lm/W	Chapter 1
L	Radiance or luminance		Chapter 1
L_e	e-Subscript: radiance	$W/m^2/sr$	
L_v	v-Subscript: luminance	$lm/m^2/sr = nt$	
L, M, N	Direction cosines		Chapter 2
M	Exitance		Chapter 1
M_e	e-Subscript: radiant exitance	W/m^2	
M_v	v-Subscript: luminous exitance	$lm/m^2 = lm$	
m	Lightpipe bend ratio		Chapter 6
M	Ray bundle		Chapter 3
MF	Merit function		Chapter 6
n	Index of refraction		Chapter 1
P	Flux or power		Chapter 1
p, q, r	Optical direction cosines		Chapter 2
P_e	e-Subscript: radiant flux/power	W	
P_v	v-Subscript: luminous flux/power	lm	
Q	Energy		Chapter 1
Q_e	e-Subscript: radiant energy	J	
Q_n	v-Subscript: luminous energy	T	

r	Radius	m	Chapter 1
$\bar{r}$	Fractional radius		Chapter 2
R	Reflectivity	(%)	Chapter 2
r_s	Source radius	m	Chapter 2
s	Skewness		Chapter 2
T	Transmissivity	(%)	Chapter 4
u	Energy density		Chapter 1
u_e	e-Subscript: radiant exitance	J/m^3	
u_n	v-Subscript: luminous exitance	T/m^3	
$\mathbf{v}$	Vector tangent		Chapter 3
$V(\mathbf{r})$	Point characteristic	m	Chapter 2
$V(\lambda)$	Luminous efficiency	(%)	Chapter 1
W	Wave front/normal congruence		Chapter 4
Φ	Flux or power		Chapter 1
Φ_e	e-Subscript: radiant flux/power	W	
Φ_v	v-Subscript: luminous flux/power	lm	
Σ_R	Reference surface		Chapter 3
Ω	Solid angle	sr	Chapter 1
Ξ	Physical étendue	m^3sr	Chapter 2
Ψ	Flux emission density	W/mm^3	Chapter 2
ε	Longitudinal asymmetry source factor		Chapter 2
η	Efficiency	(%)	Chapter 1
θ_a	Acceptance angle	°/rad	Chapter 2
θ_c	Critical angle	°/rad	Chapter 1
ξ	Étendue	m^2sr	Chapter 2
$\bar{\xi}$	Fractional étendue		Chapter 2

Note the following modifiers and subscripts:
- Unprimed: before refraction
- Primed: after refraction
- 2D subscript: two-dimensional or bi-parametric
- 3D subscript: three-dimensional
- 4D subscript: tetra-dimensional
- max subscript: maximal value
- 0: initial or central value

ACRONYMS

Term	Description	First Use
2D	Two-dimensional	Chapter 1
3D	Three-dimensional	Chapter 1
BEF	Brightness-enhancing film	Chapter 6
BRDF	Bidirectional reflectance distribution function	Chapter 1
BSDF	Bidirectional surface distribution function	Chapter 1
BTDF	Bidirectional transmittance distribution function	Chapter 1
CAD	Computer-aided design	Chapter 1

CCFL	Cold cathode fluorescent lamp	Chapter 6
CGS	Centimeter-gram-second unit system	Chapter 1
CIE	Commission Internationale de L'Éclairage	Chapter 1
CPC	Compound parabolic concentrator	Chapter 1
CPV	Concentrating photovoltaic	Chapter 4
D	Downward direction	Chapter 4
DCPC	Dielectric compound parabolic concentrator	Chapter 6
DNI	Direct normal irradiance	Chapter 4
DSMTS	Dielectric single-mirror two-stage concentrator	Chapter 5
DTIRC	Dielectric total internal reflection concentrators	Chapter 5
ECE	Economic Commission of Europe	Chapter 1
ED	Edge-ray design	Chapter 1
ES	Étendue squeezing	Chapter 3
FMVSS	Federal motor vehicle safety standards	Chapter 1
FOV	Field of view	Chapter 1
FWHM	Full width, half maximum	Chapter 6
H	Horizontal direction	Chapter 3
	It also designates the origin in the horizontal direction	Chapter 4
HCPV	High-concentration photovoltaic	Chapter 4
IR	Infrared	Chapter 1
L	Leftward direction	Chapter 4
LCD	Liquid crystal display	Chapter 2
LED	Light-emitting diode	Chapter 1
MKS	Meter-kilogram-second unit system	Chapter 1
NERD	Nonedge-ray design	Chapter 1
NURBS	Nonuniform rational b-spline	Chapter 4
PMMA	Poly(methyl methacrylate) or acrylic plastic	Chapter 6
POE	Primary optical element	Chapter 3
PV	Photovoltaic	Chapter 4
R	Rightward direction	Chapter 4
RGB	Red–green–blue (color diagrams)	Chapter 6
RMS	Root mean square	Chapter 1
RXI	Refraction–reflection-TIR concentrator	Chapter 3
SAE	Society of Automotive Engineers	Chapter 1
SI	Système Internationale	Chapter 1
SLM	Spatial light modulator	Chapter 2
SMS	Simultaneous multiple surfaces	Chapter 1
SOE	Secondary optical element	Chapter 3
TED	Tailored edge-ray design	Chapter 1
TERC	Tailored edge-ray concentrator	Chapter 5
TIR	Total internal reflection	Chapter 1
U	Upward direction	Chapter 4
UHP	Ultra-high pressure (arc lamp)	Chapter 3
V	Vertical direction	Chapter 3
	It also designates the origin in the vertical direction	Chapter 4
XR	Reflection–refraction concentrator	Chapter 4
XX	Reflection–reflection concentrator	Chapter 3

CHAPTER 1

INTRODUCTION AND TERMINOLOGY

John Koshel

This chapter introduces the reader to a number of terms and concepts prevalent in the field of illumination optics. I establish the units basis that is used throughout this book. The fields of nonimaging and illumination optics have a fundamental basis on these units; therefore, it is demanded that the reader be well versed in units and how to design, analyze, and measure with them. Next, I give an overview of the field and important parameters that describe the performance of an illumination system. The next chapter on étendue expands upon this treatment by introducing terms that are primarily focused on the design of efficient illumination systems.

1.1 WHAT IS ILLUMINATION?

Until recently the field of optical design was synonymous with lens or imaging system design. However, within the past decade, the field of optical design has included the subfield of illumination design. Illumination is concerned with the transfer of light, or radiation in the generic sense,* from the source(s) to the target(s). Light transfer is a necessity in imaging systems, but those systems are constrained by imaging requirements. Illumination systems can ignore the "imaging constraint" in order to transfer effectively the light. Thus, the term nonimaging optics is often used. In the end, one may classify optical system design into four subdesignations:

- *Imaging Systems.* Optical systems with the imaging requirement built into the design. An example is a focal-plane camera.
- *Visual Imaging Systems.* Optical systems developed with the expectation of an overall imaging requirement based upon integration of an observation system. Examples include telescopes, camera viewfinders, and microscopes that require human observers (i.e., the eye) to accomplish imaging.

* Light is defined as that within the visible spectrum, but often it loosely includes the ultraviolet and infrared parts of the electromagnetic spectrum. The proper term is electromagnetic radiation.

Illumination Engineering: Design with Nonimaging Optics, First Edition. R. John Koshel.
© 2013 the Institute of Electrical and Electronics Engineers. Published 2013 by John Wiley & Sons, Inc.

- *Visual Illumination Systems.* Optical systems developed to act as a light source for following imaging requirements. Examples include displays, lighting, and extend to the illuminator for photocopiers.
- *Nonvisual Illumination Systems.* Optical systems developed without the imaging criterion imposed on the design. Examples include solar concentrators, optical laser pump cavities, and a number of optical sensor applications.

The latter two systems comprise the field of illumination engineering. Imaging systems can be employed to accomplish the illumination requirements, but these systems are best suited for specific applications. Examples include critical and Köhler illumination used, for example, in the lithography industry, but as this book shows, there are a number of alternative methods based on nonimaging optics principles. This book focuses on these nonimaging techniques in order to transfer light effectively from the source to the target, but imaging principles are used at times to improve upon such principles. Additionally, I place no requirement on an observer within the system, but as you will discover, most illumination optics are designed with observation in mind, including the human eye and optoelectronic imaging, such as with a camera. To neglect the necessary visualization and its characteristics often has a detrimental effect on the performance of the illumination system. This last point also raises the subjective perception of the illumination system design. This factor is not currently a focus of this book, but it is discussed in order to drive the development of some systems.

I use the remainder of this chapter to discuss:

- A short history of the illumination field
- The units and terminology for illumination design and analysis
- The important factors in illumination design
- Standard illumination optics
- The steps to design an illumination system
- A discussion of the difficulty of illumination design, and
- The format used for the chapters presented herein.

Note that I typically use the terms illumination and nonimaging interchangeably, but, in fact, illumination is a generic term that includes nonimaging and imaging methods for the transfer of light to a target.

1.2 A BRIEF HISTORY OF ILLUMINATION OPTICS

The history of the field of illumination and nonimaging optics is long, but until recently it was mostly accomplished by trial and error. Consider Figure 1.1, which shows a timeline of the development of sources and optics for use in the illumination field [1]. Loosely, the field of illumination optics starts with the birth of a prevalent light source on earth—the Sun. While the inclusion of the Sun in this timeline may at first appear facetious, the Sun is becoming of increasing importance in the illu-

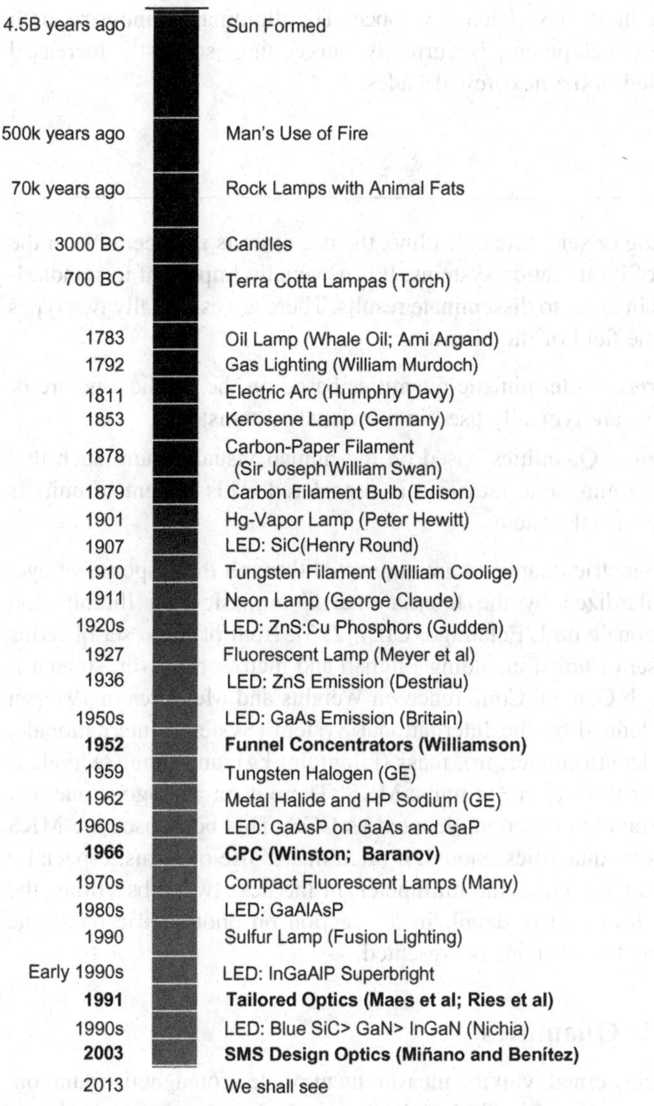

Figure 1.1 A timeline of the history of illumination and nonimaging optics. On the left is the approximate date of inception for the item on the right. Items in bold are illumination optic design concepts, those leading with "LED" are new solid-state lighting sources, and the remainder is other types of sources.

mination and nonimaging optics communities. This importance is borne out of daylighting systems, solar thermal applications, and solar energy generation. The use, modeling, and fabrication of sources is one of the largest components of the field of illumination design. Increasingly, LED sources are supplanting traditional sources since LEDs provide the potential for more efficient operation, color selection, long lifetimes, and compact configurations. It is only in the past 60 years that nonimaging optical methods have been developed. The illumination industry, both in design and source development, is currently burgeoning, so vastly increased capabilities are expected in the next few decades.

1.3 UNITS

As with any engineering or scientific discipline, the use of units is imperative in the design and modeling of illumination systems. It is especially important to standardize the system of units in order to disseminate results. There are essentially two types of quantities used in the field of illumination:

- *Radiometric Terms.* Deterministic quantities based on the physical nature of light. These terms are typically used in nonvisual systems; and
- *Photometric Terms.* Quantities based on the human visual system such that only visible electromagnetic radiation is considered. This system of units is typically used in visual systems.

Radiometric and photometric quantities are connected through the response of eye, which has been standardized by the International Commission on Illumination (Commission Internationale de L'Éclairage; CIE) [2, 3]. Both of these set of terms can be based on any set of units, including English and metric; however, standardization at the Fourteenth General Conference on Weights and Measures in 1971 on the metric system is defined by the International System (Système Internationale; SI) [4]. The units for length (meter; m), mass (kilogram; kg), and time (second; s) provide an acronym for this system of units: MKS. There is an analogous one that uses the centimeter, gram, and second, denoted as CGS. This book uses the MKS standard for the radiation quantities, though it often makes use of terms, especially length, in non-MKS units, such as the millimeter. In the next two subsections, the two set of terms are discussed in detail. In the section on photometric units, the connection between the two systems is presented.

1.3.1 Radiometric Quantities

Radiometry is a field concerned with the measurement of electromagnetic radiation. The radiometric terms as shown in Table 1.1 are used to express the quantities of measurement* [5]. The term radiant is often used before a term, such as radiant flux, to delineate between like terms from the photometric quantities; however, the accepted norm is that the radiometric quantity is being expressed if the word radiant is omitted. Additionally, radiometric quantities are often expressed with a subscript "e" to denote electromagnetic. Omission of this subscript still denotes a radiometric quantity.

* Note that I follow the convention of R.W. Boyd, *Radiometry and the Detection of Optical Radiation* for the convention of the symbols. Be cautious and forewarned that there are different conventions for the symbols for the radiometric and photometric quantities, which can cause confusion for both beginning and advanced readers.

TABLE 1.1 Radiometric Terms and their Characteristics

Term and description	Symbol	Functional form	SI units
Radiant energy	Q_e		J
Radiant energy density Radiant energy per unit volume	u_e	$\frac{dQ_e}{dV}$	J/m^3
Radiant flux/power Radiant energy per unit time	Φ_e or P_e	$\frac{dQ_e}{dt}$	J/s or W
Radiant exitance Radiant flux per unit source area	M_e	$\frac{d\Phi_e}{dA_{source}}$	W/m^2
Irradiance Radiant flux per unit target area	E_e	$\frac{d\Phi_e}{dA_{target}}$	W/m^2
Radiant intensity Radiant flux per unit solid angle	I_e	$\frac{d\Phi_e}{d\Omega}$	W/sr
Radiance Radiant flux per unit projected area per unit solid angle	L_e	$\frac{d^2\Phi_e}{dA_{s,proj}d\Omega}$	$W/m^2/sr$

Radiometric quantities are based on the first term in the table, radiant energy (Q_e), which is measured in the SI unit of joules (J). The radiant energy density (u_e) is radiant energy per unit volume measured in the SI units of J/m^3. The radiant flux or power (Φ_e or P_e) is the energy per unit time, thus it is measured in the SI unit of J/s or watts (W). There are two expressions for the radiant surface flux density, the radiant exitance and the irradiance. The radiant exitance (M_e) is the amount of flux leaving a surface per unit area, while the irradiance (E_e) is the amount of flux incident on a surface per unit area. Thus, the exitance is used for source emission, scatter from surfaces, and so forth, while irradiance describes the flux incident on detectors and so forth. Both of these terms are measured in SI units of watts per square meter (W/m^2). Radiant intensity (I_e) is defined as the power radiated per unit solid angle, thus it is in the SI units of watts per steradian (W/sr). Note that many describe the radiant intensity as power per unit area (i.e., radiant flux density), but this is expressly incorrect.* Most texts use the term of intensity when irradiance is the correct terminology, but some excellent texts use it correctly, for example, *Optics* by Hecht [6]. Modern texts in the field of illumination use it correctly [7]. For a thorough discussion on this confusion, see the article by Palmer [8]. Finally, the radiance (L_e) is the power per unit projected source area per unit solid angle, which is in the SI units of watts per meter squared per steradian ($W/m^2/sr$). In Sections 1.4–1.6, the quantities of irradiance, intensity, and radiance, respectively, are discussed in more detail.

* The use of the term intensity when irradiance is implied is unfortunate and causes considerable confusion, especially to those learning optics or radiometry. Some authors denote that "intensity" is the power per unit area, and that "radiant intensity" is power per unit solid angle. Others indicate that in physical optics intensity is defined as the magnitude of the Poynting vector. These positions are inaccurate due to the SI definition, and in my opinion should be corrected.

When one of the quantities listed in Table 1.1 is provided as a function of wavelength, it is called a spectral quantity. For example, when the intensity has a spectral distribution, it is called the spectral radiant intensity. The notation for the quantity is modified with either a λ subscript (e.g., $I_{e,\lambda}$) or by denoting the quantity is a function of wavelength (e.g., $I_e(\lambda)$). The units of a spectral quantity are in the units listed in Table 1.1, but are per wavelength (e.g., nm or μm). In order to compute the value of the radiometric quantity over a desired wavelength range, the spectral quantity is integrated over all wavelengths

$$f_e = \int_0^\infty h(\lambda) f_e(\lambda) d\lambda, \tag{1.1}$$

where $h(\lambda)$ is the filter function that describes the wavelength range of importance, f_e is the radiant quantity (e.g., irradiance), and $f_e(\lambda)$ is the analogous spectral radiant quantity (e.g., spectral irradiance).

1.3.2 Photometric Quantities

The photometric terms are applied to the human visual system, so only the visible spectrum of 360–830 nm adds to the value of a term. Due to the variability of the human eye, a standard observer is used, which is maintained by the CIE. Table 1.2 shows the analogous quantities to that of the radiometric terms of Table 1.1. The term luminous is used before the term, such as luminous flux, to delineate between radiometric and photometric quantities. Additionally, radiometric quantities are expressed with a subscript "v" to denote visual.

Luminous energy (Q_v) is measured in the units of the talbot (T), which is typically labeled as lumen-s (lm-s). The luminous energy density (u_v) is in the units of

TABLE 1.2 Photometric Terms and Their Characteristics

Term and description	Symbol	Functional form	Units
Luminous energy	Q_v		T
Luminous energy density Luminous energy per unit volume	u_v	$\frac{dQ_v}{dV}$	T/m^3
Luminous flux/power Luminous energy per unit time	Φ_v or P_v	$\frac{dQ_v}{dt}$	lm
Luminous exitance Luminous flux per unit source area	M_v	$\frac{d\Phi_v}{dA_{source}}$	lx
Illuminance Luminous flux per unit target area	E_v	$\frac{d\Phi_v}{dA_{target}}$	lx
Luminous intensity Luminous flux per unit solid angle	I_v	$\frac{d\Phi_v}{d\Omega}$	cd
Luminance Luminous flux per unit projected area per unit solid angle	L_v	$\frac{d^2\Phi_v}{dA_{s,proj}d\Omega}$	nt

T/m³. Once again, lm-s is typically used for the talbot. The luminous flux is provided in the units of lumens (lm). The two luminous flux surface density terms are for a source, the luminous exitance (M_v), and for a target, the illuminance (E_v). Both terms have the units of lux (lx), which is lumens per meter squared (lm/m²). The luminous intensity (I_v) is measured in candela (cd), which is lumens per unit steradian (lm/sr). Note that the candela is one of the seven SI base units [9]. The definition for the candela was standardized in 1979, which per Reference [9] states that it is "the luminous intensity, in a given direction, of a source that emits monochromatic radiation of frequency 540×10^{12} hertz and that has a radiant intensity in that direction of 1/683 watt per steradian." This definition may appear arbitrary, but it is established on previous definitions, thus provides a degree of consistency over the lifetime of the candela unit. The standardization of the candela for luminous intensity provides further reason of the need to correct the misuse of the terms intensity and irradiance/illuminance. The luminance (L_v) is the photometric analogy to the radiometric radiance. It is in the units of the nit (nt), which is the lumens per meter squared per steradian (lm/m²/sr). There are several other photometric units for luminance and illuminance that have been used historically. Table 1.3 provides a list of these quantities. Note that these units are not accepted SI units, and are in decreasing use. In Sections 1.4–1.6, the quantities of illuminance, intensity, and luminance, respectively, are discussed in more detail.

As with radiometric terms, spectral photometric quantities describe the distribution of the quantity as a function of the wavelength. By integrating the spectral luminous quantity over wavelength with a desired filter function, one finds the total luminous quantity over the desired spectral range

$$f_v = \int_0^\infty h(\lambda) f_v(\lambda) d\lambda. \tag{1.2}$$

Conversion between radiometric and photometric units is accomplished by taking into account the response of the CIE standard observer. The functional form is given by

$$f_v(\lambda) = K(\lambda) f_e(\lambda), \tag{1.3}$$

TABLE 1.3 Alternate Units for Illuminance and Luminance

Unit	Abbreviation	Form
Illuminance		
Foot-candle	fc	lm/ft²
Phot	ph	lm/cm²
Luminance		
Apostilb	asb	cd/ π/m²
Foot-lambert	fL	cd/ π/ft²
Lambert	L	cd/π/cm²
Stilb	sb	cd/cm²

where $f_v(\lambda)$ is the spectral photometric quantity of interest, $f_e(\lambda)$ is the analogous spectral radiometric term, and $K(\lambda)$ is the luminous efficacy, which is a function of wavelength, λ, and has units of lm/W. The luminous efficacy describes the CIE observer response to visible electromagnetic radiation as a function of wavelength. The profile of $K(\lambda)$ is dependent on the illumination level, because of the differing response of the eye's detectors. For example, for light-adapted vision, that is, photopic vision, the peak in the luminous efficacy occurs at 555 nm. For dark-adapted vision, that is, scotopic vision, the peak in the luminous efficacy is at 507 nm.* Equation (1.3) is often rewritten as

$$f_v(\lambda) = CV(\lambda) f_e(\lambda), \tag{1.4}$$

where $V(\lambda)$ is the luminous efficiency,† which is a unitless quantity with a range of values between 0 and 1, inclusive, and C is a constant dependent on the lighting conditions.‡ For photopic vision, $C = C_p = 683$ lm/W, and for scotopic vision $C = C_s = 1700$ lm/W. Note that the constant for photopic conditions is in agreement with the definition of the candela as discussed previously. The difference between the two lighting states is due to the response of the cones, which are not saturated for typical light-adapted conditions, and rods, which are saturated for light-adapted conditions. The lumen is realistically defined only for photopic conditions, so for scotopic cases the term "dark lumen" or "scotopic lumen" should be used. Light levels between scotopic and photopic vision are called mesopic, and are comprised of a combination of these two states. Figure 1.2 shows the luminous efficiency of the standard observer for the two limiting lighting conditions as a function of wavelength. Note that while similar, the response curves do not have the same shape. Light-adapted vision is broader than dark-adapted vision.

To calculate the luminous quantity when the spectral radiant distribution is known, one must integrate over wavelength using the luminous efficacy and the filter function as weighting terms. Using Equations (1.1) and (1.3), we arrive at

$$f_v = \int_0^{\infty} h(\lambda) K(\lambda) f_e(\lambda) d\lambda. \tag{1.5}$$

where the limits of integration are set to 0 and ∞, since the functional forms of $K(\lambda)$ and $h(\lambda)$ take into account the lack of response outside the visible spectrum and the wavelength range(s) of interest, respectively. Alternatively, one can determine the radiometric quantity with knowledge of the analogous photometric spectral distribution; however, only the value over the visible spectrum is calculated.

* For photopic vision, the luminous efficacy may be labeled as $K(\lambda)$ or $K_p(\lambda)$, and for scotopic vision, the luminous efficacy may be labeled as $K'(\lambda)$ or $K_s(\lambda)$. In Equation (1.3), $K(\lambda)$ is used in generic form to indicate either photopic or scotopic conditions.

† For photopic vision, the luminous efficiency may be labeled as $V(\lambda)$ or $V_p(\lambda)$, and for scotopic vision, the luminous efficiency may be labeled as $V'(\lambda)$ or $V_s(\lambda)$. In Equation (1.4), $V(\lambda)$ is used in generic form to indicate either photopic or scotopic conditions.

‡ The constant C may be labeled as C or C_p for photopic vision and as C' or C_s for scotopic vision. In Equation (1.4), C is used in generic form to indicate either photopic or scotopic conditions.

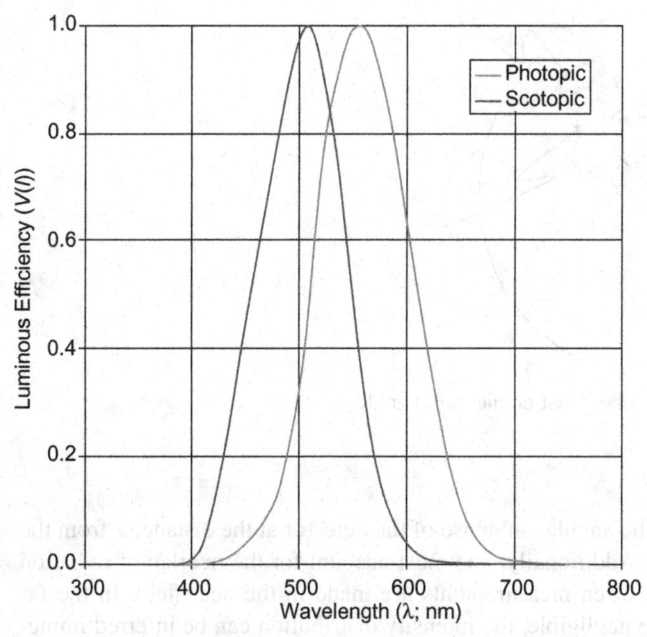

Figure 1.2 Luminous efficiency for photopic and scotopic conditions.

1.4 INTENSITY

Luminous or radiant intensity describes the distribution of light as a function of angle—more specifically a solid angle. The solid angle, $d\Omega$, in units of steradians (sr) is expressed by a cone with its vertex at the center of a sphere of radius r. As shown in Figure 1.3, this cone subtends an area dA on the surface of the sphere. For this definition, the solid angle is given by

$$d\Omega = \frac{dA}{r^2} = \sin\theta d\theta d\phi, \tag{1.6}$$

where, in reference to Figure 1.3, θ is the polar angle labeled as θ_0 and ϕ is the azimuthal angle around the central dotted line segment. By integrating the right-hand side of Equation (1.6) over a right-circular cone, one finds

$$\Omega = 4\pi \sin^2 \frac{\theta_0}{2}, \tag{1.7}$$

where θ_0 is the cone half angle as shown in Figure 1.3. A cone that subtends the entire sphere (i.e., $\theta_0 = \pi$) has a solid angle of 4π and one that subtends a hemisphere (i.e., $\theta_0 = \pi/2$) has a solid angle of 2π. Intensity is a quantity that cannot be directly measured via a detection setup, since detectors measure flux (or energy). A direct conversion to flux density is found by dividing through by the area of the detector while assuming the distribution is constant over the detector surface, but intensity

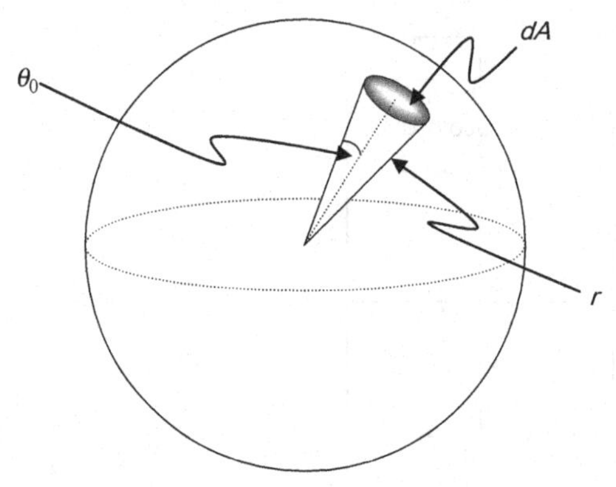

Figure 1.3 Geometry of sphere that defines solid angle.

requires knowledge of the angular subtense of the detector at the distance r from the source point of interest. Additionally, one must account for the overlap of radiation for an extended source when measurements are made in the near field. In the far field, where caustics are negligible, the intensity distribution can be inferred immediately from the flux density.

1.5 ILLUMINANCE AND IRRADIANCE

Illuminance and irradiance are the photometric and radiometric quantities respectively for the surface flux density on a target.* The material presented in this section is also applicable to the radiant and luminous exitances. These terms describe the spatial distribution of light since they integrate the luminance or radiance over the angular component. Detectors operate in this mode in the sense that the power incident on the detector surface is dependent on its area.† Figure 1.4 depicts the measurement of the flux density at a distance of r from a uniform point source by a detector of differential area dA. The subtense angle, θ, is between the normal to the detector area and the centroid from the source to the projected area of the detector area. The latter is simply the line segment joining the source (S) and the center of the detector (T), $\overline{ST}$. Using Equation (1.6), the detector area subtends an elemental solid angle from the point S of

$$d\Omega = \frac{dA_{\text{proj}}}{r^2} = \frac{dA\cos\theta}{r^2}, \tag{1.8}$$

* Use of the terms irradiance or illuminance denote the generic surface flux density term, except when units are specifically listed. Henceforth, no subscripts (i.e., "e" or "v") are used on any illumination quantities in order to denote the generic nature of the equations.

† Thus, the reason that a number of detectors have an active area of 1 cm^2. The measured power then matches the flux density with the assumption that the flux is constant over the area of the detector.

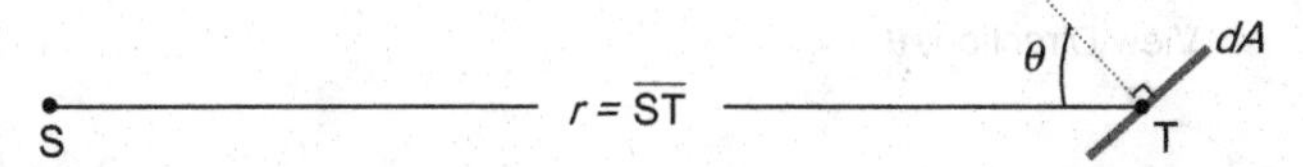

Figure 1.4 Measurement of the irradiance on a target area of dA at a distance r from a point source of flux output $d\Phi$. The target is oriented at θ with respect to the line segment joining points S and T.

where the differential projected area, dA_{proj}, will be discussed in detail in Section 1.6, Equation (1.10). Note that for the detector oriented at $\pi/2$ with respect to the line segment $\overline{ST}$, the elemental solid angle is 0. Conversely, for an orientation along the line joining the two entities, the elemental solid angle is at a maximum since the cosine term is equal to 1.

To find, for example, the irradiance due to a point source, we substitute for dA in its expression from Table 1.1 with that from Equation (1.8). The resulting equation contains an expression for the intensity, $I = d\Phi/d\Omega$, which can be substituted for,

$$E = \frac{d\Phi}{dA} = \frac{d\Phi\cos\theta}{r^2 d\Omega} = \frac{I\cos\theta}{r^2}. \tag{1.9}$$

Thus, for a point source, the flux density incident on the target decreases as a function of the inverse of the distance squared between the two objects, which is known as the inverse-square law. The cosine factor denotes the orientation of the target with respect to the line segment $\overline{ST}$.

Unlike experimental measurement of the intensity, it is easy to measure the flux areal density. With knowledge of the detector area and the power measured, one has determined the flux density at this point in space. This detection process integrates over all angles, but not all detectors measure uniformly over all angles, especially due to Fresnel reflections at the detector interface. Thus, special detector equipment, called a cosine corrector, is often included in the detection scheme to compensate for such phenomena. By measuring the flux density at a number of points in space, such as over a plane, the flux density distribution is determined. Additionally, unlike the intensity distribution, which remains constant with the distance from the optical system, the irradiance distribution on a plane orthogonal to the nominal propagation direction evolves as the separation between the optical system and plane is changed. Simply said, near the optical system (or even extended source), the rays are crossing each other such that local spatial (i.e., irradiance) distribution evolves with distance. In the far field, where the crossing of rays (e.g., caustics) are negligible, the irradiance distribution has the form of the intensity distribution. Thus, one is in the far field when these two distributions have little difference between their shapes. This point is discussed in more detail in Chapter 7.

1.6 LUMINANCE AND RADIANCE

Luminance and radiance are fundamental quantities of any emitter. As is shown in the next chapter, these terms are conserved in a lossless system, which drives the

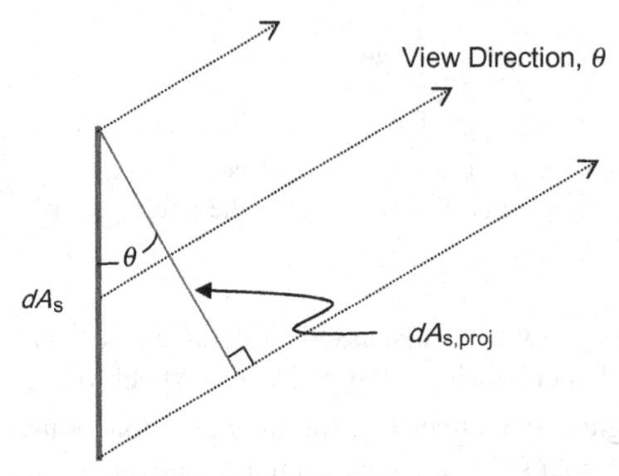

Figure 1.5 Depiction of $dA_{s,proj}$ of source element dA_s along a view direction θ.

ultimate limit of the performance of system design. Another term often used for luminance or radiance is brightness; however, within the vision community, brightness assumes the response for an actual observer to a prescribed area of an object* [10]. The radiance distribution† from a source describes the emission from each point on its surface as a function of angle. Thus, knowing the radiance distribution, one can ascertain the propagation of radiation through a known optical system. The result is that the intensity and flux density distributions, and for that matter, the radiance distribution can be determined at arbitrary locations in the optical system. In conclusion, radiance provides the best quantity to drive the design process of an illumination system for two reasons:

- An accurate source model implies that an accurate model of an illumination system can be done. The process of developing an accurate source model is a focus of the next chapter.
- Radiance is conserved, which provides the limit on system performance while also providing a comparison of the performances of different systems. The implications of conservation of radiance are presented in detail in Chapter 2.

The radiance distribution is measured as a function of source flux per projected unit area (i.e., spatial considerations), $dA_{s,proj}$, per unit solid angle (i.e., angular considerations), $d\Omega$. The projected area takes into account the orientation of the target surface with respect to the source orientation, which is shown in Figure 1.5.‡ Pursuant to this figure, the elemental projected area is given by

$$dA_{s,proj} = dA_s \cos\theta, \tag{1.10}$$

* Though the difference between brightness and luminance is minor, caution indicates that using the accepted terms of luminance and radiance is preferred.

† For simplicity, I use radiance to mean both luminance and radiance.

‡ Note that some express the radiance as the flux per unit area per projected solid angle. While the latter definition does not agree with the standard, it does provide for the same quantity.

where dA_s is the actual area of the source and θ is the view angle. Substituting Equation (1.10) into the expression for the radiance

$$L = \frac{d^2\Phi}{\cos\theta dA_s d\Omega}. \tag{1.11}$$

In connection with radiance, there are two standard types of distributions: Lambertian and isotropic. These radiance distributions are described in more detail in the next two subsections.

1.6.1 Lambertian

When a source or scatterer is said to be Lambertian, it implies its emission profile does not depend on direction,

$$L(\mathbf{r}, \theta, \phi) = L(\mathbf{r}) = L_s, \tag{1.12}$$

where $\mathbf{r}$ is the vector describing the points on the surface, θ is the polar emission angle, and ϕ is the azimuthal emission angle. Thus, only the dependence on the areal projection modifies the radiance distribution. Substituting Equation (1.12) into Equation (1.11) while using Equation (1.6) for the definition of the elemental solid angle and then integrating over angular space, one obtains

$$\int \frac{d^2\Phi}{dA_s} = \int dM = \int_{4\pi} L_s \cos\theta d\Omega. \tag{1.13}$$

The solution of this gives

$$\begin{aligned} M &= L_s \int_0^{2\pi} \int_0^{\pi/2} \cos\theta \sin\theta d\theta d\phi \\ &= M_{\text{lam}} = \pi L_s, \end{aligned} \tag{1.14}$$

where the "lam" subscript denotes a Lambertian source. The exitance for a Lambertian source or scatterer is always the product of π and its radiance, L_s. Similarly, one can integrate over the area to determine the intensity

$$\int \frac{d^2\Phi}{d\Omega} = \int dI = \int_{\mathrm{D}} L_s \cos\theta dA.$$

Integrating provides

$$\begin{aligned} I &= \cos\theta \int_D L_s dA \\ &= I_{\text{lam}} = I_s \cos\theta, \end{aligned} \tag{1.15}$$

where D is the surface of the source and

$$I_s = \int_D L_s dA. \tag{1.16}$$

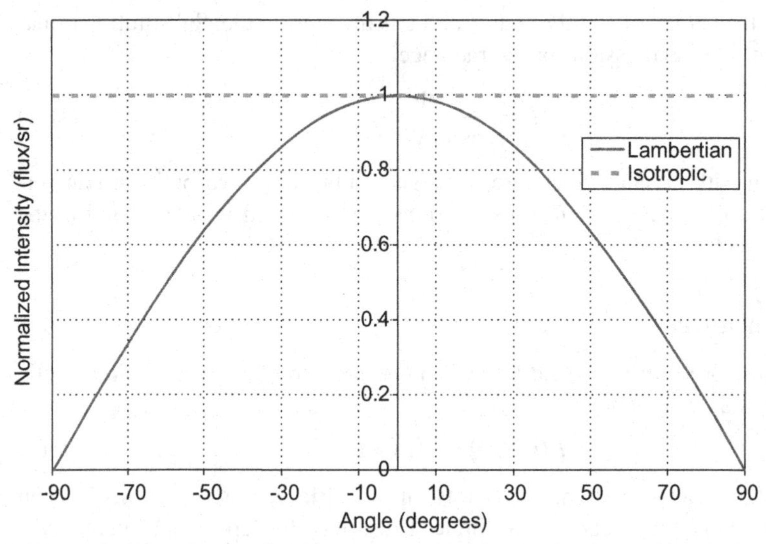

Figure 1.6 Plot of normalized intensity profiles for Lambertian and isotropic spatially uniform planar sources.

Thus, the intensity for a Lambertian source changes as a function of the cosine of the view angle. For a spatially uniform planar emitter of area A_s, the intensity distribution is

$$I_{lam} = A_s L_s \cos\theta. \tag{1.17}$$

Figure 1.6 shows the normalized intensity profile for an emitter described by Equation (1.17). Note that as the angle is increased with respect to the normal of the surface (i.e., $\theta = 0$ degrees), less radiation is seen due to the projection characteristics. In other words, as the polar angle increases, the projected area decreases, resulting in a reduction in intensity. It is an important point that while Lambertian means independent of direction, it does not preclude the reduction in intensity due to the projection of the emission area.

1.6.2 Isotropic

An isotropic source or scatterer is also said to be uniform, but such a source accounts for its projection characteristics. Thus, the cosine dependencies shown in Equations (1.15) and (1.17) are compensated within the radiance function by

$$L(\mathbf{r}, \theta, \phi) = \frac{L(\mathbf{r})}{\cos\theta} = \frac{L_s}{\cos\theta}. \tag{1.18}$$

The exitance and intensity for such a source are

$$M_{\text{iso}} = 2\pi L_s, \tag{1.19}$$

and

$$I_{\text{iso}} = I_s, \tag{1.20}$$

where the "iso" subscripts denote isotropic and I_s is as defined in Equation (1.16). For the spatially uniform planar emitter of area A_s, the intensity distribution is given by

$$I_{\text{iso}} = A_s L_s. \tag{1.21}$$

Therefore, an isotropic source has twice the exitance compared with an analogous Lambertian source. It also has a constant intensity profile as a function of angle for a planar emitter. The normalized isotropic intensity profile of Equation (1.21) is also shown in Figure 1.6. Realistically, only point sources can provide isotropic emission. There are no independent sources that provide isotropic illumination, but the integration of tailored optics and an emitter can provide such.

1.7 IMPORTANT FACTORS IN ILLUMINATION DESIGN

The transfer of light from the source to the target typically has two important parameters: transfer efficiency and the distribution at the target. Transfer efficiency is of particular importance due to increasing needs of energy efficiency—that is, electricity costs are increasing, and concerns over environmental effects are rising. However, often counter to efficiency requirements is the agreement of the achieved distribution to the desired one. For example, the desire for uniformity can easily be achieved by locating a bare source a great distance from the target plane, albeit at the expense of efficiency. Therefore, in the case of uniformity, there is a direct trade between these two criteria such that the illumination designer must develop methods to provide both through careful design of the optical system. Lesser criteria include color, volume requirements, and fabrication cost. As an example, the design of a number of illumination optics is driven by the reduction of cost though the process time to manufacture the overall system. So not only must the designer contend with trades between efficiency and uniformity, but also cost constraints and other important parameters. The end result is a system that can be quickly and cheaply fabricated with little compromise on efficiency and uniformity. The selection and functional form of the important criteria for an illumination design provide a merit function that is used to compare one system with another. The development of a merit function based on the material presented here is discussed in detail in Chapter 7.

1.7.1 Transfer Efficiency

Transfer efficiency, η_t, is defined as the ratio of flux at the target, Φ_{target}, to that at the input, which is most often the flux emitted from the source, Φ_{source}:

$$\eta_t = \frac{\Phi_{\text{target}}}{\Phi_{\text{source}}}. \tag{1.22}$$

This simple definition includes all emission points, all angles, and the entire spectrum from the source. Criteria based on spatial positions, angular domain, and spectrum can be used to select limited ranges of the emission from the source. The spectral radiance, $L_{e,\lambda}$, or spectral luminance, $L_{v,\lambda}$, describe terms* that replace the flux components in Equation (1.22),

$$\eta_t = \frac{\iiint h_{\text{target}}(\mathbf{r},\Omega,\lambda)L_{\text{target},\lambda}(\mathbf{r},\Omega,\lambda)d\mathbf{r}d\Omega d\lambda}{\iiint h_{\text{source}}(\mathbf{r},\Omega,\lambda)L_{\text{source},\lambda}(\mathbf{r},\Omega,\lambda)d\mathbf{r}d\Omega d\lambda}, \tag{1.23}$$

where h_{target} and h_{source} are filters denoting the functional form for position (**r**, which has three spatial components such as x, y, and z), angle (Ω, which is the solid angle and composed of two components such as θ, the polar angle, and ϕ, the azimuthal angle), and spectrum (λ) for the target and source, respectively. $L_{\text{target},\lambda}$ and $L_{\text{source},\lambda}$ denote the spectral radiance or luminance for the target and source, respectively. Note that h_{target} and h_{source} almost always have the same form for the spectrum, but the position and angular aspects can be different. These filter and radiance functions can be quite complex analytically, thus they are often approximated with experimental measurements or numerical calculations. The end result is that solving Equation (1.23) can be rather cumbersome for realistic sources and target requirements, but it provides the basis of a driving term in illumination design, étendue, and the required conservation of this term. It is shown in Chapter 2 that the étendue is related to radiance, thus Equation (1.23) expresses a fundamental limit and thus efficiency for the transfer of radiation to the target from the source.

1.7.2 Uniformity of Illumination Distribution

The uniformity of the illumination distribution defines how the modeled or measured distribution agrees with the objective distribution. The illumination distribution is measured in at least one of three quantities:

- Irradiance or illuminance: measured in flux/unit area
- Intensity: measured in flux/steradian, or
- Radiance or luminance: measured in flux/steradian/unit area.

For the radiometric quantities (i.e., irradiance, radiant intensity, and radiance), the unit of flux is the watt (see Table 1.1). For the photometric quantities (i.e., illuminance, luminous intensity, and luminance) the unit of flux is the lumen (see Table 1.2). These quantities are described in depth in Sections 1.3–1.6. Other quantities have been used to express the illumination distribution, but they are typically hybrids or combinations of the three terms provided above.

Uniformity is determined by comparing the sampled measurement or model with the analogous one of the goal distribution. Note that the irradiance/illuminance and intensity distributions can be displayed with two orthogonal axes, while the

* The spectral radiance is used for radiometric calculations, while the spectral luminance is used for photometric calculations.

radiance and luminance quantities require a series of depictions to show the distribution. There are a multitude of methods to determine the uniformity, including:

- Peak-to-valley variation of the distribution, Δf:

$$\Delta f = \max\left[f_{\text{model}}(i, j) - f_{\text{goal}}(i, j)\right] - \min\left[f_{\text{model}}(i, j) - f_{\text{goal}}(i, j)\right]. \quad (1.24)$$

- Variance of the distribution compared with the goal, σ^2:

$$\sigma^2 = \frac{1}{(mn-1)} \sum_{j=1}^{n} \sum_{i=1}^{m} \left[f_{\text{model}}(i, j) - f_{\text{goal}}(i, j)\right]^2. \quad (1.25)$$

- Standard deviation of the distribution, σ: the square root of the variance provided in Equation (1.25).

where the f terms denote the selected quantity that defines uniformity (e.g., irradiance or intensity). The model and goal subscripts denote the sampled model and goal distributions, respectively. The terms i and j are the counters for the m by n samples, respectively, over the two orthogonal axes. The term root mean square (RMS) deviation is found by taking the square root of an analogous form of the bias-corrected variance expressed in Equation (1.25). For the RMS variation, σ_{rms}, the factor in front of the summation signs in Equation (1.25) is replaced with $1/nm$.* In all cases, a value of 0 for the uniformity term (e.g., Δf or σ^2) means that the uniformity, or agreement with the target distribution, is perfect for the selected sampling. The choice of the uniformity metric is dependent on the application, accuracy of the modeling, and ease of calculation. The RMS variation of the selected quantity is the standard method of calculating the uniformity of a distribution.

1.8 STANDARD OPTICS USED IN ILLUMINATION ENGINEERING

There is a multitude of types of optics that can be used in the design of an illumination system. These types can essentially be broken down into five categories: refractive optics (e.g., lenses), reflective optics (e.g., mirrors), total internal reflection (TIR) optics (e.g., lightpipes), scattering optics (e.g., diffusers), and hybrid optics (e.g., catadioptric Fresnel elements or LED pseudo collimators). In the next five subsections, I discuss each of these optic types in more detail. Each of these optic types has different utility in the field of illumination design, and this book uses separate chapters to discuss and delineate these. As a general rule of thumb, reflective optics provide the most "power" to spread light, but at the expense of tolerance demands and typical higher absorption losses. Refractive optics allow more compact systems to be built, but at the expense of dispersion, an increased number of elements, and higher fabrication costs due to alignment issues and postproduction manufacturing

* See MathWorld, hosted and sponsored by Wolfram Research; specifically, see http://mathworld.wolfram.com/StandardDeviation.html for further discussion of this topic.

demands. TIR optics provide in theory the optimal choice, except when one includes potential leakage due to both scattering and a lack of fulfilling the critical-angle condition at all interfaces. Scatter is not typically employed as a method to provide a critical illumination distribution, but, rather, a means to make the distribution better match uniformity goals at the expense of efficiency. Scatter is also used for subjective criteria, such as look and feel of the optical system. Hybrid optics that employ reflection, refraction, scatter, and TIR are becoming more prevalent in designs since they tend to provide the best match in regards to optimizing goals.

1.8.1 Refractive Optics

Refractive optics are a standard tool in optical design, primarily used for imaging purposes, but they can also be used in illumination systems. There is a variety of refractive optics used in illumination systems. In fact, the types of refractive optics are too numerous to discuss here, but they range from standard imaging lenses to arrays of pillow optics to protective covers. Examples of refractive optics in use in illumination systems include: (i) singlet lens for projector headlamp, (ii) pillows lens array for transportation applications, (iii) nonimaging Fresnel lens for display purposes, and (iv) a protective lens for an automotive headlight. The first three examples are using the refraction to assist in obtaining the target illumination distribution, but (iv) has minimal, if not negligible, impact on the distribution of light at the target. The primary purpose of the latter lens is to protect the underlying source and reflector from damage due to the environment.

Image-forming, refractive optics are not optimal for illumination applications in the sense that they do not maximize concentration, which is defined in the next chapter. The reason for the limitation is due to aberrations as the *f*-number is decreased. This topic for imaging systems has been investigated in detail [11]. In Reference [11], it is pointed out that the theoretical best is the Luneberg lens, which is a radial gradient lens shaped as a perfect sphere and has a refractive index range from 1.0 to 2.0 inclusive. This lens is impossible to manufacture due to the index range spanning that of vacuum to high-index flint glass, such as Schott LaSF35. Realistic lenses, such as an *f*/1 photographic objective or an oil-immersed, microscope objective, are examples of imaging lenses that provide the best performance from a concentration point of view. These two examples do not provide optimal concentration, and they also suffer in that their sizes are small, or in other words, while their *f*/# might be small, accordingly the focal length and thus diameter of the clear aperture are also small. However, there are cases when nonoptimal imaging, refractive systems are preferred over those that provide optimal concentration. An example is a pillow lens array to provide homogeneity over a target. Each of the pillow lenses creates an aberrated image of the source distribution on the target plane, thus the overlap of these distributions due to the lens array can be used to create an effective uniform distribution. Additionally, nonimaging lenses are an option, but this type of lens almost always involves other phenomena, especially TIR, to accomplish their goals. These optics are treated in Section 1.8.5 on hybrid optics. Refractive, nonimaging lenses use, in part, imaging principles to great effect to transfer the light from the source to the target. Two examples are the nonimaging

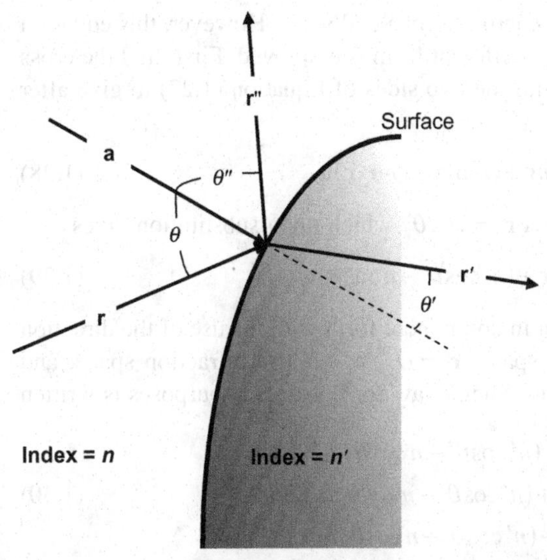

Figure 1.7 Schematic of refraction, reflection, and total internal reflection at an optical surface. The incident ray vector is shown by **r** in index n. The refracted ray vector is $\mathbf{r}'$ in index n'. The reflected and TIR ray vectors are $\mathbf{r}''$ in index n. For TIR to occur $n > n'$, and $\theta \geq \theta_c$.

Fresnel lens, which has been discussed in the literature [12], and the catadioptric lens used to collimate as best as possible the output from an LED. The latter is a focus of Chapter 7 on optimization and tolerancing of nonimaging optics. Nonimaging, refractive designs are making great use of tailoring, where tailoring is defined as the designation of the optical surface to provide a prescribed distribution at the target in order to meet design goals. Tailoring is discussed further in the next chapter and the succeeding chapters in this book. The systems and principles presented here and others are described in more depth in Chapters 4–6.

Refractive optics use materials of differing indices of refraction, n, to alter the propagation path of light from the source to the target. Refraction, displayed in Figure 1.7, is governed by the law of refraction, often called Snell's law,

$$n \sin\theta = n' \sin\theta', \tag{1.26}$$

where θ and θ' are the angles of incidence and refraction, respectively, and n and n' are the indices of refraction in source (i.e., object) and target (i.e., image) spaces, respectively. In other words, the unprimed notation denotes the quantities prior to refraction, and the primed notation indicates the quantities after refraction. In illumination systems, three-dimensional (3D) ray tracing is needed, thus the vector form is preferred,

$$n\mathbf{r} \times \mathbf{a} = n'\mathbf{r}' \times \mathbf{a}, \tag{1.27}$$

where **a** is the surface normal into the surface as shown in Figure 1.7, and **r** and $\mathbf{r}'$ are the unit vectors depicting the rays. Equation (1.27) also implies that **r**, $\mathbf{r}'$, and **a** are coplanar, which is a necessary condition for Snell's law. However, this equation is not tractable to ray tracing, thus a different form is employed. First, find the cross product of the surface normal **a** with the two sides of Equation (1.27) to give after rearrangement of the terms

$$n'\mathbf{r}' = n\mathbf{r} + (n'\mathbf{a}\cdot\mathbf{r}' - n\mathbf{a}\cdot\mathbf{r})\mathbf{a}. \tag{1.28}$$

Next, we note that $\mathbf{a}\cdot\mathbf{r} = \cos\theta$ and $\mathbf{a}\cdot\mathbf{r}' = \cos\theta'$, which upon substitution gives

$$n'\mathbf{r}' = n\mathbf{r} + (n'\cos\theta' - n\cos\theta)\mathbf{a}. \tag{1.29}$$

Finally, this expression is rewritten in component form with the use of the direction cosines $\mathbf{r} = (L, M, N)$ in incident space, $\mathbf{r}' = (L', M', N')$ in refraction space, and $\mathbf{a} = (a_L, a_M, a_N)$. In component form, Snell's law for ray-tracing purposes is written

$$\begin{aligned} n'L' &= nL + (n'\cos\theta' - n\cos\theta)a_L, \\ n'M' &= nM + (n'\cos\theta' - n\cos\theta)a_M, \text{ and} \\ n'N' &= nN + (n'\cos\theta' - n\cos\theta)a_N. \end{aligned} \tag{1.30}$$

Note that the refraction angle θ' must be found prior to the determination of the direction cosines after refraction. Snell's law as given in Equation (1.26) is used for this purpose

$$\theta' = \arcsin\left(\frac{n}{n'}\sin\theta\right). \tag{1.31}$$

Substitution of Equation (1.31) into Equation (1.30) and diving through by n' provides the ray path after refraction for the purposes of numerical ray tracing. The only caveat to the implementation of Equation (1.30) is that $\mathbf{r} = (L, M, N)$, $\mathbf{r}' = (L', M', N')$, and $\mathbf{a} = (a_L, a_M, a_N)$ are all unit vectors.

1.8.2 Reflective Optics

Reflective optics are also a standard tool in optical design, but admittedly imaging design uses them less than refractive ones. In illumination design, reflective optics have been more prevalent than refractive ones. The primary reasoning is that it is easier to design the optics that provide optimal concentration between the input and output apertures, and the tolerance demands for nonimaging systems are less in comparison to imaging ones. There is a wealth of reflective optics in use in illumination designs from conic reflectors, such as spherical, parabolic, elliptical, and hyperbolic, to edge-ray designs (EDs) [13], such as the compound parabolic concentrator (CPC), to tailored edge-ray designs (TEDs) [13, 14]. The conic reflectors typically involve imaging requirements, and illumination designs use these properties to capture the light emitted by a source. Nonimaging designs use the edge-ray principle to transfer optimally the light from source to target. EDs are optimal in two dimensions (2D), called troughs, but 3D designs, called wells, do not transfer some skew rays from the source to target. ED assumes a constant acceptance angle with uniform angular input, but TED uses a functional acceptance angle. Thus, TED accommodates realistic sources while providing tailoring of the distribution at the target dependent on requirements and the source characteristics. Systems employing the edge-ray principle are based on their parent, conic designs, thus there are two classes: elliptical and hyperbolic.

Tailored designs are currently at the forefront of reflector design technology in the nonimaging optics sector. Note that tailored designs started within the realm of reflective illumination optics, but they are now present in all sectors of nonimaging optics, from refractive to hybrid. Tailored reflectors use one of two methods: discrete faceted or continuous/freeform reflectors. The former is comprised of pseudo-independent, reflective segments to transfer the light flux. The freeform design can be thought of as a faceted design but made up of an infinite number of individual segments. The designs are typically smooth, although discontinuities in the form of cusps (i.e., the first derivative is discontinuous) are sometimes employed, but steps are not allowed since that is the feature of faceted designs. While tailored designs employ optics principles in order to define their shape, the chief goal is to maintain a functional acceptance angle profile. Recently, applications that are not based entirely on the optics of the problem have dictated that the edge-ray can limit performance of the overall system. These systems use interior rays to motivate the reflector development, thus this domain is called nonedge-ray reflector design (NERD) [15]. One application is optical pumping of gain material, which introduces some imaging properties in order to obtain more efficient laser output at the expense of transfer efficiency [16].

Examples of reflective illumination designs include: (i) luminaire for room lighting, (ii) faceted headlight, (iii) faceted reflector coupled to a source for projection and lighting applications (i.e., MR16), and (iv) a reflector coupled to an arc source for emergency warning lighting. For luminaires for architectural lighting, the shape of the reflector is either a conic or an arbitrary freeform surface. In both of these cases, there is minimal investment in the design of the reflector, but rather a subjective look and feel is the goal of the design process. In recent years, luminaire designs for specific architectural applications, such as wall-wash illumination, has integrated tailoring in order to provide better uniformity and a sharp cutoff in the angular distribution. Increasingly, especially with the development of solid-state lighting, luminaire reflectors employ some level of tailoring in order to meet the goals of uniformity while also maintaining high transfer efficiency. They must also meet marketing requirements that drive their appearance in both the lit and unlit states. For example, the headlight shown in (ii) must adhere to stringent governmental illumination standards (e.g., ECE [European] or FMVSS/SAE [US]), but must also conform to the shape of the car body and provide a novel appearance.

Specular reflectors employ either employ reflective materials, such as polished aluminum or silver; reflective materials deposited on substrates; or dielectric thin film coating stacks deposited on polished substrates. For reflective deposition, the coatings are typically aluminum, chromium, or even silver and gold. The substrate is typically an injection-molded plastic. For dielectric coatings, the substrates can be absorbing, reflective, or transmissive. Dielectric coatings on reflective substrates protect the underlying material but also can be used to enhance the overall reflectivity. Dielectric coatings placed on absorbing or transmissive substrates can be used to break the incident spectrum into two components so that the unwanted light is removed from the system. An example is a hot mirror that uses a dichroic placed on an absorbing substrate. The visible light is reflected by the coating, while the infrared (IR) light is absorbed by the substrate. In all cases, the law of reflection can be used to explain the behavior of the reflected rays. For reflection, which is also shown in Figure 1.7, the incident ray (**r**) angle, θ, is equal to that of the reflected ray (**r**″) angle, θ'',

$$|\theta| = |\theta''|. \tag{1.32}$$

Note that we update the prime notation used for refraction to a double prime notation to denote reflection. The development employed in the previous section can be used to determine the direction cosine equations for the propagation of a reflected ray by using the law of refraction with $n' = n'' = -n$ and $\mathbf{r}' = -\mathbf{r}''$.* Using this formalism, Snell's law for reflection gives $\theta = -\theta''$, and Equations (1.28) and (1.29) upon reflection give

$$\mathbf{r}'' = \mathbf{r} - (\mathbf{a}\cdot\mathbf{r}'' + \mathbf{a}\cdot\mathbf{r})\mathbf{a} = \mathbf{r} - 2(\mathbf{a}\cdot\mathbf{r})\mathbf{a}, \tag{1.33}$$

and

$$\mathbf{r}'' = \mathbf{r} - 2\mathbf{a}\cos\theta. \tag{1.34}$$

The direction cosines are then written,

$$\begin{aligned} L'' &= L - 2a_{L'}\cos\theta \\ M'' &= M - 2a_{M'}\cos\theta, \text{ and} \\ N'' &= N - 2a_{N'}\cos\theta. \end{aligned} \tag{1.35}$$

Thus, Equations (1.30) and (1.31) are essentially the same, and often optical ray-tracing software implements Equation (1.20) with the caveat that a negative index of refraction is used with an antithetical reflection vector.

1.8.3 TIR Optics

TIR optics use the "frustration" of refraction in order to propagate light within the higher index material that traps the light. Examples of TIR illumination optics include (i) large-core plastic optical fibers, (ii) lightpipes, (iii) lightguides for display applications, and (iv) brightness enhancement film. TIR is not a standard in imaging design, used sparingly in components like prisms and fibers. In illumination optics, TIR optics are gaining in popularity, especially in hybrid form as discussed in Section 1.8.5. This popularity gain is due to real and perceived benefits, including higher transfer efficiency, assistance with homogenization, compact volume, and the component provides guiding along its length. Unfortunately, to turn these benefits into realistic TIR optics, it requires a sizable investment in design and fabrication.

* A number of authors say that reflection is a special case of refraction, but they only point out that $n' = n'' = -n$, while neglecting to point out that $\mathbf{r}' = -\mathbf{r}''$.

The higher transfer efficiency is due to the 100% Fresnel reflection as long as the critical angle, θ_c, or greater, is satisfied. The critical angle is governed by Snell's law

$$\theta_c = \arcsin\left(\frac{n'}{n}\right), \tag{1.36}$$

where this angle describes the point at which light that would refract from the higher index material (n) into a lower-index material (n') at an angle of $\theta' = 90°$ to the surface normal (i.e., along the tangent to the surface). With TIR, the light will be trapped within the optic, propagating along the reflection ray vector $\mathbf{r}''$, as shown in Figure 1.7. The only perceived losses are at the input and output apertures of the TIR medium due to Fresnel reflections. With angles of incidence in the range of ±45°, the Fresnel reflection, typically considered a loss, is around 4% per uncoated interface. However, with only the two surfaces, input and output, a transfer efficiency of greater than 90% is still difficult to obtain because of scatter from both volume and surface imperfections that frustrate the TIR condition, leakage due to lack of holding to the critical angle condition, and back reflection from other surfaces of the optic due to shape and guiding requirements for the TIR optic. For example, some lightpipes used in automotive, dashboard applications have been known to have transfer efficiency around 10% to the output.

Homogenization requires careful development of the TIR optic to ensure proper mixing of the source(s) within the medium. A standard rule-of-thumb is that to obtain homogenization, the length of the TIR optic must be ten times or greater than its width. This length-to-width ratio thus demands a rather large volume in comparison to non-TIR methods, such as pillow lens arrays. Additionally, input and output locations within the optic may make increased demands on the system design, resulting in a trade on other benefits. These trade studies are often driven by volume limitations, thus while the final design can remain compact, one must give up uniformity and efficiency.

Notwithstanding, TIR optics are often the best choice for the design of illumination systems, especially those requiring remote or distributed lighting. The most prevalent TIR optics in the illumination sector are lightpipes and lightguides. A standard use of lightpipes is for discrete lighting applications, such as indicators and buttons, while lightguides are used for distributed lighting seen, for example, in backlit displays. This distinction is arbitrary, thus the terms lightpipe and lightguide are often used interchangeably. There are three aspects to the design of the lightpipes: the coupling of the source to it, the efficient propagation of the captured flux, and the outcoupling with the desired pattern. The incoupling usually involves secondary optics, proximity, and widening of the optic after the input aperture. The design of the guide after coupling, as stated previously, is driven by volume limitations, output demands, and reduction of leakage. The design of the lightpipe propagation sections can be tedious and complex, requiring intense user interaction due to the limited number of algorithms. The outcoupling is achieved through four standard methods: refraction at the output surface; narrowing of the guide optic; bumps, holes, or scatter sites on the surface of the lightpipe; and shaping of the optic to allow controlled leakage. These topics and others are considered in detail in Chapter 6 of this book.

1.8.4 Scattering Optics

Scatter optics purposely use surface roughness or volume scattering in order to approach the desired distribution. They do such by trading off transfer efficiency because of scattering outside the desired distribution or even backscattering. Additionally, scatter centers placed on or within optical components provide a method to lessen hot spots (i.e., regions resulting in high illumination levels) or cold spots (i.e., regions resulting in low illumination levels); to reduce or smooth the gradient between neighboring illuminated regions, especially since human observation is drawn to the gradient rather than the level; to provide a desired appearance, lit and/or unlit; frustrate TIR; and provide mixing to hide individual sources, especially those from different spectral sources. Scatter is also unintentionally present in all illumination optics because nothing can be made perfectly. Surfaces will never be completely smooth, and volumes will have some level of inclusions that cause scattering.

Scattering sites are used in all types of components: reflective, refractive, TIR, and absorbing/opaque. Examples include: (i) a tailored diffuser, (ii) a cosine-corrected diffuser, (iii) a roughened housing for a transmissive, automotive optic, and (iv) a rough, metallic luminaire. The first two examples use scatter to provide a distinct distribution of light, where the first is using surface scattering and the second is using volume scattering. For a number of illumination optics incorporating roughened surfaces, the scatter is placed on the optic to not only assist in meeting required standards, but to provide a smoother appearance to the illumination distribution while also hiding to some extent the interior operation of the illumination system. For example, a metal luminaire reflector incorporates a nonspecular surface to provide a fairly uniform and soft illumination of the lit environment.

Surface scattering, both intentional and unintentional, is explained by the bidirectional surface distribution function (BSDF). The BSDF is the generic form to describe the scatter distribution, but the bidirectional reflectance distribution function (BRDF) and bidirectional transmittance distribution function (BTDF) denote the reflection and transmission distributions, respectively. A special case of the BSDF for weakly scattering surfaces, such as those found in optical systems, is the Harvey model. These terms describe how the light is scattered as a function of incident and observation angles. Volume scattering is described by model distributions, such as Mie scatter, Henyey–Greenstein [17], and the Gegenbauer model [18]. Finally, there is Lambertian and isotropic (uniform) scatter for both surfaces and volumes. Both provide equal scatter as a function of angle, but isotropic scatter compensates for the cosine falloff due to the projection of the scattering surface. The reader is directed to the book *Optical Scattering* by Stover for more information about scattering theory and measurement [19].

1.8.5 Hybrid Optics

Hybrid illumination optics, also called combination optics, are those that integrate by design at least two of the four optical phenomena described in the previous sections. With this definition, one could construe that a source coupled to a reflector in close proximity to a lightguide is a hybrid optic; however, this is an example of a system with distinct optical components each doing a separate function. Rather, the definition for hybrid illumination optic is further refined by demanding that the optical phenomena must be integrated into a single optical component. Note that a simple lightpipe that has planar surfaces at the input and output apertures performs two refractions; however, these refractions are just incidental, and the optic is considered solely a TIR structure. Alternatively, if the input or output surfaces were shaped to provide a desired distribution, then the optic would be a hybrid since it integrates by design two of the four optical phenomena previously presented.

Examples of hybrid illumination optics include (i) a catadioptric Fresnel element, (ii) a backlighting lightguide using scatter sites for output, and (iii) an integrated coupler-illuminator for LED sources. In (i), the light first uses refraction at the entrance aperture of the Fresnel element, followed by TIR at the next intercept, and then refraction at the output aperture. For (ii), the lightguide uses TIR to propagate the light along its length and then scatter to eject the light from the lightguide. The integrated coupler-illuminator for LED sources is the leading example of the utility of hybrid illumination optics. This optic integrates the source coupling and then the illumination into the desired target distribution. This method ensures compactness, higher efficiency, and required uniformity. Design algorithms are being proposed for the development of these hybrid optics. One in particular, the simultaneous multiple surfaces (SMS) method, can incorporate reflection, TIR, and refraction into the design of a single optic to provide desired wave fronts upon emission [20]. The SMS design method is described in detail in Chapter 4. The ray propagation algorithms explained in the previous sections are used to explain the behavior of the individual ray paths, but the final design of the optic is based on the combination of the many ray paths. For all of these reasons and others, hybrid illumination optics are at the forefront of illumination engineering, and will likely prove for the near future to be the selected optic for demanding illumination requirements. Hybrid illumination optics are discussed in more depth in Chapter 7.

1.9 THE PROCESS OF ILLUMINATION SYSTEM DESIGN

There are many steps that go into the design and fabrication of an optical system, in particular an illumination system. Figure 1.8 shows a flowchart for the steps of the process. The primary steps are divided into two phases: design and fabrication. The solid arrows inline with the steps shows the natural progression of the steps, while the exterior solid lines show potential iteration paths in order to ensure that the built system meets requirements. The path on the left-hand side of the primary path shows the potential iterations involved in the design phase, while the path on the right-hand side of the primary path shows the potential iterations initiated in the fabrication phase. The dashed lines show optional paths to circumvent a step in the primary path. A description of each step is provided here:

Design Phase

- *Concept.* The concept for the system is determined, including the type of optics that can potentially be used, goals, and a description of the system.
- *Baseline.* The design specifications, including efficiency, distribution, color, cost, and volume. Additionally, an étendue analysis is done in order to determine what source is required to meet the goals. Note that it is acceptable practice to set design specifications higher than required in order to alleviate tolerance and fabrication concerns.
- *Literature.* The designer investigates the literature taking into account the goals. Sometimes, the type of optics to be employed can be determined during this step, for example, reflector optics coupled to a source. This step can essentially be ignored if the designer has experience with the application being pursued.
- *Initial Study.* This step is also called conceptual study. If the type of optics has not already been determined, this step studies the effectiveness of various system configurations. At the end of this step, the chosen path and configuration for future design is selected.
- *Design.* The illumination optics are developed. In this step, the important optimization and tolerance parameters are determined. At the end of this step, the untoleranced system is at a minimum close to meeting design goals.
- *Optimization.* An optimization or perturbation routine is employed to improve the performance of the system. The optimization parameters are determined during the previous design steps. At the end of this step, the untoleranced system exceeds design specifications.

Fabrication Phase

- *Tolerance.* The important tolerance parameters are determined during the design steps. Two types of tolerancing are done: parameter sensitivity and Monte Carlo analysis. For parameter sensitivity, each of the important parameters is perturbed over a prescribed range to ascertain system performance. For Monte Carlo analysis, all of the important parameters are perturbed randomly in order to map out the performance of realistic systems. Unfortunately, this step is often ignored by many designers due to time constraints and a perception that the fabricated system will easily meet requirements. However, this neglect often leads to the major reason that manufactured illumination systems do not meet specifications.
- *Fabrication.* The components of the illumination system are integrated into a CAD layout for the complete system, and then the individual components are manufactured.
- *Testing.* There are two potential aspects to this step: component testing and system testing. For component testing, which is an optional substep, individual parts of the complete illumination system are tested to ensure they conform to the design specification. The parts are then integrated into the systems.

Finally, the system is tested. Comparison between the measurements and design results ensures the system meets requirements.

Note that the design iteration path only includes steps within the design phase, but the fabrication iteration path includes steps within both phases. The reasoning is that during the fabrication phase, it may be determined that the base design is not suitable for the application due to tolerance or fabrication demands. In this event, the design phase may have to be reopened to obtain a system that is able to meet the goals.

Figure 1.8 provides a fairly complete description of the steps involved in the successful completion of an illumination system; however, certain designs may require less or more steps. Ultimately, this determination is based on the complexity of the system, experience of the designer, level of system requirements, and evolving

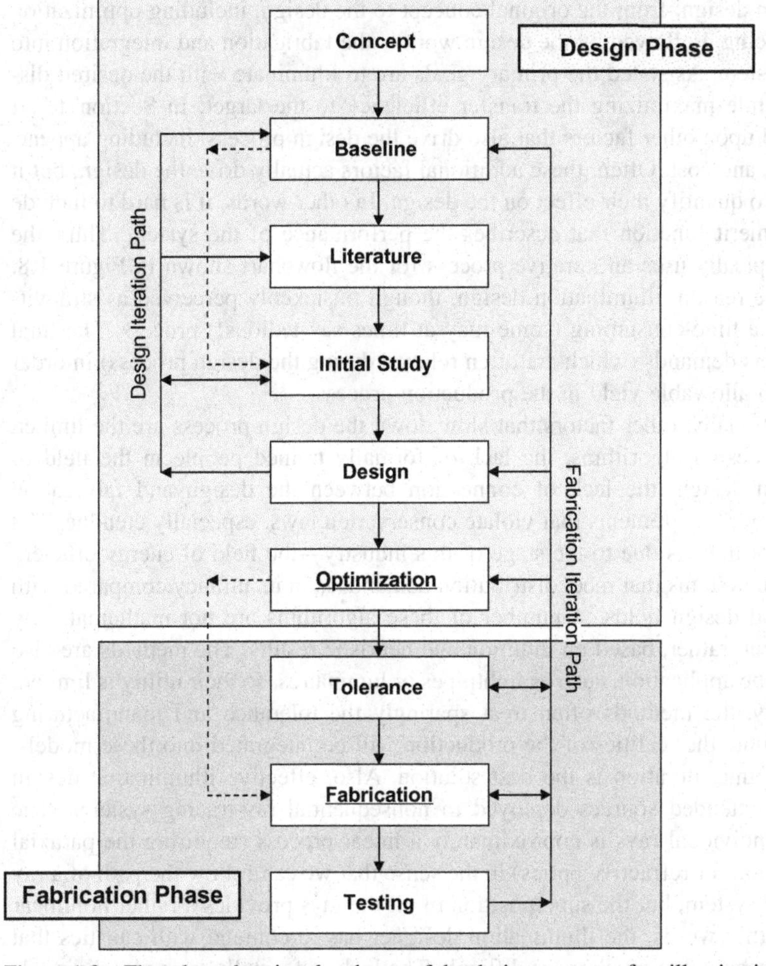

Figure 1.8 Flowchart showing the phases of the design process of an illumination system. The Optimization and Fabrication steps indicate a path that many designers do at their peril.

metrics and goals. Additionally, in order to address fabrication realities, a novel method of integrating design, optimization, and tolerancing into a single large step is employed. Essentially, this implies three phases of the system development process: concept phase, design phase, and fabrication phase. The concept phase would include the concept, baseline, literature, and initial study steps.

Now we introduce a number of terms, especially the radiometric and photometric units, in order to establish a basis for illumination optics design, analysis, and measurement. The understanding of the units is imperative in order to make more effective illumination system designs.

1.10 IS ILLUMINATION ENGINEERING HARD?

As presented in Sections 1.1, 1.8, and 1.9, there are a number of factors that comprise illumination design: from the original concept to the design, including optimization and tolerancing. Followed by the design work is the fabrication and integration into the final system. As stated the primary goals are to illuminate with the desired distribution while maximizing the transfer efficiency to the target. In Section 1.7, I commented upon other factors that also drive the design process, including appearance, color, and cost. Often, these additional factors actually drive the design, but it is difficult to quantify their effect on the design. In other words, it is hard to include them in a merit function that describes the performance of the system. Thus, the designer typically uses an iterative process for the flowchart shown in Figure 1.8. For this one reason, illumination design, though mistakenly perceived as straightforward, is a time-consuming (some may at times say tedious!) process. The final system meets demands (which are often relaxed during the design process) in order to obtain an allowable yield in the production process.

Additionally, other factors that slow down the design process are the limited number of design algorithms, the lack of formally trained people in the field of illumination design, the lack of connection between the design and fabrication industries, and requirements that violate conservation laws, especially étendue. The lack of algorithms is due to the stage of this industry—the field of energy efficient illumination systems that meet distribution demands is in its infancy compared with other optical design fields. A number of these algorithms are not mathematically intensive, but, rather, based on intuition and heuristic results. The methods are also specific to the application, such as lightpipes or luminaires, so their utility is limited. Additionally, the methods often treat sparingly the tolerance and manufacturing issues. In time, the realities of the production will be integrated into these models, but at this time, iteration is the best solution. Also, effective illumination design deals with extended sources deployed in nonsequential ray-tracing systems. The tracing of individual rays is approximately a linear process (requiring the paraxial approximation for refractive optics) in the sense that we can follow the path of a ray through the system, but the superposition of all the rays provides a rather nonlinear result. In other words, the illumination designer has to contend with caustics that make the process that much more difficult. The lack of formally trained individuals in the field is primarily due to the infancy and perception of the field. The perception is based on the world around us in which lighting is seen to be a commodity and thus inexpensive. However, a sizable amount of energy is wasted in the lighting industry, from unwanted light such as glare, inefficient sources, and solutions that provide look rather than form. As energy costs continue to rise and an increasing need to tailor illumination, more individuals will be formally trained in the field of illumination design. All industries contend with a disconnect between the designers and the fabricators. The illumination industry is especially susceptible to this due to the lack of formally trained engineers, the infancy of the industry, and the lack of terminology to describe tolerance and fabrication demands. Continued investment in the connection between these two arms of the illumination industry will reduce the system design costs. Finally, every illumination designer must at one time in their career try to violate the conservation of étendue, which is explained in the next chapter. Étendue and its conservation is the driving physics in the design of illumination systems, and by attempting to violate it, the designer will actually learn its properties and consequences. The increasing demands on illumination systems mean that we are often approaching the limit of étendue, but note that it is hard to calculate the étendue limit for a system because of the realistic emission characteristics of the source. Also, it is difficult to convey the concept of conservation of étendue to those not trained in the field of illumination design. One goal of this book is to assist the new illumination designer to the concept of étendue and its limitations while also providing some information to educating the untrained, especially your boss or customer!

I end this section with the observation that it does not fail to surprise me the amount of time in the optical design phase that must be dedicated to the appearance of the system, the volume limitations, and the yield requirements. On every project, I learn something new that increases my experience in the field of illumination design. So, at this time, the field of illumination design tends to be a rather hard discipline requiring the user to gain experience through trial and error, intuition, and the growing amount of literature in the field.

1.11 FORMAT FOR SUCCEEDING CHAPTERS

In Chapter 2, the important illumination topics of étendue, its conservation, concentration, acceptance angle, and skew invariant are introduced. In Chapter 3, the concept of étendue is expanded by discussing the "squeezing" of it. Étendue squeezing is a technique to address dilution and mixing within the optical design process. In Chapter 4, the SMS method is presented. Examples are presented to illustrate this design method. Chapter 5 looks at the application of nonimaging optics in the field of solar energy concentrators. Novel methods including those of Chapter 4 are used to design efficient, limited tracking solar concentrators. Chapter 6 gives a detailed treatment of lightpipe design, analysis, and modeling methods. Chapter 7 provides information about the successful use of optimization and tolerancing methods in the design of nonimaging optics.

REFERENCES

1. R.J. Koshel, Why illumination engineering? in *Nonimaging Optics and Efficient Illumination Systems IV, SPIE Proc.* 6670, 667002 (2007). Bellingham, WA.
2. CIE, *Commission Internationale de l'Éclairage Proceedings 1924*, Cambridge University Press, Cambridge, UK (1926).
3. CIE, *Bureau Central de la CIE* **1**, Section 4, **3**, 37 (1951).
4. W. Crummett and A.B. Western, *University Physics: Models and Applications*, Wm. C. Brown Publishers, Dubuque, IA (1994).
5. R.W. Boyd, *Radiometry and the Detection of Optical Radiation*, John Wiley and Sons, New York (1983).
6. E. Hecht, *Optics*, 4th ed., Addison Wesley, Reading, MA (2001).
7. R. Winston, J.C. Miñano, and P. Benítez, *Nonimaging Optics*, Elsevier Academic Press, Burlington, MA (2005).
8. J.M. Palmer, Getting Intense about Intensity, *Optics and Photonics News* **6**(2), 6 (1995) or see the website http://www.optics.arizona.edu/Palmer/intenopn.html (last accessed 9 September 2012).
9. NIST, *The International System of Units (SI)*, NIST Special Publication 330, 2001 ed., B.N. Taylor, ed., NIST, Gaithersburg, MD (2001).http://physics.nist.gov/Pubs/pdf.html or http://physics.nist.gov/cuu/Units/bibliography.html for downloadable version of this and other NIST publications in the area of units and measurements.
10. D. Malacara, *Color Vision and Colorimetry: Theory and Applications*, SPIE, Bellingham, WA (2002).
11. W.T. Welford and R. Winston, *High Collection Nonimaging Optics*, Academic Press, Inc., San Diego, CA (1989).
12. R. Leutz and A. Suzuki, *Nonimaging Fresnel Lenses: Design and Performance of Solar Concentrators*, Springer, Berlin (2001).
13. R. Winston, J.C. Miñano, and P. Benítez, *Nonimaging Optics*, Elsevier Academic Press, Burlington, MA (2005).
14. J. Chaves, *Introduction to Nonimaging Optics*, CRC Press, Boca Raton, FL (2008).
15. R.J. Koshel, Non-edge ray reflector design (NERD) for illumination systems, *Novel Optical Systems Designs and Optimization II, SPIE Proc.* 4092, (2000), p. 71. Bellingham, WA.
16. R.J. Koshel and I.A. Walmsley, Non-edge ray design: improved optical pumping of lasers, *Opt. Eng.* **43**, 1511 (2004).
17. S.L. Jacques and L.-H. Wang, Monte Carlo modeling of light transport in tissues, in A.J. Welch and M.J.C. van Gemert, eds., *Optical Thermal Response of Laser Irradiated Tissue*, Plenum Press, New York (1995), pp. 73–100.
18. A.N. Yaroslavsky, I.V. Yaroslavsky, T. Goldbach, and H.-J. Schwarzmaier, Influence of the scattering phase functionapproximation on the optical properties of blood determined from the integrating sphere measurements, *J. Biomed. Opt.* **4**, 47–53 (1999).
19. J.C. Stover, *Optical Scattering*, 2nd ed., SPIE Press, Bellingham, WA (1995).
20. P. Benítez, J.C. Miñano, J. Blen, R. Mohedano, J. Chaves, O. Dross, M. Hernández, and W. Falicoff, Simultaneous multiple surface optical design method in three dimensions, *Opt. Eng.* **43**, 1489–1502 (2004).

CHAPTER 2

ÉTENDUE

John Koshel

Étendue is one of the most basic yet important concepts in the design of nonimaging and illumination optics. First, it explains the flux transfer characteristics of the optical system, and, second, it plays an integral role in the ability to shape the distribution of radiation at the target. Interestingly, the concept of étendue was only formally accepted in the 1970s through a series of letters to a journal [1–4]. Reference [1] asked for input to a proposal to be submitted to the Nomenclature Sub-Committee of the International Commission for Optics in order to standardize the name for what ultimately became étendue [5]. Initially, the author of this journal letter favored the term "optical extent," with the term étendue not likely due to "the typing disadvantage of é, (and) the pronunciation difficulty of the French u" [1]. Through the series of replies, both in the journal and by private communications, the term étendue gained favor "since the accent will be dropped anyway" [4]. The dropping of the accent is not applicable today since modern computer technology makes it simple to include. This recent history of varied opinions and potential confusion illustrates the complexity inherent in the term étendue, indicating that it has the potential to be interpreted in many ways. In fact, after defining étendue and illustrating its innate conservation in optical systems, we present a series of terms that are analogous or often confused with étendue.

In this chapter, a series of proofs and examples show that the flux transfer characteristic is an integral characteristic of étendue through its inherent conservation. The shaping of the distribution from the source to the target is also an inherent part of étendue, but it is often better viewed through the first derivative of étendue—the skewness. Like étendue, skewness is also invariant through an optical system of rotational symmetry. We end this chapter by presenting a discussion of the limitations of étendue, especially in the physical domain of optics, and thus present an alternate expression for development.

Note that for this chapter, the radiometric terms of irradiance, radiance, and radiant intensity or simply intensity are used. The photometric quantities of illuminance, luminance, and luminous intensity, respectively, can be substituted. Additionally, we use imaging terms such as "entrance pupil" or "exit pupil" in order to remain in agreement with the imaging community. We find that using such expressions from the imaging community can alleviate potential confusion to those new to the field

Illumination Engineering: Design with Nonimaging Optics, First Edition. R. John Koshel.
© 2013 the Institute of Electrical and Electronics Engineers. Published 2013 by John Wiley & Sons, Inc.

of nonimaging optics. Alternate terms of entrance aperture or exit aperture can be used instead.

2.1 ÉTENDUE

Étendue is a French word that means, as a verb, extended, and, as a noun, reach [6]. In the field of optics étendue is a quantity arising out of the geometrical characteristics of flux propagation in an optical system. As will be seen, the use of étendue in the field of optics is in good agreement with the French root. It describes both the angular and spatial propagation of flux through the system, so it obviously relates to the radiance propagation characteristics of a system. In a lossless system, which is one without absorption, scatter, gain, or Fresnel reflection losses, all flux that is transmitted by the entrance pupil of the system is emitted from the exit pupil. The étendue of a system is defined as [7, 8]

$$\xi = n^2 \iint_{\text{pupil}} \cos\theta dA_s d\Omega, \tag{2.1}$$

where n is the index of refraction in source space, and the integrals are performed over the entrance pupil. The total flux that propagates through this system is found from $L(\mathbf{r}, \hat{\mathbf{a}})$, which is the source radiance at the point $\mathbf{r}$ in the direction of the unit vector $\hat{\mathbf{a}}$. By integrating, we obtain the flux,*

$$\begin{aligned} \Phi &= \iint_{\text{pupil}} L(\mathbf{r}, \hat{\mathbf{a}}) dA_{s,\text{proj}} d\Omega \\ &= \iint_{\text{pupil}} L(\mathbf{r}, \hat{\mathbf{a}}) \cos\theta dA_s d\Omega, \end{aligned} \tag{2.2}$$

where the spatial integral is over the source area, which is in view of the entrance pupil, and the angular integral is over the source emission that is within the field of view of the pupil, that is, that subtended by the pupil from the source point $\mathbf{r}$. If a spatially uniform, Lambertian source is assumed, then the source radiance function is given by L_s, which upon substitution in Equation (2.2) gives

$$\Phi = L_s \iint_{\text{pupil}} \cos\theta dA_s d\Omega. \tag{2.3}$$

So, the total flux that is transmitted by an optical system with a spatially uniform, Lambertian emitter in terms of the étendue is given by

$$\Phi = \frac{L_s \xi}{n^2}. \tag{2.4}$$

* The notation of Reference [7] is used for this development.

Note that the étendue is a geometric quantity that describes the flux propagation characteristics for a lossless system. The term lossless puts constraints on the interpretation of a system étendue. The term implies that absorption, scatter, and reflection losses are not considered, but, more importantly, it restricts the integration of Equations (2.1–2.3) to that of the entrance pupil criterion. In realistic systems, there are not only absorption, aberration, diffraction, and reflection losses, but also losses associated to manufacturing tolerances and lack of capture of source flux by the entrance pupil. Therefore, the geometric étendue provides a theoretical limit to flux transfer capability of a real-world optical system.

As an example, consider a planar emitter oriented such that it is viewed directly by the entrance pupil, and it is of a suitably small size, A_s, such that the solid angle subtended at the entrance pupil does not vary appreciably over its extent. The solid angle subtense is given by the half-cone angle of θ_a. This angle is called the acceptance angle and denotes the half-angle accepted by the entrance pupil and transmitted with no loss out of the exit pupil. The acceptance angle is presented in more depth in later sections. Using the expression for the differential solid angle given in Chapter 1, the étendue for this system is

$$\xi = n^2 A_s \int_0^{2\pi}\int_0^{\theta_a} \cos\theta \sin\theta d\theta d\phi = \pi n^2 A_s \sin^2 \theta_a. \tag{2.5}$$

To this point, the étendue has been described as a property of an optical system, such as a camera lens or concentrator; however, the definition can be extended to consider the étendue of a source. The source étendue describes the geometric flux emission characteristics of the emitter by itself, and it provides a fundamental limit on the ability of a following optical system to collect and transmit this flux to the intended target. As discussed in Chapter 1, imaging systems are étendue limited due primarily to aberration considerations as the f/# is reduced. In other words, for realistic sources, such as tungsten bulbs and LEDs, imaging systems are constrained in their capability to capture the flux from the source and transmit it without loss to the exit pupil. Nonimaging systems, or more generically, illumination systems, use the quantity of étendue to drive their design such that loss is greatly reduced. In fact, as will be seen in the following chapters, there are theoretical constructs that provide no loss, thus conserving étendue, which is broached in generic terms in the next subsection.

2.2 CONSERVATION OF ÉTENDUE

We have alluded to the fact that étendue is conserved in a lossless system. In this section, a few proofs to the veracity of this claim are made. First, we show conservation of étendue through conservation of radiance and energy. Second, we show conservation of the geometric quantity, generalized étendue, with the use of the eikonal. Third, a proof based on the law of thermodynamics is presented [7]. The reasoning behind presenting multiple proofs is not only to show that étendue and conservation are fundamental concepts of physics, but also that different

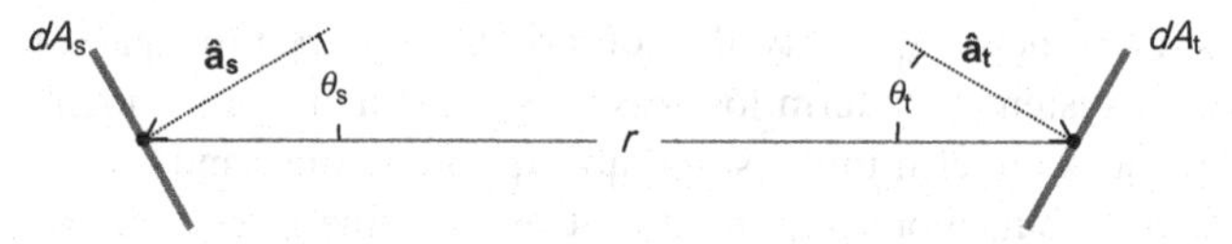

Figure 2.1 Representation of the source (dA_s) and target (dA_t) in a lossless optical system to show the conservation of radiance upon propagation in a homogeneous medium.

interpretations of the physics provides the same result. Simply said, it is imperative that the readers understand the nuances of étendue and its conservation in order to be able to effectively design illumination systems. It is anticipated that at least one of these proofs will be more familiar to a new practitioner in the fields of nonimaging and illumination optics, thus making their comprehension better. Finally, a number of other authors have presented these proofs in disparate publications, so the presentation herein brings these three treatments into one place. Note that additional proofs can be developed using Liouville's theorem in statistical mechanics [9] and Stokes' theorem applied to Hamiltonian systems [10].

A simple interpretation of failing to conserve étendue means that by "giving up" étendue, one reduces the flux-transmission capabilities of an optical system. The important point is that conservation of étendue provides the theoretical limit on the capabilities of a system to transfer the flux from a source to the target. Additionally, the étendue conservation equation can be used to find the shapes of optical systems that do conserve flux for a given source.

2.2.1 Proof of Conservation of Radiance and Étendue

First, we investigate the conservation of radiance with propagation in a uniform, lossless medium. Figure 2.1 represents the transfer of flux from the source (dA_s) to the target (dA_t). The source and target are oriented at θ_s and θ_t, respectively, to the centroid joining the two elemental areas. The differential solid angle from the source subtended by the target area is given by $d\Omega_s$, and the analogous target solid angle is given by $d\Omega_t$. The ratio of the two radiances, L_s and L_t, is

$$\frac{L_s}{L_t} = \frac{d^2\Phi_s / \cos\theta_s dA_s d\Omega_s}{d^2\Phi_t / \cos\theta_t dA_t d\Omega_t}. \tag{2.6}$$

Note that $\Phi_s = \Phi_t$ in order to maintain the lossless condition. Canceling the flux differentials and substituting for the differential solid angles as defined in Chapter 1, we obtain

$$\frac{L_s}{L_t} = \frac{\cos\theta_t dA_t d\Omega_t}{\cos\theta_s dA_s d\Omega_s} = \frac{\cos\theta_t dA_t \left(dA_s \cos\theta_s / r^2\right)}{\cos\theta_s dA_s \left(dA_t \cos\theta_t / r^2\right)} = 1, \tag{2.7}$$

where r is the separation of the elemental objects. Equation (2.7) is then rewritten as

$$L_s = L_t, \tag{2.8}$$

which proves that radiance is conserved as it propagates through a homogeneous medium.

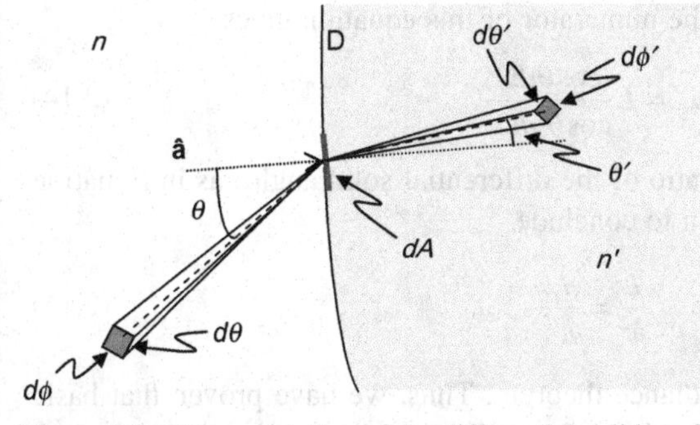

Figure 2.2 Representation of the differential solid angles at an index of refraction transition in order to validate that basic radiance is invariant with index change.

We next investigate what is called the basic radiance [7], L/n^2, as it propagates through a lossless system composed of uniform media, with distinct boundaries separating these media. Equation (2.8) shows that radiance is conserved in any single medium, but what happens as the flux is incident on an index boundary? Figure 2.2 shows the layout of the propagation of a beam element of radiance L and differential solid angle subtense of $d\Omega$ that is incident on an elemental area of dA located on the boundary between the two media. This beam is incident at a polar angle of θ and azimuthal angle of ϕ with respect to the unit normal to dA, $\hat{\mathbf{a}}$. The source (object) medium has index n, while the target (image) medium has index n'. The differential flux carried by this beam is

$$d^2\Phi = L\cos\theta dA d\Omega. \tag{2.9}$$

Upon refraction at this surface, the centroid polar and azimuthal angles of the differential solid angle subtense, $d\Omega'$, are given by θ' and ϕ' respectively. The differential parts of the polar and azimuthal angles are prior to refraction, $d\theta$ and $d\phi$, and after refraction, $d\theta'$ and $d\phi'$. A condition of Snell's law is that a refracted ray stays in its respective plane of incidence, so

$$d\phi = d\phi'. \tag{2.10}$$

Differentiating Snell's law with respect to the polar angle, θ, gives

$$n\cos\theta d\theta = n'\cos\theta' d\theta'. \tag{2.11}$$

Substituting Equations (1.5), (2.10), and (2.11) into the ratio of the two differential solid angle, $d\Omega$ and $d\Omega'$, results in

$$\frac{d\Omega}{d\Omega'} = \frac{\sin\theta d\theta d\phi}{\sin\theta' d\theta' d\phi'} = \frac{n'^2\cos\theta'}{n^2\cos\theta}. \tag{2.12}$$

The radiance emanating from the elemental area dA after refraction with the condition of no loss (i.e., flux is conserved) is

$$L' = \frac{d^2\Phi}{\cos\theta' dA d\Omega'}. \tag{2.13}$$

Substituting Equation (2.9) into the numerator of this equation gives

$$L' = L\frac{\cos\theta d\Omega}{\cos\theta' d\Omega'}. \tag{2.14}$$

Finally, we substitute in for the ratio of the differential solid angles as in Equation (2.12) and reorganize the equation to conclude

$$\frac{L}{n^2} = \frac{L'}{n'^2}. \tag{2.15}$$

Equation (2.15) is labeled the radiance theorem. Thus, we have proven that basic radiance is invariant upon the transition between two mediums of indices n and n'. This equation is equally valid for reflection since, as noted in Chapter 1, reflection is a special case of Snell's law when $n' = -n$. Note that radiance is conserved within gradient-index media by using a piecewise approach. In each propagation step, the index variation is infinitesimally small, such that radiance is conserved over the small step as per Equation (2.8). Then, the basic radiance is conserved with a change in the index of refraction as per Equation (2.15).

Finally, now that we have proven that radiance is conserved through propagation in arbitrary media and at index transitions, we can study the conservation of étendue. Using Equations (2.4) and (2.9) to express the differential flux in terms of the differential étendue,

$$d^2\Phi = \frac{L_s}{n^2} d^2\xi. \tag{2.16}$$

We have shown with Equation (2.15) that the basic radiance, L_s/n^2, is invariant, and that in a lossless system, the differential flux, $d^2\Phi$, must be conserved due to the law of conservation of energy. Therefore, the differential étendue, $d^2\xi$, must also be invariant. The total étendue for an optical system is found by integrating over the pupil,

$$\xi = \iint_{\text{pupil}} d^2\xi, \tag{2.17}$$

which must be conserved. Equations (2.16) and (2.17) prove that conservation of étendue is realized in a lossless system that transmits radiation accepted by the entrance pupil of the optical system.

Additionally, another way to think about the two equations, Equations (2.8) and (2.15), are that they represent the étendue equivalents of the transfer and refraction equations, respectively, of ray tracing in an imaging system. Equation (2.8) transfers the input radiation from one surface to the next, while Equation (2.15) handles the refraction at a boundary.

2.2.2 Proof of Conservation of Generalized Étendue

The previous subsection used the conservation of radiance to prove that étendue is conserved; however, as noted previously, étendue is a geometric quantity, thus there

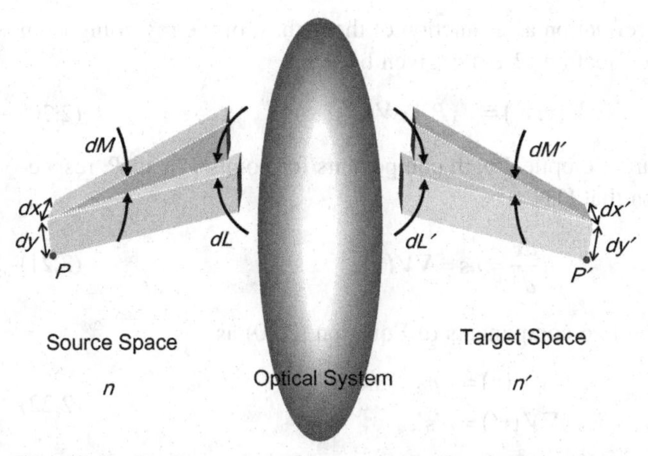

Figure 2.3 Representation of beam propagation from source space (**r**) to target space (**r′**) via an intermediate optical system.

must be a proof that étendue in the geometric or general sense is conserved. The proof of generalized étendue, which is attributable to arbitrary systems not constrained by the paraxial approximation, symmetric to asymmetric optical systems, and any type of component (e.g., refractive, reflective, and gradient), is shown here. As with the previous proof, it can be applied to sources.

Figure 2.3 shows the configuration of a generic system, with the source space of index n labeled in the coordinates $\mathbf{r} = (x, y, z)$ and the target space of index n' denoted in coordinates $\mathbf{r'} = (x', y', z')$. We investigate the propagation of a ray that goes from point P in source space to P' in target space via the intermediate optical system. We then extend our study to include the propagation of a beam of differential area dA to differential area dA'. Additionally, the directions of the source and image beams are extended to subtend that of a differential angular extent of $dLdM$ and $dL'dM'$, respectively. Note that the L and M terms are the standard direction cosines. The quantity that is invariant through the optical system of Figure 2.3 is

$$dxdydpdq = dx'dy'dp'dq', \tag{2.18}$$

where the dp and dq terms are the differential optical direction cosine terms, which are equivalent to ndL and ndM, respectively. This equation expresses the (optical) phase space volume and its conservation within the optical system shown in Figure 2.3. Equation (2.18) thus provides the final form for the invariance of the generalized étendue with propagation through an optical system. The purpose of this section is to prove the veracity of Equation (2.18).

Starting with the point characteristic, V [11], which describes the optical path length of the ray from P to P', we have the form as a function of the coordinates

$$V(\mathbf{r};\mathbf{r'}) = \int_{P}^{P'} nds, \tag{2.19}$$

where n is the index of refraction as a function of the path, s, of the ray going from P to P'. The solution to Equation (2.19) is given by

$$V(\mathbf{r};\mathbf{r}') = V(P') - V(P), \tag{2.20}$$

where the V functions are the optical path components for points P' and P, respectively. It has been shown that [11]

$$n\frac{d\mathbf{r}}{ds} = n\mathbf{s} = \nabla V(\mathbf{r}). \tag{2.21}$$

Thus, we can express the two components of Equation (2.20) as

$$\begin{aligned} \nabla V(\mathbf{r}) &= -n\mathbf{s} \quad \text{and} \\ \nabla V(\mathbf{r}') &= n'\mathbf{s}', \end{aligned} \tag{2.22}$$

where the vectors $\mathbf{s}$ and $\mathbf{s}'$ denote the respective direction cosines in source space and target space. The individual optical direction cosine components in the transverse directions of Equation (2.22) may then be written

$$\begin{aligned} nL = p &= -\frac{\partial V}{dx} = -V_x, \\ nM = q &= -\frac{\partial V}{dy} = -V_y, \\ n'L' = p' &= \frac{\partial V}{dx'} = V_{x'}, \quad \text{and} \\ n'M' = q' &= \frac{\partial V}{dy'} = V_{y'}. \end{aligned} \tag{2.23}$$

The subscripts on the V terms denote partial derivatives with respect to the listed variables. Differentiating the individual equations in Equation (2.23) with respect to the propagation path, we find expressions for the differential optical direction cosines

$$\begin{aligned} dp &= -V_{xx}dx - V_{xy}dy - V_{xx'}dx' - V_{xy'}dy', \\ dq &= -V_{yx}dx - V_{yy}dy - V_{yx'}dx' - V_{yy'}dy', \\ dp' &= V_{x'x}dx + V_{x'y}dy + V_{x'x'}dx' + V_{x'y'}dy', \quad \text{and} \\ dq' &= V_{y'x}dx + V_{y'y}dy + V_{y'x'}dx' + V_{y'y'}dy'. \end{aligned} \tag{2.24}$$

Rewriting Equation (2.24) to separate the source space terms ($\mathbf{r}$) and the target space terms ($\mathbf{r}'$), and then using matrices to denote the set of equations

$$\begin{pmatrix} V_{xx} & V_{xy} & 1 & 0 \\ V_{yx} & V_{yy} & 0 & 1 \\ V_{x'x} & V_{x'y} & 0 & 0 \\ V_{y'x} & V_{y'y} & 0 & 0 \end{pmatrix} \begin{pmatrix} dx \\ dy \\ dp \\ dq \end{pmatrix} = \begin{pmatrix} -V_{xx'} & -V_{xy'} & 0 & 0 \\ -V_{yx'} & -V_{yy'} & 0 & 0 \\ -V_{x'x'} & -V_{x'y'} & 1 & 0 \\ -V_{y'x'} & -V_{y'y'} & 0 & 1 \end{pmatrix} \begin{pmatrix} dx' \\ dy' \\ dp' \\ dq' \end{pmatrix}. \tag{2.25}$$

If **A** is the square matrix on the left-hand side, **B** is the square matrix on the right-hand side, and **w** and **w′** are the column vectors on the left- and right-hand sides, respectively, then

$$\mathbf{A}\mathbf{w} = \mathbf{B}\mathbf{w}'. \tag{2.26}$$

We desire to find **w′** as a function of **w**, thus multiplying each side by the inverse of **B**, $\mathbf{B}^{-1}$,

$$\mathbf{w}' = \mathbf{B}^{-1}\mathbf{A}\mathbf{w}. \tag{2.27}$$

Equation (2.27) represents a change of variables from the source space coordinates $(x, y; p, q)$ to the target space coordinates $(x', y'; p', q')$. The Jacobian matrix, **J**, provides the equality during this transformation, such that

$$\mathbf{w}' = \mathbf{J}\mathbf{w}, \tag{2.28}$$

where

$$\mathbf{J} = \mathbf{B}^{-1}\mathbf{A}. \tag{2.29}$$

The change of variables then provides,

$$dx'dy'dp'dq' = |\mathbf{J}|\,dxdydpdq = \frac{|\mathbf{A}|}{|\mathbf{B}|}dxdydpdq, \tag{2.30}$$

where $|\mathbf{J}|$ represents the determinant of the Jacobian matrix, or simply called the Jacobian. In order to prove Equation (2.18), we need to show that $|\mathbf{J}| = 1$ in Equation (2.30). We find the determinants of the component matrices of **J**:

$$|\mathbf{A}| = \begin{vmatrix} V_{xx} & V_{xy} & 1 & 0 \\ V_{yx} & V_{yy} & 0 & 1 \\ V_{x'x} & V_{x'y} & 0 & 0 \\ V_{y'x} & V_{y'y} & 0 & 0 \end{vmatrix} = V_{x'x}V_{y'y} - V_{x'y}V_{y'x}, \tag{2.31}$$

and,

$$|\mathbf{B}| = \begin{vmatrix} -V_{xx'} & -V_{xy'} & 0 & 0 \\ -V_{yx'} & -V_{yy'} & 0 & 0 \\ -V_{x'x'} & -V_{x'y'} & 1 & 0 \\ -V_{y'x'} & -V_{y'y'} & 0 & 1 \end{vmatrix} = V_{xx'}V_{yy'} - V_{xy'}V_{yx'}. \tag{2.32}$$

The order of the differentiation in Equations (2.31) and (2.32) can be changed, thus the determinants for **A** and **B** are equal, such that $|\mathbf{J}| = 1$ and Equation (2.18) is proven. Thus, étendue in the generalized sense is invariant upon propagation through a lossless system. This equality is based upon the use of Clairaut's theorem [12], which says that the second-order mixed partial derivatives are equal if these partials of the function, V in this case, are continuous in the region of interest. If there is a discontinuity in the second partial derivatives at a point within the neighborhood of interest, then the matrix is said to be nonsymmetric, which means the second-order

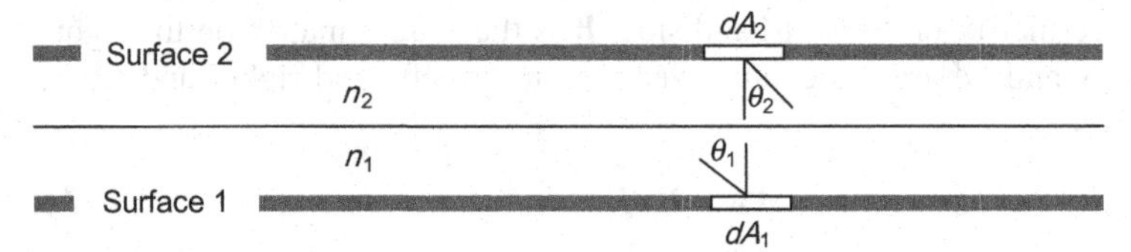

Figure 2.4 Two infinite, parallel plates in thermal equilibrium to illustrate the conservation of étendue using the laws of thermodynamics.

partial derivatives are not commutative. In this scenario, conservation of generalized étendue is not ensured; however, counterexamples to the validity of conservation are nongeometric. Such examples introduce loss since it essentially means there are multiple paths for the point characteristic V through the optical system (i.e., there is a discontinuity or degeneracy).

2.2.3 Conservation of Étendue from the Laws of Thermodynamics

In this section, we present conservation of étendue from the basic laws of thermodynamics, in particular the zeroth law (e.g., thermal equilibrium) and the first law (conservation of energy). Starting with Figure 2.4, we have two infinite, planar surfaces with the lower one (surface 1) in index n_1 and the upper one (surface 2) in index n_2 with the stipulation that $n_1 < n_2$ [7, 13]. Additionally, all surfaces, including the dielectric transition, are parallel to one another. The radiance of the bottom surface is L_1 and the radiance of the upper surface is L_2. An elemental area of dA_1 emits into an elemental annular solid angle of $d\theta_1$ at angle θ_1. Therefore, the radiated flux is given by

$$d^2\Phi_1 = 2\pi L_1 dA_1 \cos\theta_1 \sin\theta_1 d\theta_1. \tag{2.33}$$

Likewise, this treatment is done for surface 2, such that

$$d^2\Phi_2 = 2\pi L_2 dA_2 \cos\theta_2 \sin\theta_2 d\theta_2. \tag{2.34}$$

These equations are simply expansions of Equation (2.9); however, we will incorporate the laws of thermodynamics to arrive at our answer.

A fraction of the flux from surface 1 is reflected by the dielectric interface and then absorbed by surface 1. This assumption implies that there is perfect absorption by surface 1, which is attainable by a blackbody. An outcome of this blackbody assumption means that the radiance L_1 is independent of angle. The amount of reflection is $R_1(\theta_1)$, where R_1 as a function of the angle of incidence is the reflectivity found with the Fresnel equations. Thus, flux proportional to $1 - R_1(\theta_1)$ goes from surface 1 to surface 2. As with surface 1, there is an assumption that all of the incident flux on surface 2 is absorbed, so the second surface is also a blackbody. In the antithetical direction, we have a reflectivity as a function of angle of $R_2(\theta_2)$, so the flux amount absorbed at the first surface is proportional to $1 - R_2(\theta_2)$.

Using the zeroth and first laws of thermodynamics, we have a system in equilibrium, such that the flux transferred from one surface to the other must be equal to the flux flow in the other direction. For thermal equilibrium, one uses $\Phi_{1,\text{in}} = \Phi_{1,\text{out}}$ and $\Phi_{2,\text{in}} = \Phi_{2,\text{out}}$. Using the reflectivity and transmissivity terms, one obtains for both surfaces, upon collecting like terms

$$[1-R_1(\theta_1)]d^2\Phi_1 = [1-R_2(\theta_2)]d^2\Phi_2. \tag{2.35}$$

Equation (2.35) is an expression of equilibrium, while the fact that both surfaces see the same net energy flow means that there is conservation of energy. Next, substitute in Equations (2.33) and (2.34) and integrate over the areal and angular variables. Upon integration and using the fact that $A_1 = A_2 = \infty$ such that the areal components can be removed

$$L_1 \int_0^{\pi/2} [1-R_1(\theta_1)]\sin\theta_1 \cos\theta_1 d\theta_1 = L_2 \int_0^{\arcsin(n_1/n_2)} [1-R_2(\theta_2)]\sin\theta_2 \cos\theta_2 d\theta_2. \tag{2.36}$$

Note that the upper limit for the right-hand side is restricted by the condition of total internal reflection at the critical angle. Snell's law, $n_1\sin\theta_1 = n_2\sin\theta_2$, is used on the right-hand side to arrive at

$$L_1 \int_0^{\pi/2} [1-R_1(\theta_1)]\sin\theta_1 \cos\theta_1 d\theta_1 = \frac{n_1^2}{n_2^2} L_2 \int_0^{\pi/2} [1-R_2(\theta_2)]\sin\theta_1 \cos\theta_1 d\theta_1. \tag{2.37}$$

There is reversibility for $R_1(\theta_1)$ and $R_2(\theta_2)$, such that $R_1(\theta_1) = R_2(\theta_2)$, which means rays from one direction to the other will follow the same path for rays going the opposite direction. Rewriting Equation (2.37),

$$L_1 \int_0^{\pi/2} [1-R_1(\theta_1)]\sin\theta_1 \cos\theta_1 d\theta_1 = \frac{n_1^2}{n_2^2} L_2 \int_0^{\pi/2} [1-R_1(\theta_1)]\sin\theta_1 \cos\theta_1 d\theta_1. \tag{2.38}$$

The integrands in Equation (2.38) do not vanish, thus we arrive at the radiance theorem of Equation (2.15). Conservation of étendue follows as per Equations (2.16) and (2.17).

2.3 OTHER EXPRESSIONS FOR ÉTENDUE

In the literature, there are many terms or expressions in use that imply étendue and its conservation. These many terms can cause confusion to the lay reader, so here we offer insight into the plethora of terms used to express étendue. We also discuss the applicability and nuances of each term.

2.3.1 Radiance, Luminance, and Brightness

First, due to the invariant nature of radiance through an optical system as shown in Section 2.2, the very term radiance is often synonymous with étendue. Though this use is understandable—we did show the conservation of étendue via conservation of radiance in Section 2.2.1—this use is incorrect. This limitation is easily evident due to the units of each quantity: étendue is in the units of area-solid angle (e.g.,

m^2-sr), and radiance has units of flux per projected area per solid angle (e.g., W/m^2-sr). The subtle difference is that étendue does not contain knowledge of the distribution of flux within the optical system, but rather, it solely represents the flux transmission capabilities over the entrance pupil, or, for that matter, any transverse cross section, of the optical system. Additionally, the radiance is described for a source, while the étendue describes the transmission characteristics of an optical system. A simple and convenient way to think of étendue and radiance for the purposes of designing illumination systems is:

- Radiance describes the flux emission characteristics of a source in space and angle.
- Étendue describes the aperture transmission function through an optical system.
- Upon transmission through the optical system, the radiance function can be modified in space and position by the étendue function, which provides a new radiance distribution for the propagated radiation.
- This new radiance distribution can then be propagated to the next aperture, such that an iterative procedure of a radiance function filtered by an étendue function to produce a new radiance distribution is performed until a target is struck, where the desired metric can be determined, such as flux transfer as per Equation (2.4).

This procedure lends itself well to first-order calculations to design optical systems based on the transfer of radiance through an étendue filter, especially those that conserve étendue. Because of this symbiotic relationship of étendue and radiance, the terms are often used interchangeably, but it is imperative that users of such terms understand the nuances described here. Thus, in order to reduce potential confusion, one should not use the terms for one another.

Luminance is the photometric analog of radiance; therefore, the same cautions that were provided previously are continued here. In Chapter 1, we discussed the term brightness, which is analogous to luminance and therefore radiance; however, it is dependent on an actual observer, not a standardized one. The inclusion of an actual observer brings perceptual issues into the discussion. Therefore, our cautions are magnified since perceptual issues increase the disconnect between étendue and brightness. The reader must ensure that the differences are understood, and that in order to alleviate any potential confusion, the terms not be used interchangeably.

2.3.2 Throughput

Throughput, often with the term optical placed in front of it, is a term often heard in the imaging design community. It describes the flux transmission capabilities of such a system, so throughput is directly tantamount with étendue. The only caveat is that throughput is almost always used in conjunction with imaging (i.e., lens) systems rather than with illumination systems. Its connection with imaging systems often constrains it to the paraxial domain. Additionally, some use throughput to describe the radiance distribution, rather than the étendue of the lens system. Note that the "optical throughput" typically indicates some form of the radiance, while the "throughput" signifies the étendue. In conclusion, throughput without inclusion of the radiance profile is an analog for étendue, and if the paraxial domain is implied, then it is the paraxial version of étendue.

2.3.3 Extent

In agreement with the French definition of étendue, the terms optical extent, geometrical extent, or extent by itself are used to describe the étendue of the system. As with throughput, the precursor of "optical" is usually synonymous with the radiance, while the geometric extent usually signifies some form of the étendue. Thus, the geometric extent is typically analogous to étendue, but as with the other cases described, caution on the use of these various terms is mandated. The reader must ensure what quantity is being expressed, if a paraxial approximation is implied, and if there are perceptual components.

2.3.4 Lagrange Invariant

The conservation of each of the aforementioned terms is also a common occurrence. However, specifically in the lens design community the optical or Lagrange invariant, H, is used for conservation within paraxial systems [14]

$$H = nhu = n'h'u', \tag{2.39}$$

between the object and image spaces, respectively, or, more generally,

$$H = n(\bar{y}u - \bar{u}y) = n'(\bar{y}u' - \bar{u}'y), \tag{2.40}$$

where the n terms are the refractive indicia, the y terms are the paraxial transverse displacements, and the u terms are the paraxial angles with respect to the optical axis. The "barred" terms (e.g., $\bar{u}$) represent chief-ray terms, while the "un-barred" terms (e.g., y) represent marginal-ray terms. Also, the "unprimed" terms (e.g., n) represent prior to refraction, while the "primed" terms (e.g., u') represent after refraction. Finally, h and h' in Equation (2.39) are the paraxial object and image heights, respectively. The invariance of this term through an imaging system is only one facet of its importance, since its square, H^2, is proportional to the flux transmission and information-carrying capability, that is, number of resolvable spots in an image [15]. Like étendue, the Lagrange invariant expresses the flux transmission capabilities of an optical system, and it is invariant, but it is a paraxial expression. Simply, it is the paraxial form of étendue, but its utility is limited because nonimaging optics are typically using angles well outside the limits of the paraxial approximation (i.e., input or output angles of 90°).

2.3.5 Abbe Sine Condition

In nonparaxial notation the Abbe sine condition is used [16],

$$nh\sin\theta = n'h'\sin\theta', \tag{2.41}$$

where the terms in this equation follow the syntax used for Equations (2.39) and (2.40), expect $u = \sin\theta$ and $u' = \sin\theta'$. The Abbe sine condition has other properties in imaging design since it describes a stigmatic system that is devoid of coma and spherical aberrations. Note the similarity of Equation (2.41) to those of the previous section, where the h terms represent the differential spatial terms (e.g., dx or dy in Eq. 2.18) and the $n\sin\theta$ terms represent the differential direction cosine terms (e.g., dp or dq in Eq. 2.18). Therefore, the Abbe sine condition is a direct expression for conservation of étendue, but it has the added meaning of no coma and spherical aberrations.

2.3.6 Configuration or Shape Factor

In the field of radiative heat transfer, the power transfer from one surface to another is given in integral form by [17–20]

$$\Phi_{1\to 2} = \int_{A_1}\int_{A_2} \frac{L_1(\theta,\phi)\cos\theta_1\cos\theta_2}{d^2} dA_1 dA_2 \underset{\text{Lambertian}}{\Rightarrow} L_1 \int_{A_1}\int_{A_2} \frac{\cos\theta_1\cos\theta_2}{d^2} dA_1 dA_2, \tag{2.42}$$

Where L_1 is the radiance, ϕ is the azimuthal angle, θ_1 and θ_2 are the orientations of the two surfaces with respect to the axis joining them, d is the distance between the two surfaces, and A_1 and A_2 are the areas of the two surfaces. The second form of the equation is for the special case when the radiance is Lambertian. In the field of radiative heat transfer, it is typical to assume the radiator emits as a blackbody or with a Lambertian radiance profile. The right-hand side of Equation (2.42) has two components: the radiometric part given by L_1 and the geometric part given by the double integral. Equation (2.42) is simply a different form of the étendue given by Equation (2.1), which is made by substituting for the differential solid angle $d\Omega = dA \cos\theta/d^2$ (see Chapter 1). The configuration or shape factor is defined as

$$F_{1\to 2} = \frac{\Phi_{1\to 2}}{\Phi_1}, \tag{2.43}$$

where $\Phi_{1\to 2}$ is as per Equation (2.42), and Φ_1 is the total power emitted by the first surface. Thus, the configuration factor is the ratio of the radiation emitted by one surface that is incident on the second surface. In the case of Lambertian emission from the first surface $\Phi_1 = M_1 A_1 = \pi L_1 A_1$, where M_1 is the radiant exitance of the first surface. Substituting the total power of the Lambertian first surface and Equation (2.42) into Equation (2.43), one obtains

$$F_{1\to 2} = \frac{1}{\pi A_1}\int_{A_1}\int_{A_2} \frac{\cos\theta_1\cos\theta_2}{d^2} dA_1 dA_2 = \frac{\xi_{1\to 2}}{n^2 \pi A_1} = \frac{\Omega_{1\to 2}}{\pi}, \tag{2.44}$$

where $\xi_{1\to 2}$ and $\Omega_{1\to 2}$ are the étendue and solid angle subtended by the first surface from the second surface, respectively. In the étendue expression, the denominator is the étendue of the hemispherical, Lambertian emission from the first surface. In the solid angle expression, the denominator is the solid angle emission from the first surface. Therefore, in all cases, the configuration factor is a ratio of the radiation, étendue, or solid angle that impinges on the second surface in comparison to the like quantity from the radiator. The computation of configuration factors can be tedious, especially for more complex geometric arrangements, so there are literature sources that provide electronic format for the integrals [20, 21].

In conclusion, the configuration factor directly expresses the étendue of the optical system, but it does such by comparing it to the étendue of the emitter. This method of expressing étendue is useful since it also provides the transfer efficiency of the optical system. This latter utility is a primary focus of discussion in the next section.

2.4 DESIGN EXAMPLES USING ÉTENDUE

A number of examples are shown here. Each uses the radiance distribution of a source that is propagated through a theoretical optical system that perfectly couples over an arbitrary spatial radius (R) and acceptance angle (θ_a). The examples show the flux-transmission characteristics of the optical system as a function of the étendue. The first example is for a Lambertian, uniform spatial disk. The second changes the angular emission profile to that of isotropic. The third case further adds a nonuniform spatial emission pattern. Finally, the fourth example is more realistic with a tubular source.

2.4.1 Lambertian, Spatially Uniform Disk Emitter

Using Figure 2.5a, one starts with Equation (2.4) to find the flux-transmission performance of the optical system. Since the optical system is perfect, there are no losses within it, so we can use conservation of étendue to determine the flux at the target. One only needs to solve for Equation (2.1) to determine the amount of flux transferred from the source to the target side of the optical system. The solution to Equation (2.1) over the full area of the source (A_s) and the angular range of $\theta \in [0, \theta_a]$ is

$$\xi_s(\theta_a) = n^2 A_s \int_0^{2\pi} \int_0^{\theta_a} \cos\theta \sin\theta d\theta d\varphi = \pi n^2 A_s \sin^2 \theta_a, \tag{2.45}$$

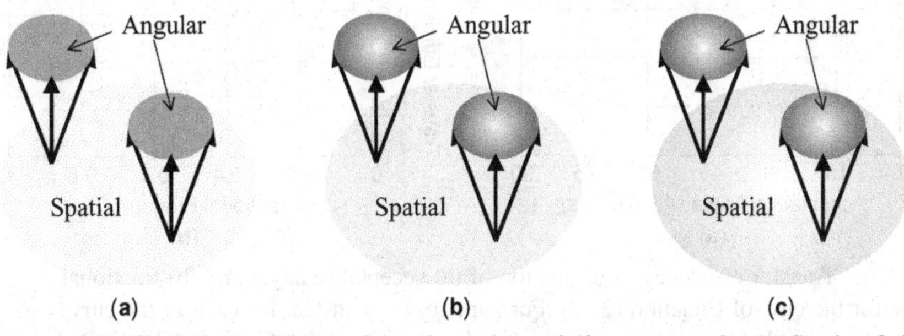

Figure 2.5 Depictions of the spatial and angular emission patterns for (a) Section 2.4.1 (Lambertian and spatially uniform source), (b) Section 2.4.2 (isotropic and spatially uniform source), and (c) Section 2.4.3 (isotropic and spatially non-uniform source).

where we have substituted in the integrand for the differential solid angle with $d\Omega = \sin\theta d\theta d\phi$, and the s-subscript on the étendue terms denotes that the source extent perfectly matches the field of the view of the optical system. Therefore, after the optical system, the flux at the target, with a source of radius r_s that emits with a Lambertian, spatially uniform source radiance of L_s, is

$$\Phi = \pi L_s A_s \sin^2\theta_a = \pi^2 L_s r_s^2 \sin^2\theta_a. \tag{2.46}$$

If the optical system captures all of the source emission, then $\theta_a = \pi/2$, giving a theoretical best flux at the target of

$$\Phi(\theta_a = \pi/2) = \Phi_{opt} = \pi L_s A_s = \pi^2 L_s r_S^2. \tag{2.47}$$

By comparing a system of nonoptimal acceptance angle (Eq. 2.46) to the optimal case (Eq. 2.47) [22], we find the transfer efficiency, η, of the system as a function of the acceptance angle

$$\eta(\theta_a) = \frac{\Phi(\theta_a)}{\Phi_{opt}} = \frac{\xi_s(\theta_a)}{\xi_{opt}} = \sin^2\theta_a. \tag{2.48}$$

The theoretical limit of perfect transfer efficiency occurs when $\theta_a = \pi/2 = 90°$. There are two ways to plot Equation (2.48): transfer efficiency as a function of acceptance angle or transfer efficiency as a function of fractional étendue. The fractional étendue is given by $\bar{\xi}_s(\theta_a) = \xi_s(\theta_a)/\xi_{opt}$. The fractional étendue for this case is the configuration factor as described in Section 2.3.6. Equation (2.48) is plotted in Figure 2.6a for the first case and Figure 2.6b for the second case ($r = r_s$). A fractional étendue of 1 on the horizontal axis of Figure 2.6b is equal to the optimal case étendue of $\pi^2 n^2 r_s^2$ in the units of area-sr. Figure 2.6a shows that increasing the acceptance angle has an increasing effectiveness until the acceptance angle of the optical system is 45°, upon which it has decreasing effectiveness. This nature is due to the Lambertian angular emission property. Figure 2.6b shows that the transfer efficiency is linear with étendue because of the Lambertian, spatially uniform source condition, as shown in Equation (2.4).

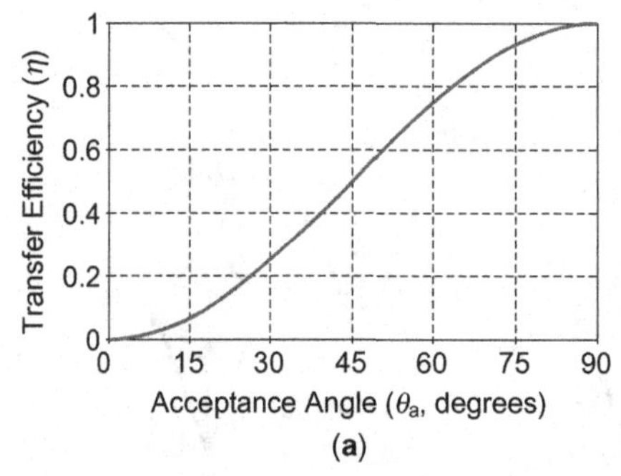

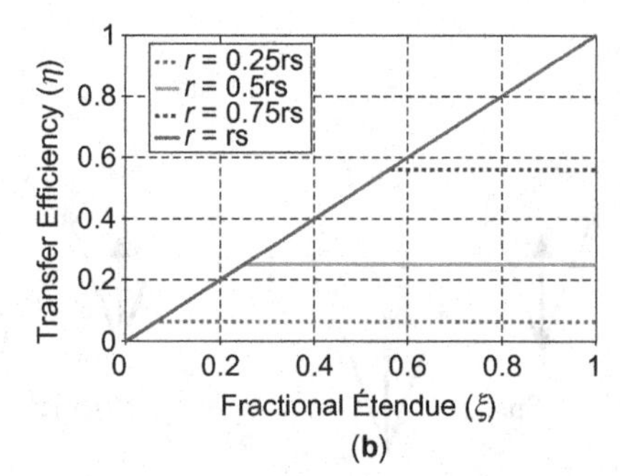

Figure 2.6 Transfer efficiency as a function of (a) acceptance angle and (b) fractional étendue for the case of Equation (2.50). For panel b, r is constant for each of the curves so the acceptance angle is varied to maintain étendue. A fractional étendue of 1 is equal to étendue of $\pi^2 n^2 r_s^2$ in units of area-sr.

Many would use Figure 2.6a to describe the flux transfer characteristics of the optical system in conjunction with the source; however, there are a number of reasons that make the form of Figure 2.6b more attractive to designers. First, we have assumed that the full spatial extent of the source perfectly matches the field of view of the optical system. Second, we have used a Lambertian angular distribution and uniform spatial distribution, which do not happen readily in real systems. Third, there are hereunto undeveloped limitations in the transfer efficiency of systems whose cross-sectional shape evolves with propagation. We remove the second case limitation in the next two examples (see Sections 2.4.2 and 2.4.3), and the third limit in Section 2.4.4, and further refine in Section 2.6 (skew invariant). For the first case, we expand the presentation by including the spatial extent of the source that can be coupled by the optical system. The efficiency is now given by

$$\eta(r,\theta_a) = \frac{\Phi(r,\theta_a)}{\Phi_{opt}} = \frac{\xi(r,\theta_a)}{\xi_{opt}} = \bar{\xi}(r,\theta_a) = \frac{r^2 \sin^2\theta_a}{r_s^2} = \bar{r}^2 \sin^2\theta_a, \tag{2.49}$$

where r is the radius of the source coupled by the optical system, and $\bar{r} = r/r_s$ is the normalized source radius coupled by the optical system. Once again, the transfer efficiency is linear with the fractional étendue (note that the s-subscript is dropped on the fractional étendue term since we do not assume that source perfectly matches the field of view of the optical system); however, if one limits the coupled radius of the source or acceptance angle, you cannot couple all of the radiation from the whole source. For the case where we prescribe the coupled radius of the source, then the acceptance angle is allowed to vary. However, the maximum fractional étendue occurs when $\theta_a = \pi/2$, after which there is no way to increase the flux transfer of the optical system. In mathematical form, this result is given by

$$\eta(\bar{\xi};\theta_a) = \begin{cases} \bar{\xi}, & \bar{\xi} \in [0, \bar{r}^2] \\ \bar{r}^2, & \bar{\xi} \in (\bar{r}^2, 1] \end{cases}. \tag{2.50}$$

The lack of r as a term in the efficiency function on the left-hand side indicates that it is a constant. Cases for Equation (2.50) are shown in Figure 2.6b for $r = mr_s$, $m \in [0.25, 0.5, 0.75, 1]$. Note that restricting the coupled radius of the source, as expected, restricts the transfer efficiency of the optical system to less than 1. Likewise, one can examine the transfer efficiency for a prescribed acceptance angle. This case allows the coupled source radius to vary; however, one cannot couple beyond $r = r_s$. In mathematical form, this result is given by

$$\eta(\bar{\xi};r) = \begin{cases} \bar{\xi}, & \bar{\xi} \in [0, \sin^2\theta_a] \\ \sin^2\theta_a, & \bar{\xi} \in (\sin^2\theta_a, 1] \end{cases}. \tag{2.51}$$

The lack of θ_a as a term in the efficiency function on the left-hand side indicates that it is a constant. Cases for Equation (2.51) are shown in Figure 2.7 for $\theta_a = m\pi/12$, $m \in [1, 2, 3, 6]$. Note that restricting the acceptance angle of the optical system, as expected, restricts the transfer efficiency of the optical system to less than 1.

Limiting the coupling of a spatially uniform, Lambertian source does not show unexpected behavior since there is a direct tradeoff of solid angle (i.e., $d\Omega$ term) for

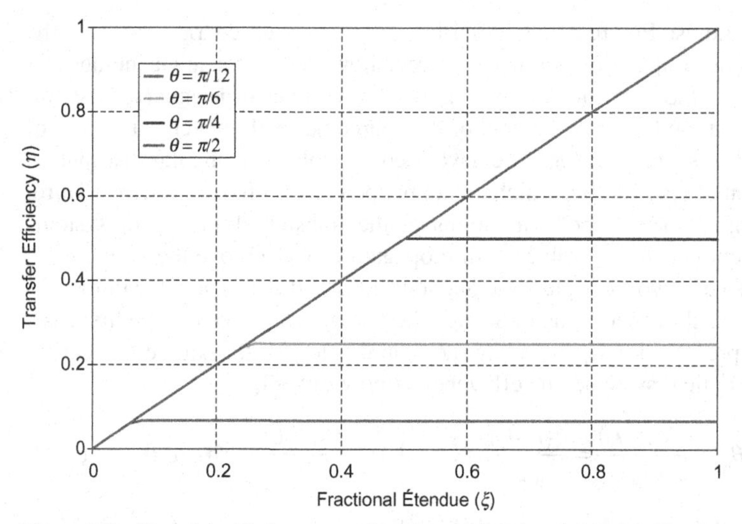

Figure 2.7 Transfer efficiency as a function of fractional étendue for the case of Equation (2.51). θ_a is held constant for each of the curves so the coupled source radius is varied to obtain conservation of étendue. Note that a fractional étendue of 1 is equal to an étendue of $\pi^2 n^2 r_s^2$ with the units of area-sr.

projected area (i.e., $\cos\theta\, dA$ term). Fractional étendue plots show the conservation of étendue behavior by joining together the spatial and angular extents. This is a result of the source radiance function being independent of both angle and position. Thus, the only outcome is that by restricting the coupling of spatial extent (see Eq. 2.50 and Fig. 2.6b) or the acceptance angle (see Eq. 2.51 and Fig. 2.7) results in a reduced transfer efficiency of the total source flux.

2.4.2 Isotropic, Spatially Uniform Disk Emitter

Per Figure 2.5b, we have the same spatial geometry, but in this case, the disk emits with an isotropic angular distribution. As shown in Chapter 1, the radiance of the source is inversely related to the cosine of the emission angle. Following the method from the previous section, the transfer efficiency is

$$\eta(R,\theta_a)=\frac{\Phi(R,\theta_a)}{\Phi_{opt}}=\frac{r^2}{r_s^2}(1-\cos\theta_a)=\bar{r}^2(1-\cos\theta_a)=\bar{\xi}\frac{(1-\cos\theta_a)}{\sin^2\theta_a}. \tag{2.52}$$

It is not practical to use configuration factor terminology for this case because a cosine component from the isotropic nature is included in the geometric configuration factor integral. This result means that the radiometric and geometric terms are not separable; therefore, we elect to use the term fractional étendue rather than configuration factor for all cases. For a given acceptance angle θ_a, we allow r to vary from $[0, r_s]$, which gives

$$\eta(\bar{\xi};r)=\begin{cases}\bar{\xi}\dfrac{(1-\cos\theta_a)}{\sin^2\theta_a}, & \bar{\xi}\in[0,\sin^2\theta_a]\\ 1-\cos\theta_a, & \bar{\xi}\in(\sin^2\theta_a,1]\end{cases}. \tag{2.53}$$

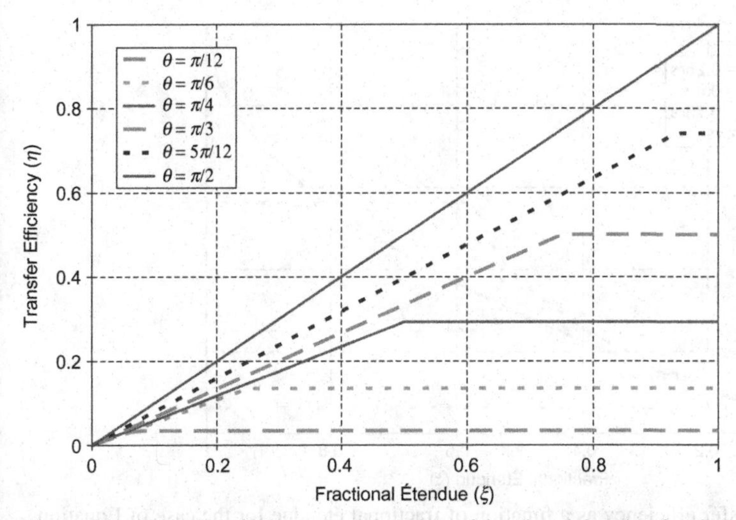

Figure 2.8 Transfer efficiency as a function of fractional étendue for the case of Equation (2.53). θ_a is held constant for each of the curves so the coupled source radius is varied to obtain conservation of étendue. Note that a fractional étendue of 1 is equal to an étendue of $\pi^2 n^2 r_s^{\,2}$ with the units of area-sr.

This result is shown in Figure 2.8 for $\theta_a = m\pi/12$, $m \in [1, 2, 3, 4, 5, 6]$. Note that for each curve, the transfer efficiency increases as the fractional étendue increases until certain values, which all correspond to a source radius of $r = r_s$ (i.e., the lower part of the right-hand side of Eq. 2.53). Beyond these étendue values there is no additional source extent that can be coupled by the optical system. Thus, increasing the spatial field of view of these systems will dilute the étendue transfer characteristics of the system. This term dilution indicates that there is étendue that is not filled with flux, or in other words, it is unused étendue. Additionally, the transfer efficiency increases as the prescribed acceptance angle increases. This result is due to the isotropic emission from the source. Therefore, in comparison with the Lambertian case of the previous section, there is more flux with an increase in the acceptance angle.

Likewise, for a given coupled radius, $r \le r_s$, one finds the transfer efficiency as a function of the varying acceptance angle

$$\eta(\bar{\xi};\theta_a) = \begin{cases} \bar{\xi}\dfrac{(1-\cos\theta_a)}{\sin^2\theta_a}, & \bar{\xi} \in [0, \bar{r}^2] \\ \bar{r}^2, & \bar{\xi} \in (\bar{r}^2, 1] \end{cases}. \tag{2.54}$$

This equation is not easy to plot as a function of fractional étendue because the first part of the right-hand side is a function of the additional variable, θ_a. However, due to conservation of étendue, we can use

$$\sin^2\theta_a = \frac{\bar{\xi}}{\bar{r}^2}. \tag{2.55}$$

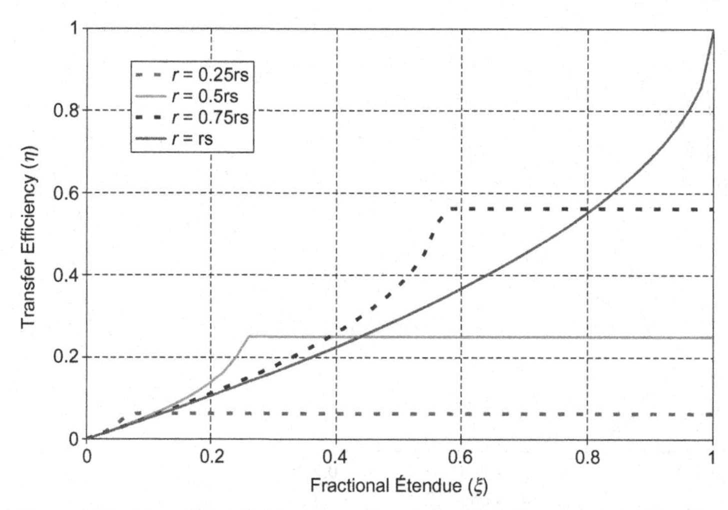

Figure 2.9 Transfer efficiency as a function of fractional étendue for the case of Equation (2.56). r is held constant for each of the curves so the acceptance angle is varied to obtain conservation of étendue. Note that a fractional étendue of 1 is equal to an étendue of $\pi^2 n^2 r_s^2$ with the units of area-sr.

The ability to use conservation of étendue as written in Equation (2.55) allows one to do a simple change of variables to obtain expressions with the fractional étendue term and constants. Updating Equation (2.54) with Equation (2.55), the transfer efficiency is

$$\eta(\bar{\xi};\theta_a)=\begin{cases}\bar{r}^2\left(1-\sqrt{1-\bar{\xi}/\bar{r}^2}\right), & \bar{\xi}\in[0,\bar{r}^2]\\ \bar{r}^2, & \bar{\xi}\in(\bar{r}^2,1]\end{cases}. \tag{2.56}$$

Figure 2.9 shows the results for Equation (2.56) for $r = mr_s$, $m \in [0.25, 0.5, 0.75, 1]$. Once again, for each coupled source radius, there is a threshold upon which the optical system cannot improve the transfer efficiency. This point occurs for each case when $\theta_a = \pi/2$. However, unlike the previous cases, the transfer efficiency improves more rapidly for a smaller coupled source radius. This result is due to the nonlinear $\cos\theta_a$ dependence of Equation (2.55), or, analogously, the square root dependence on the fractional étendue. Physically, this outcome is once again due to the isotropic nature of the source emission. There is increased transfer efficiency compared to the Lambertian case as the acceptance angle is increased.

2.4.3 Isotropic, Spatially Nonuniform Disk Emitter

Per Figure 2.5c, we have the same spatial geometry, but in this case, the disk emits with an isotropic angular distribution and a spatial distribution that peaks at the center of the disk and drops to zero at its edge. The source radiance function is

$$L(r,\theta_a)=L_s\frac{(1-r/r_s)}{\cos\theta_a}=L_s\frac{(1-\bar{r})}{\cos\theta_a}, \tag{2.57}$$

where $r \in [0, r_s]$ and $\theta_a \in [0, \pi/2]$. The optimal (i.e., total) flux and as a function of coupled source radius and acceptance angle are given by

$$\Phi(r,\theta_a) = \frac{2}{3}\pi^2 L_s r^2 (3-2\bar{r})(1-\cos\theta_a), \tag{2.58}$$

and

$$\Phi_{opt} = \frac{2}{3}\pi^2 L_s r_s^2. \tag{2.59}$$

The transfer efficiency is then

$$\eta(r,\theta_a) = \frac{\Phi(r,\theta_a)}{\Phi_{opt}} = \bar{r}^2(3-2\bar{r})(1-\cos\theta_a) = \bar{\xi}(3-2\bar{r})\frac{(1-\cos\theta_a)}{\sin^2\theta_a}. \tag{2.60}$$

For a given acceptance angle θ_a, we allow r to vary within the range $[0, r_s]$, which gives

$$\eta(\bar{\xi}; r) = \begin{cases} \bar{\xi}(3-2\bar{r})\dfrac{(1-\cos\theta_a)}{\sin^2\theta_a}, & \bar{\xi} \in [0, \sin^2\theta_a] \\ (1-\cos\theta_a), & \bar{\xi} \in (\sin^2\theta_a, 1] \end{cases}. \tag{2.61}$$

With the use of conservation of étendue (Eq. 2.55), the normalized radius term is removed,

$$\eta(\bar{\xi}; r) = \begin{cases} \bar{\xi}\left(3\sin\theta_a - 2\sqrt{\bar{\xi}}\right)\dfrac{(1-\cos\theta_a)}{\sin^3\theta_a}, & \bar{\xi} \in [0, \sin^2\theta_a] \\ (1-\cos\theta_a), & \bar{\xi} \in (\sin^2\theta_a, 1] \end{cases}. \tag{2.62}$$

This equation is depicted in Figure 2.10 for $\theta_a = m\pi/12$, $m \in [1, 2, 3, 4, 5, 6]$. The plot shows that increasing the coupled angle increases the transfer efficiency for a given fractional étendue. As the fractional étendue increases, the transfer efficiency increases, but at a decreasing rate until the coupled source radius is equal to r_s. At this point, the transfer efficiency is constant. The decreasing rate of improvement in transfer efficiency is due to the nonuniform emittance, which results in decreasing emitted flux as the coupled source radius increases.

Likewise, for a given coupled radius, $r \le r_s$, one finds the transfer efficiency as a function of the varying acceptance angle

$$\eta(\bar{\xi}; r) = \begin{cases} \bar{r}^2(3-2\bar{r})\left(1-\sqrt{1-\bar{\xi}/\bar{r}^2}\right), & \bar{\xi} \in [0, \bar{r}^2] \\ \bar{r}^2(3-2\bar{r}), & \bar{\xi} \in (\bar{r}^2, 1] \end{cases} \tag{2.63}$$

Figure 2.11 shows the results for Equation (2.63) for $r = mr_s$, $m \in [0.25, 0.5, 0.75, 1]$. In comparison with the uniform spatial emission profile, as shown in Figure 2.9, there is increased performance with $r < 1$. This result is due to decreasing return on opening up the spatial field of view of the optical system due to the nonuniform emittance. As r increases, there is decreasing flux available since the distribution is peaked at the center of the disk emitter. This example is a good illustration where it does not benefit the designer to fully couple the spatial extent of the source if there

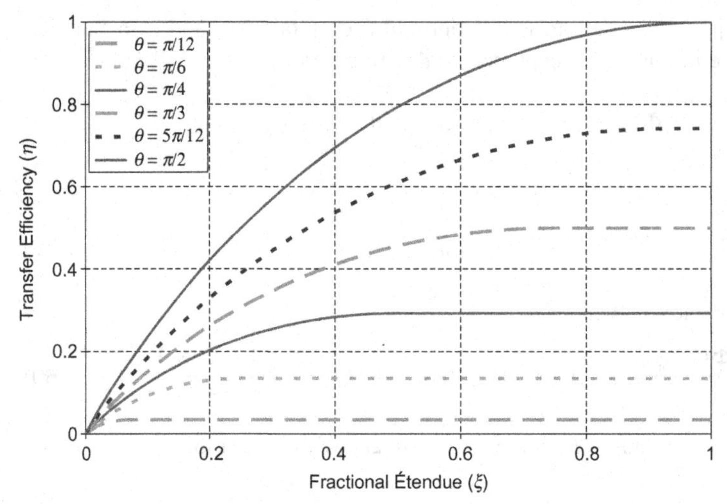

Figure 2.10 Transfer efficiency as a function of fractional étendue for the case of Equation (2.62). θ_a is held constant for each of the curves so the coupled source radius is varied to obtain conservation of étendue. Note that a fractional étendue of 1 is equal to an étendue of $\pi^2 n^2 r_s^2$ with the units of area-sr.

is little flux in these added regions. Therefore, the designer would know that flux transfer is better realized by addressing the angular coupling (Fig. 2.10) in comparison with spatial coupling (Fig. 2.11).

2.4.4 Tubular Emitter

The previous examples are illustrative of the utility of étendue plots to better understand and design optical systems for prescribed sources. However, planar emitters are not realistic in comparison with actual source geometries. For example, LEDs are better represented by the three-dimensional (3D) shape of the dies, filaments by helixes, and arcs, as expected, by arcs. However, these shapes can be difficult to develop analytic models; therefore, one typically approximates the shape of the emission region with simpler 3D shapes. For example, a tubular emitter can be used for filaments and arcs. We illustrate this by showing an example that is similar to that from the literature [23].

Consider an illumination system, such as a projector, making use of an arc or filament source to illuminate the spatial light modulator(s) (SLM[s]) in the system (i.e., three-color LCD panels or a micromirror array). In order to develop a working model of the throughput characteristics of this illumination system, typically, a tubular volume emitter approximates the spatial emission geometry of the source. The angular emission distribution is Lambertian from each point in the tubular volume. For determination of the étendue of the source, we look at the surface area of the tubular emitter and treat it as uniform; however, the spatial emission distribution is both a function of the radial position and the spatial position. This source is

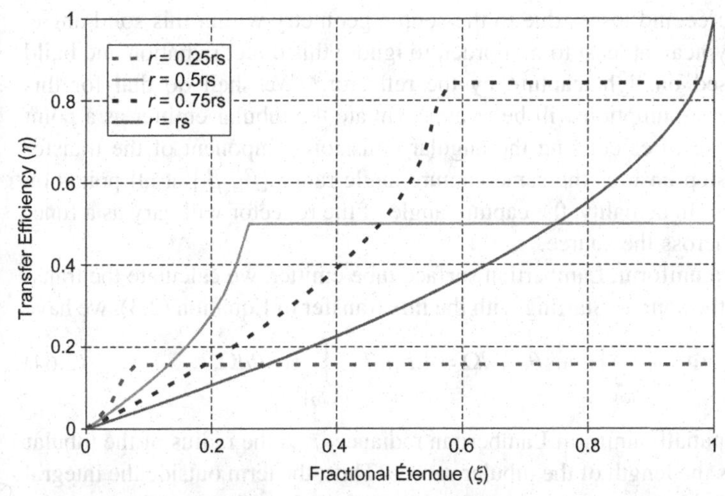

Figure 2.11 Transfer efficiency as a function of fractional étendue for the case of Equation (2.63). r is held constant for each of the curves so the acceptance angle is varied to obtain conservation of étendue. Note that a fractional étendue of 1 is equal to an étendue of $\pi^2 n^2 r_s^2$ with the units of area-sr.

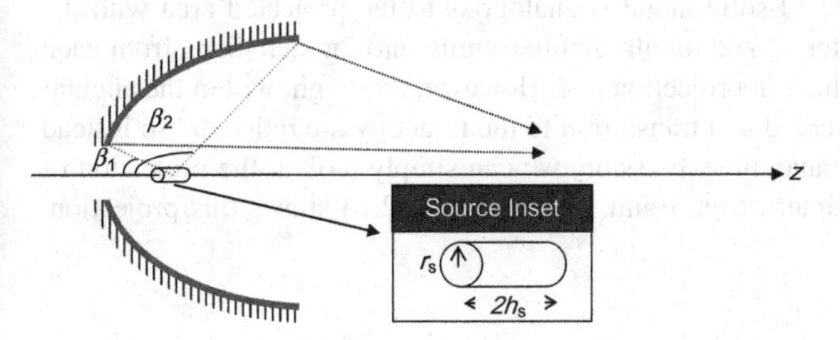

Figure 2.12 Geometry layout for a tubular emitter of radius r_s and length $2h_s$ centered at the nominal focus of a reflector that captures light within the angular range [β_1, β_2], as shown in cross section.

placed near the nominal focus of an optical component, such as a reflector.* As per Figure 2.12, the light from the source from angle β_1 to β_2 is captured by the reflector and transferred to the SLM. The light from the negative z-axis to β_1 is lost through the source hole (i.e., through which the bulb is mounted to be located at the nominal focus location), while light emitted from β_2 to the positive z-axis does not strike the reflector and is considered lost. Some of this radiation directly illuminates the SLM(s) in the system; however, the amount of radiation is low due to the Lambertian

* Note that until recently, conic reflectors, such as ellipses, were solely used to transfer the light from the source to the SLMs; however, more complex shapes, such as faceted ones or continuous ones based on, for example, nonuniform rational B-splines (NURBS), have been employed. These types of optics do not have the concept of "focus" inherent to them, but, rather, have a caustic region in which the source rays emanate. These more complex optic shapes and their designs are discussed in following chapters.

nature of the source and losses due to the source geometry within this solid angle. Therefore, it is typical, at least to first order, to ignore this direct radiation and build your system based on light capture by the reflector.* We shall do that for this example. Another assumption will be to approximate the tubular emitter as a point source for the sake of calculating the angular emission component of the transfer efficiency. This step ensures the same capture angle range $[\beta_1, \beta_2]$ at all points on the tubular source. In actuality, the capture angle of the reflector will vary as a function of position across the source.

Assuming a uniform, Lambertian surface tube emitter, we calculate the transfer efficiency of the source. Starting with the flux transfer of Equation (2.3), we have

$$\Phi = L_s \iint_{\text{reflector}} \cos\theta dA_s d\Omega = 4\pi r_s h_s L_s \int_{[\beta_1,\beta_2]} \cos\theta d\Omega, \tag{2.64}$$

where L_s is the spatially uniform Lambertian radiance, r_s is the radius of the tubular source, and $2h_s$ is the length of the tubular source. Thus, the term outside the integral in the right-hand side represents the surface area of the lateral sides of the tubular source. The circular ends do not emit. The integral in the right-hand side of Equation (2.64) is over the projected solid angle, which can be solved by looking at the geometry.

Consider the kernel of the integral, $\cos\theta\, d\Omega$, which is the differential, projected solid angle. The projected solid angle is analogous to the projected area with the inclusion of the $\cos\theta$ term. The tubular emitter emits into 2π steradians from each point on its surface, which in projection is π. However, only light within the angular range of $[\beta_1, \beta_2]$ is accepted and transferred to the target by the reflector. So instead of solving the tedious analytic expression, we can simply look at the projection of the angular region of interest on a unit sphere. Figure 2.13 shows this projection,

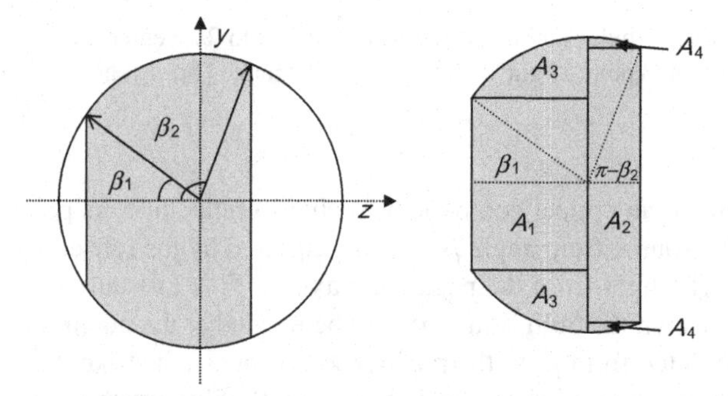

Figure 2.13 Geometry of the projected solid angle on the meridional plane of the optical system. The z-axis is as in Figure 2.12, the y-axis is the up direction of this same figure, and the unit-radius projected sphere onto this plane appears as a circle of radius 1. The integral in the left-hand side of Equation (2.64) is the "area" of the shaded region.

* The chapter on freeform optics shows a method that considers this direct radiation called the simultaneous multiple surfaces (SMS) method.

where the z-axis in this figure agrees with that of Figure 2.12. The projection of this region on the unit sphere is an internal segment delineated by latitude lines. This projection is valid for each point on the tubular emitter since we have approximated it as a point source for the sake of calculating the angular range of interest. Therefore, we need to only calculate the "area" of the shaded region in Figure 2.13. There are six regions as denoted by areas A_1 and A_2 and the doubled regions of A_3 and A_4. The areas of these four regions are:

$$\begin{aligned} A_1 &= \sin 2\beta_1, \\ A_2 &= -\sin 2\beta_2, \\ A_3 &= \frac{1}{4}(\pi - 2\beta_1 - \sin 2\beta_1), \text{ and} \\ A_4 &= \frac{1}{4}(2\beta_2 - \pi + \sin 2\beta_2). \end{aligned} \tag{2.65}$$

All of the areas of these projected solid angle regions are found with standard geometry methods. The regions A_1 and A_2 are the areas of the two respective rectangles, while one can a mathematical reference book to find the half circular segment regions A_3 and A_4.

Taking the individual components of Equation (2.65), we can compute the integrated projected solid angle "area"

$$\begin{aligned} A &= A_1 + A_2 + 2A_3 + 2A_4 \\ &= \beta_2 - \beta_1 - \frac{\sin 2\beta_2 - \sin 2\beta_1}{2}. \end{aligned} \tag{2.66}$$

The "area" A is the solution to the integral on the left-hand side of Equation (2.64). Substituting this into Equation (2.64) and assuming that our optical system may not couple the full spatial extent, $r_s \to r$ and $h_s \to h$, we obtain

$$\Phi = 2\pi r h L_s (2\beta_2 - 2\beta_1 - \sin 2\beta_2 + \sin 2\beta_1) = \frac{L_s \xi_{system}}{n^2}, \tag{2.67}$$

where the system étendue, including the combination of the source and capture angles of the reflector, is given by

$$\xi_{system} = 2\pi n^2 r h (2\beta_2 - 2\beta_1 - \sin 2\beta_2 + \sin 2\beta_1). \tag{2.68}$$

This latter step can be done because of our Lambertian assumption. If the source were not Lambertian, there would be additional terms within the integral of Equation (2.64) such that étendue could not be explicitly calculated for the system. The optimal étendue is that of the source, which is given when $\beta_1 = 0$, $\beta_2 = \pi$, $r = r_s$, and $h = h_s$,

$$\xi_{opt} = 4\pi^2 n^2 r_s h_s, \tag{2.69}$$

and the optimal power of the system emitted by a uniform, Lambertian source is

$$\Phi_{opt} = 4\pi^2 r_s h_s L_s, \tag{2.70}$$

Next, we look at the fractional étendue

$$\frac{\xi_{\text{system}}}{\xi_{\text{opt}}} = \bar{\xi}_{\text{system}} = \frac{\bar{r}\bar{h}}{2\pi}(2\beta_2 - 2\beta_1 - \sin 2\beta_2 + \sin 2\beta_1) = \bar{r}\bar{h}\xi_0, \tag{2.71}$$

where $\xi_0 = (2\beta_2 - 2\beta_1 - \sin 2\beta_2 + \sin 2\beta_1)/2\pi$, $\bar{r} = r/r_s$, and $\bar{h} = h/h_s$. We want to plot the transfer efficiency as a function of Equation (2.71) as we have done in previous sections. Now, we turn our attention to the actual emission profile of the tubular volume emitter by using the flux emission density function, Ψ, in cylindrical coordinates

$$\Psi(\bar{r}, \bar{h}, \theta) = \begin{cases} \Psi_0(1-\bar{r})(\varepsilon + 2(1-\varepsilon)|\bar{h}|), & \bar{r} \le 1 \text{ and } |\bar{h}| \le 1 \\ 0, & \text{otherwise} \end{cases}, \tag{2.72}$$

where Ψ_0 is the flux emission density (i.e., in units lumens/mm^3) on the central axis of the tubular emitter, $\varepsilon \in [0, 2]$ is the longitudinal asymmetry emission factor, the normalized capture radius is given by $\bar{r} = r/r_s$, and the normalized capture length is given by $\bar{h} = h/h_s$ where h is the position on the z-axis with the source centered at the origin and oriented along this axis. The term ε represents uniform emission along the length of the tubular emitter for a value of 1, a distribution peaked at the center occurs with a value of 2, and a distribution peaked at the ends of the tube (i.e., the electrodes) occurs with a value of 0. The last term in Equation (2.72) is a linear distribution with respect to the normalized position along the axis, and its functional form ensures that the flux emission density is normalized over the valid range of $h = [-h_s, h_s]$.* Typical arc lamps can be approximated by the case of $0 \le \varepsilon \le 1$, that is, the positions around the electrodes emit more radiation than the center. Refined models for the source emission can be used, in particular those based on direct measurement of the source output [24], but this simplified model allows an analytic treatment while providing the insight that we seek. Integrating Equation (2.72) over r, z, and θ, we find the total power emitted by the source within a tube of fractional parameters $\bar{r}$ and $\bar{h}$,

$$\Phi = \begin{cases} \Phi_0(3\bar{r}^2 - 2\bar{r}^3)(\varepsilon\bar{h} + (1-\varepsilon)\bar{h}^2), & \bar{r} \le 1 \text{ and } |\bar{h}| \le 1 \\ \Phi_0, & \text{otherwise} \end{cases}, \tag{2.73}$$

where Φ_0 is the total amount of lumens emitted by the source described in Equation (2.72) and it is equal to $2\pi r_s^2 h_s \Psi_0/3$, which is the volume of a cone of radius r_s and height h_s. Dividing through by the total source power, we find the transfer efficiency as a function of the source parameters

$$\eta(\bar{r}, \bar{h}; \gamma) = \frac{\Phi}{\Phi_0} = \begin{cases} (3\bar{r}^2 - 2\bar{r}^3)(\varepsilon\bar{h} + (1-\varepsilon)\bar{h}^2), & \bar{r} \le 1 \text{ and } |\bar{h}| \le 1 \\ 1, & \text{otherwise} \end{cases}, \tag{2.74}$$

Finally, returning to Equation (2.71) and assuming for the sake of simplified analysis that the source has geometrical parameters of $\bar{h} = \bar{r}$, we substitute into

* Note that this equation is slightly modified to that of Reference [17]. This change is due to keeping Equation (2.72) normalized over its valid range of interest.

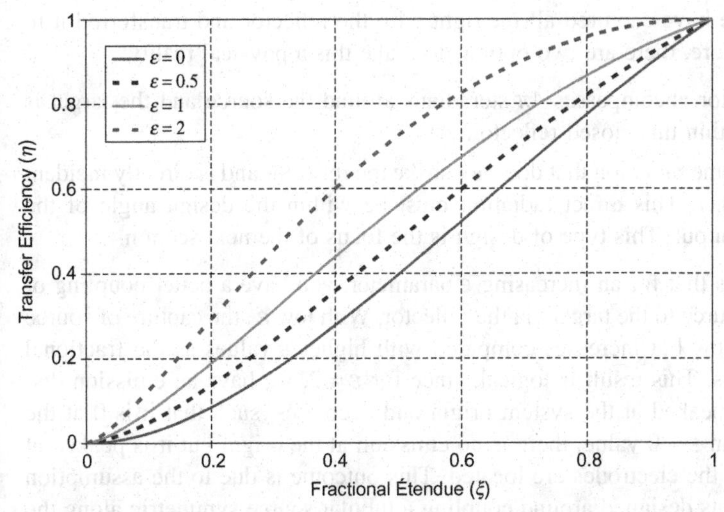

Figure 2.14 Transfer efficiency as a function of fractional étendue for Equation (2.75) for $\varepsilon \in [0, 0.5, 1, 2]$. None of the other source parameters are set; therefore, the various parameters $\overline{r}$, $\overline{h}$, β_1, and β_2 are allowed to vary. This lack of explicit definition means that these parameters are allowed to trade for one another—that is, there is conservation of étendue for various values of the parameters per Equation (2.71). Note that a fractional étendue of 1 is equal to an étendue of $4\pi^2 n^2 r_s h_s$ with the units of area-sr.

Equation (2.74) for these geometrical parameters as a function of fractional étendue

$$\eta(\overline{r}, \overline{h}; \gamma) = \begin{cases} \left(3\frac{\overline{\xi}}{\xi_0} - 2\left(\frac{\overline{\xi}}{\xi_0}\right)^{3/2}\right)\left(\varepsilon\sqrt{\frac{\overline{\xi}}{\xi_0}} + (1-\varepsilon)\frac{\overline{\xi}}{\xi_0}\right), & \overline{\xi} \le \xi_0 \\ 1, & \xi_0 < \overline{\xi} \le 1 \end{cases}, \tag{2.75}$$

where we have dropped the "system" subscript on étendue. We plot Equation (2.75) as a function of fractional étendue in Figure 2.14 for four cases $\varepsilon \in [0,0.5,1,2]$. As per conservation of étendue, there is no loss in the optical system for our selection of the source parameters. So there is no absorption loss at the reflector, no scatter loss, and flux incident on the reflector within the range of the design parameters is transferred to the target without loss. The latter point assumes that there is no "aberration" loss in the system—it is a perfect light transformer.

One interesting facet of this figure is that the various source parameters, $\overline{r}$, $\overline{h}$, β_1, and β_2, are allowed to trade for one another as long as conservation of étendue is realized per Equation (2.71). Thus, there are infinite possibilities for what these four parameters can be to realize étendue conservation. In reality this allowed trade for this figure is not true—there are issues with taking light from one distribution pattern and transforming it another pattern. This phenomenon is due to the skew invariant, which is directly related to étendue, and is the focus of Section 2.6. This example is updated in that section to include loss due to skewness. For a fractional étendue of 1, we have captured all the light with the reflector and transferred it to the target; therefore, there are two options to make this a physical reality:

- The reflector encompasses 4π steradians around the source and the target is located within this closed reflector, or
- There is some radiation that does not strike the reflector and is directly incident on the target. This direct radiation must be within the design angle of the reflector output. This type of design is the focus of the next section.

The figure shows that for an increasing ε parameter, you have a better coupling of light from the source to the target via the reflector. With low ε, the capture of source radiation starts low but increases compared with higher ε values as the fractional étendue increases. This result is logical, since for $\varepsilon = 2$, we have an emission distribution that is peaked at the system origin and decreases, such that it is 0 at the electrodes. For an $\varepsilon = 0$ value, there is no emission at the origin but it is peaked at the edges where the electrodes are located. This outcome is due to the assumption that the reflector is designed around coupling a tubular source symmetric along the z axis and centered at the origin. Therefore, for the $\varepsilon = 0$ value, there is little flux available to be coupled for low fractional étendue values, while the opposite holds true for the $\varepsilon = 2$ case. Of course, the system could be designed around a different position, such as one of the two electrode positions; however, this would end up leading to one of three possibilities:

- More loss due to the need to couple a larger source assuming the reflector is designed around a symmetric source located at the origin
- A larger reflector to couple this apparent larger symmetric source, or
- A novel design that couples around an asymmetric source, such that different locations on the reflector surface couple different regions from the source. Note that this type of design technique is called tailored design.

Note that a value of $0 \le \varepsilon \le 1$ is a typical approximation for an arc source, so the system designer must contend for a larger source or incorporate a tailored design. Ultimately, this outcome is the reason that systems that use arc lamps, such as projector displays, incorporate a source with as small an arc gap as possible. Finally, Equation (2.75) assumes that we have variable values for the source parameters. Instead, assume a fixed capture angle of $\beta_1 = 0$ and $\beta_2 \in m\pi/4$, $m \in [1, 2, 3]$, with $\varepsilon = 0.5$, then Equation (2.74) becomes

$$\eta(\bar{r}, \bar{h}; \gamma) = \begin{cases} \left(3\frac{\bar{\xi}}{\xi_0} - 2\left(\frac{\bar{\xi}}{\xi_0}\right)^{3/2}\right)\left(0.5\sqrt{\frac{\bar{\xi}}{\xi_0}} + 0.5\frac{\bar{\xi}}{\xi_0}\right), & \bar{\xi} \le \bar{r}^2\xi_0 \\ \left(3\xi_0 - 2\xi_0^{3/2}\right)\left(0.5\sqrt{\xi_0} + 0.5\xi_0\right), & \bar{r}^2\xi_0 < \bar{\xi} \le 1 \end{cases}, \tag{2.76}$$

This result is plotted in Figure 2.15 for the three β_2 values. In this case, one sees that there is an upper limit to the fractional étendue that can be coupled to the target, which occurs at $\bar{r} = \bar{h} = \sqrt{\xi_0}$. Therefore, there is a limit to the transfer efficiency.

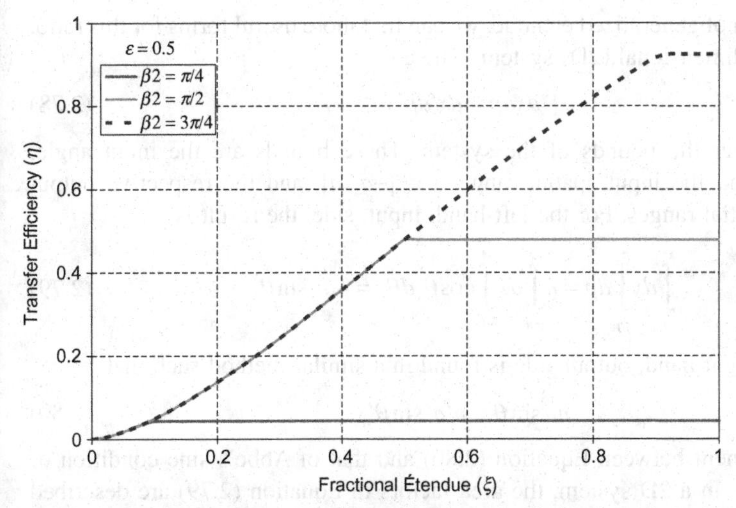

Figure 2.15 Transfer efficiency as a function of fractional étendue for Equation (2.76) for $\beta_1 = 0$ and $\beta_2 \in m\pi/4$, $m \in [1, 2, 3]$, with $\varepsilon = 0.5$. In this case, restricting the angular range sets an upper limit on the fractional étendue that can be coupled to the target; therefore, the transfer efficiency is limited. Note that a fractional étendue of 1 is equal to an étendue of $4\pi^2 n^2 r_s h_s$ with the units of area-sr.

Figure 2.15 only displays light that is captured by the reflector, and it is then assumed that there is no loss to the target that follows. As discussed previously, there is another factor, called skewness, that limits the ability to transfer light from one distribution (e.g., a circular tube such as the arc source) to another distribution (e.g., a circular plane, such as the exit aperture of the reflector). We present the skew invariant in Section 2.6, but in order to improve upon our discussion of nonimaging optics, we first investigate the concentration ratio. Associated with the concentration ratio is the acceptance angle of an optical system. We have already made use of the acceptance angle in our discussions, so in the next section, its understanding is expanded.

2.5 CONCENTRATION RATIO

Concentration is a term related to étendue and its conservation. Concentration defines the ratio of the flux from the input area (A) that is transmitted by an output aperture area A'. For this reason, it is called the concentration ratio

$$C = \frac{A}{A'}. \tag{2.77}$$

This expression is actually a limit of the laws of thermodynamics, which as previously shown is a forbearer of the invariance of radiance and étendue [8, 25]. The concentration ratio is a fundamental building block for nonimaging optical systems, due primarily to the development of solar concentrators. Using Equation (2.18) for the conservation of generalized étendue, we can find more useful forms for this ratio. First, in a two-dimensional (2D) system, where

$$dxdp = dx'dp', \tag{2.78}$$

we integrate over the bounds of the system. These bounds are the input angle, $\theta_x \in [-\theta, \theta]$, and the input spatial range, $x \in [-a, a]$, and the respective output angular and spatial ranges. For the left-hand, input, side, the result is

$$\int_{x_1}^{x_2} dx \int_{p_1}^{p_2} dp = n \int_{-a}^{a} dx \int_{-\theta}^{\theta} \cos\theta_x d\theta_x = 4an\sin\theta. \tag{2.79}$$

Likewise, the right-hand, output side is found in a similar method such that

$$na\sin\theta = n'a'\sin\theta'. \tag{2.80}$$

Note the agreement between Equation (2.80) and that of Abbe's sine condition of Equation (2.41). In a 2D system, the area factors in Equation (2.79) are described by the aperture dimensions of a and a', respectively, such that the 2D concentration ratio is given by

$$C_{2D} = \frac{a}{a'} = \frac{n'\sin\theta'}{n\sin\theta}. \tag{2.81}$$

In a three-dimensional (3D) system, the concentration ratio is found with the analogous method, except now the dy and dq terms are retained. For the left-hand side, assuming Cartesian symmetry

$$\int_{x_1}^{x_2} dx \int_{y_1}^{y_2} dy \int_{p_1}^{p_2} dp \int_{q_1}^{q_2} dq = n^2 \int_{-a}^{a} dx \int_{-b}^{b} dy \int_{-\theta}^{\theta} \cos\theta_x d\theta_x \int_{-\varphi}^{\varphi} \cos\theta_y d\theta_y = 16abn^2 \sin\theta \sin\varphi, \tag{2.82}$$

where the bounds are $\theta_y \in [-\varphi, \varphi]$ and $y \in [-b, b]$. The right-hand side is found similarly. The result is thus found to be

$$C_{3D} = \frac{A}{A'} = \frac{ab}{a'b'} = \frac{n'^2 \sin\theta' \sin\varphi'}{n^2 \sin\theta \sin\varphi}, \tag{2.83}$$

In a rotationally symmetric system, $dxdy = ydyd\beta$, where r is the radius coordinate and has the range $r \in [0, a]$, and β is the spatial angle with range $\beta \in [0, 2\pi]$, and $\theta_x = \theta_y$. Solving both sides of the equation and simplifying gives the 3D concentration ratio

$$C_{3D} = \frac{A}{A'} = \left(\frac{n'\sin\theta'}{n\sin\theta}\right)^2, \tag{2.84}$$

where $A = \pi a^2$ and $A' = \pi a'^2$.

Optimal expressions for concentrators of a prescribed acceptance angle, θ_a, are found when the output light is emitted over a hemisphere, such that $\theta' = \pi/2$. In this case, the optimal, 2D concentration ratio is

$$C_{2D,opt} = \frac{n'}{n\sin\theta_a}, \tag{2.85}$$

and the optimal, 3D concentration ratio for the rotationally symmetric case is

$$C_{3D,opt} = \left(\frac{n'}{n \sin \theta_a} \right)^2 . \tag{2.86}$$

Equations (2.85) and (2.86) provide the theoretical limit for the concentration, assuming a lossless system that transmits all radiation incident on the input aperture of area A over the acceptance angle of $\pm\theta_a$, from the output aperture of area A' in a hemispherical angular distribution.

2.6 ROTATIONAL SKEW INVARIANT

While conservation of étendue limits the transfer characteristics of optical systems, the skewness is usually more demanding. It is typical to design optical systems, including nonimaging ones, around rotational symmetry. Rotationally symmetric design has the potential to limit performance due to the skewness of the rays that are emitted from the source that are then transferred by this symmetric system to the target. This limitation is due to the fact that the rotationally symmetric optical system cannot alter the skewness of the rays; therefore, it is impossible to efficiently couple the light from a source of one shape into a target of a different shape. The designer must include rotationally asymmetry into the system, including asymmetric geometry (i.e., a system that has "twist"), anisotropic scatter, anisotropic gradient index materials, and so forth in order to change one distribution shape into another. Thus, in a rotationally symmetric system, the skewness of individual rays is invariant, and the resulting skewness distribution of all the rays from the source is also invariant.

2.6.1 Proof of Skew Invariance

In order to show the skewness (or skew invariant) of a ray in an optical system, we start with Figure 2.16 and its definition [25]

$$s = \vec{r} \cdot \left(\vec{k} \times \hat{z} \right), \tag{2.87}$$

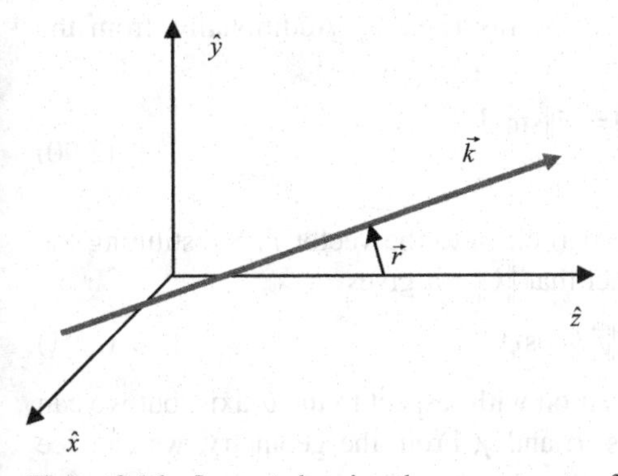

Figure 2.16 Layout showing the components of skewness per Equation (2.87).

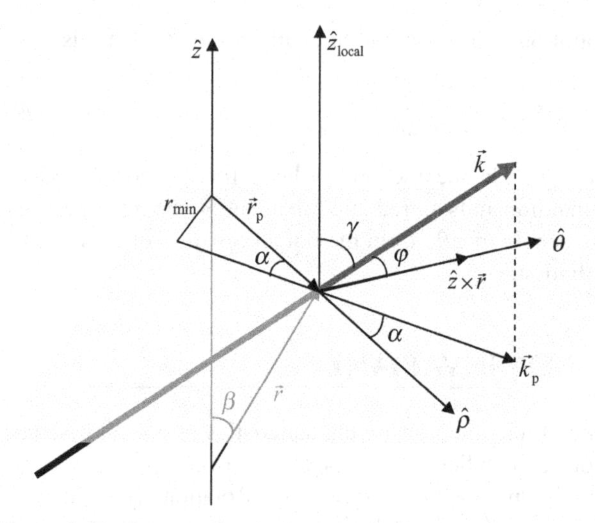

Figure 2.17 Layout of a single ray to prove skew invariance. The $\hat{z}$-axis indicates the optical axis that defines the axis of rotational symmetry. A projection surface is shown in gray. Any component that is shaded indicates that it lies below the projection surface, while those that are black are on the surface or above it. The only vectors extending beyond the projection surface are $\hat{z}$, $\hat{z}_{\text{local}}$, and $\vec{k}$.

where $\vec{r}$ is an arbitrary vector that joins the optical axis to the ray, $\vec{k}$ is the vector along the direction of the ray with magnitude $\|\vec{k}\| = n$, n is the index of refraction of the media of the ray, and $\hat{z}$ is the unit vector along the direction of the optical axis. Using Figure 2.17 for the remainder of this discussion [26], we show that the rotational skewness is invariant with propagation through an optical system. Starting with the definition of skewness in Equation (2.87), we can rewrite it based on the properties of the triple product

$$s = \vec{k} \cdot (\hat{z} \times \vec{r}). \tag{2.88}$$

Based on the geometry in Figure 2.17, we can rewrite the equation

$$s = \|\vec{k}\| \cdot \|\hat{z} \times \vec{r}\| \cos\varphi, \tag{2.89}$$

where φ is the angle from the $\hat{\theta}$-axis to the ray (i.e., $\vec{k}$). Additionally, from the geometry

$$\begin{aligned} \|\hat{z} \times \vec{r}\| &= \|\vec{r}\| \sin\beta \\ &= \|\vec{r}_p\|, \end{aligned} \tag{2.90}$$

where β is the angle from the optical axis (i.e., $\hat{z}$) to the vector $\vec{r}$. Substituting the latter expression while also using the fact that $\|\vec{k}\| = n$ gives

$$s = n\|\vec{r}_p\| \cos\varphi, \tag{2.91}$$

The $\cos\varphi$ term orients the ray into its position with respect to the $\hat{\theta}$-axis, but we can do this operation with two other angles, α and γ. From the geometry, we can see that the operations to accomplish this are

$$\cos\varphi = \sin\gamma \sin\alpha, \tag{2.92}$$

where γ is the angle from the $\hat{z}$-axis to the ray and α is the angle from the $\hat{\rho}$-axis to the projected $\vec{k}_p$ vector. Substituting Equation (2.92) into Equation (2.91), we obtain

$$s = n\|\vec{r}_p\|\sin\gamma\sin\alpha. \tag{2.93}$$

Finally, from the figure, we note that $r_{\min} = \|\vec{r}_p\|\sin\alpha$, which denotes the limit of closest approach of the ray to the optical axis. Substituting this term into Equation (2.93)

$$s = nr_{\min}\sin\gamma = r_{\min}k_t, \tag{2.94}$$

where $r_{\min}$ is the minimal magnitude of the vector $\vec{r}$, and $k_t = n\sin\gamma$ is the tangential component of $\vec{k}$.

Equation (2.94) is invariant for a ray as it propagates in medium of index n since all the terms are constant for the ray. The term $r_{\min}$ is simply the closest approach of the ray to the defined optical axis, which is a constant. The term k_t is the tangential component of the ray, which is a constant in medium of index n. Upon refraction or reflection, we use Snell's law

$$k_t = n\sin\gamma = n'\sin\gamma' = k_t'; \tag{2.95}$$

therefore,

$$s = r_{\min}k_t = r_{\min}k_t' = s', \tag{2.96}$$

Equation (2.96) proves that the skewness for a given ray is constant, or invariant, with propagation through an optical system.

2.6.2 Refined Tubular Emitter Example

In Section 2.4.4, we investigated the transfer efficiency as a function of fractional étendue for a tubular emitter to simulate the capture of radiation by an arc source integrated with a reflector. In that example, we solely investigated the base form of étendue to formulate our understanding; however, as shown in the previous section, one must also consider the skew invariance of the radiation in order to better analyze the transfer characteristics of the optical system. Greatly using Reference [25] for the development herein, we set up the geometry as shown in Figure 2.18, which is similar to that of Figure 2.17, but in this case, we project onto the meridional plane (i.e., the z–ρ plane).* The differential solid angle in the tangential direction is given by

$$d\Omega = \sin\left(\frac{\pi}{2} - \theta\right)d\theta d\phi = \cos\theta d\theta d\phi, \tag{2.97}$$

* Note that Figures 2.17 and 2.18 use some of the same angles with different variable designations. Thus, the discussions in the two sections should be treated independently of one another. This disparity may at first be confusing to the reader, but it was necessary in order to effectively develop the arguments while using standard terminology.

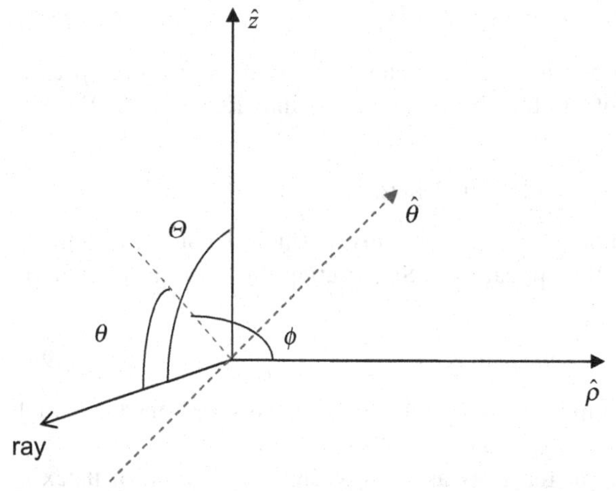

Figure 2.18 Layout of a single ray to the meridional plane ($z - \rho$) with the optical axis in the z-direction.

where θ is the tangential angle between the ray and its projection onto the meridional plane, and ϕ is the azimuthal angle (i.e., the rotation angle around the tangential axis with respect to the meridional plane or the $\hat{\theta}$-axis of Fig. 2.17). Per Equation (2.1), we also have the differential étendue

$$d^2\xi = n^2 \cos\Theta dA d\Omega = n^2 \cos\Theta \cos\theta dA d\theta d\phi, \tag{2.98}$$

where Θ is the angle between the optical axis (z) and the ray. Additionally, the last equality term has substituted for $d\Omega$ per Equation (2.97). From the geometry, as per Equation (2.92), one can accomplish the correct orientation of the ray with respect to the optical axis using two separate rotations

$$\cos\Theta = \cos\theta \sin\phi. \tag{2.99}$$

From the definition of the skew invariant and the geometry of Figure 2.18

$$s = nr\sin\theta. \tag{2.100}$$

Its derivative with respect to θ gives after rearrangement

$$\frac{ds}{r} = n\cos\theta d\theta. \tag{2.101}$$

Additionally, using trigonometry and Equation (2.100), we find

$$\cos\theta = \sqrt{1 - \frac{s^2}{n^2 r^2}}. \tag{2.102}$$

Substituting Equations (2.99), (2.101), and (2.102) into Equation (2.98), an expression for the differential étendue is found

$$d^2\xi = \frac{n}{r}\sqrt{1 - \frac{s^2}{n^2 r^2}} \sin\phi dA ds d\phi. \tag{2.103}$$

Integrating this expression, the skewness distribution is found

$$\frac{d\xi}{ds} = \int_{\text{Surface}} \int_{\phi_{\min}}^{\phi_{\max}} \frac{n}{r}\sqrt{1-\frac{s^2}{n^2r^2}} \sin\phi d\phi dA, \tag{2.104}$$

where the areal integral is over the surface of the emitter, and $[\phi_{\min}, \phi_{\max}]$ is the azimuthal acceptance range. For the tubular emitter coupled to the reflector, the emission varies as a function of the azimuthal range since the reflector captures a portion of the actual emission. Using Equations (2.99) and (2.100), we find the azimuthal acceptance range

$$\phi_{\min} = \arcsin\left(\frac{\cos\Theta_{\min}}{\sqrt{1-\frac{s^2}{n^2r^2}}}\right) \tag{2.105}$$

and

$$\phi_{\max} = \pi - \arcsin\left(\frac{\cos\Theta_{\max}}{\sqrt{1-\frac{s^2}{n^2r^2}}}\right), \tag{2.106}$$

where these equations differ from Reference [25] since the azimuthal cutoff angle may differ in the minimal and maximal directions. From Equation (2.94), it is known that

$$r_{\min} = \frac{|s|}{n\sin\Theta_{\min,\max}} \text{ and } |s| < nr\sin\Theta_{\min,\max}. \tag{2.107}$$

For the sake of simplicity, assume that $\Theta_{\max} = \Theta_{\min}$ for a tubular emitter coupled to a circular disk via a reflector, assume that $\Theta_{\max} = \Theta_{\min}$. Using Equation (2.17), we can substitute for our limits of integration in Equation (2.104), which is then integrated for ϕ

$$\frac{d\xi}{ds} = \int_{nr\sin\Theta_{\max}>|s|} \frac{2n\sin\Theta_{\max}}{r}\left(1-\frac{s^2}{n^2r^2\sin^2\Theta_{\max}}\right)^{1/2} dA. \tag{2.108}$$

For a circular disk of coupled radius R, the skewness distribution is

$$\frac{d\xi}{ds} = \begin{cases} 4\pi nR\sin\Theta_{\max}\left[\sqrt{1-\left(\frac{|s|}{nR\sin\Theta_{\max}}\right)^2} - \left(\frac{|s|}{nR\sin\Theta_{\max}}\right)\arccos\left(\frac{|s|}{nR\sin\Theta_{\max}}\right)\right], & \left(\frac{|s|}{nR\sin\Theta_{\max}}\right) \le 1 \\ 0, & \text{otherwise} \end{cases} \tag{2.109}$$

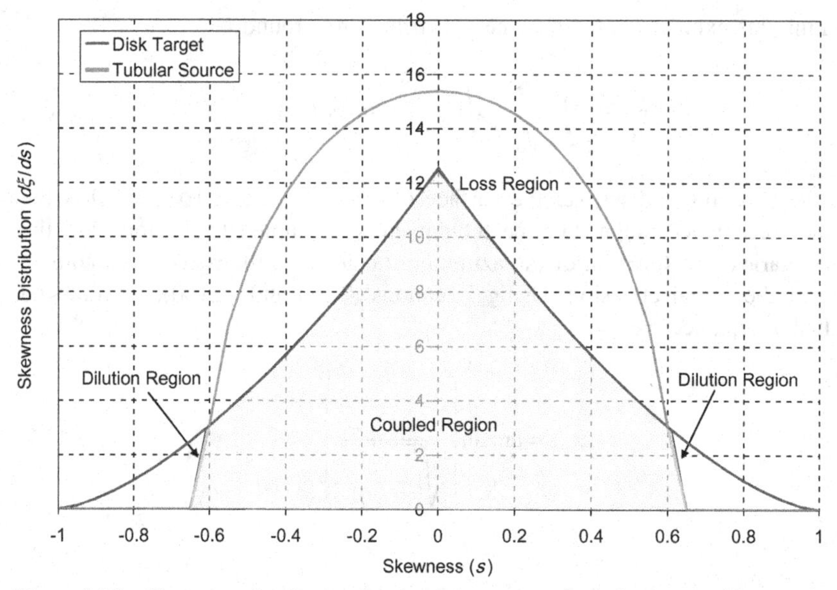

Figure 2.19 Skewness distribution for the disk target and tubular source. The overlap between the two distributions denotes the phase space where flux can be coupled from the source to the target. There is dilution in the region where the source skewness distribution is less than that of the target. There is loss for the parts where the source skewness distribution is greater than that of the target.

For a tube of coupled radius R and length $2h$, the skewness distribution is

$$\frac{d\xi}{ds} = \begin{cases} 8\pi nh\sin\Theta_{max}\sqrt{1-\left(\dfrac{|s|}{nR\sin\Theta_{max}}\right)^2}, & \left(\dfrac{|s|}{nR\sin\Theta_{max}}\right) \le 1. \\ 0, & \text{otherwise} \end{cases} \tag{2.110}$$

Finally, we are able to calculate the effect skewness mismatch for the tubular emitter coupled to a circular disk via a reflector. The reflector is rotationally symmetric, so it cannot alter the skewness of the input distribution. It is a passive device with respect to the skewness, so the computation of the skewness overlap between Equations (2.109) and (2.110) needs to be done to find the efficiency of the system. A specific case is investigated here such that the surface areas of the tube and the disk are the same: $R_{disk} = 1$, and $R_{tube} = h = 1/\sqrt{2}$. The disk allows input from 0 to π, while the tube emits into an angular range of $[\pi/3, 2\pi/3]$. Figure 2.19 shows the skewness distribution for each of the two geometries. There are a number of regions that comprise this plot:

- *Coupled Region.* For the case where $d\xi_{source}/ds \le d\xi_{target}/ds$, the flux within the phase space denoted by this region is coupled from the source to the target.
- *Loss Region.* For the case where $d\xi_{source}/ds \ge d\xi_{target}/ds$, the flux within the phase space denoted by this region is not coupled from the source to the target.

- *Dilution Region.* For the case where $d\xi_{source}/ds \leq d\xi_{target}/ds$ and no available source flux, the phase space is filled by some of the flux from the coupled region. Simply, there is a diffusion of the coupled flux into these regions.

Therefore, the coupled étendue for a system that also obeys conservation of étendue is denoted by the integral of the minimum value for the skewness distribution for the two curves: source and target. Mathematically, this is denoted as

$$\xi_{max} = \int_{-\infty}^{\infty} \min\left[\frac{d\xi_{source}}{ds}, \frac{d\xi_{target}}{ds}\right] ds. \tag{2.111}$$

The term ξ_{max} denotes the case where conservation of étendue is realized. For a system with loss, $\xi_{actual} < \xi_{max}$. The crossing points for the case shown in Figure 2.19 occur at $s \approx \pm 0.60011$. Using the values from above and a source like that of Section 2.4.4, but with only the case where $\varepsilon = 1.0$,* Equation (2.76) at the point $\bar{\xi} = \bar{r}^2 \xi_0$ indicates that the system, without including skew invariance, is 68.40% efficient at transferring the radiation from the source to the target. Including the skew invariance loss, the system is only 45.33% efficient. This disparity is due both to the large loss region, as shown in Figure 2.19. This loss region is due to nonmatching of the two shapes: cylindrical tube to that of a disk. As per the previous footnote, further complexity and nonmatching arises if the emission distribution from the nominal cylindrical tube is nonuniform.

2.7 ÉTENDUE DISCUSSION

This chapter has provided a fairly involved treatment of the term étendue and the associated properties of its conservation, concentration, and skewness. Still, the topic can be greatly expanded upon with additional examples and insights. Etendue is one of the most important aspects to describing and designing efficient nonimaging systems. It is important to remember that étendue is solely a geometric factor—it does not include any physical properties of the light. However, due to that, one can seemingly break the limits of étendue by using such optical phenomena as the spectrum, polarization, coherence, and even mixing to increase the flux in a beam. A short explanation of each method to increase flux without affecting étendue is warranted:

- *Spectral Methods.* A dichroic optical element can be used to add two separate beams with the same geometrical parameters. The two beams have spectra of $\Delta\lambda_1$ and $\Delta\lambda_2$ and fluxes of Φ_1 and Φ_2, and these spectra do not overlap. One beam is reflected by the dichroic window, while the other is transmitted, so the total power is $\Phi_{tot} = \Phi_1 + \Phi_2$ with the original beam parameters.

* The case where $\varepsilon = 1.0$ is for uniform emission from the length of the cylinder. The other cases of $\varepsilon = [0.0, 0.5, 2.0]$ are not done here because the functional forms for the rotational skew invariance are decidedly more difficult to compute. For the other values of ε, the transfer efficiency solely using the étendue while ignoring rotational skew invariance are: 0.0 (53.38%), 0.5 (60.89%), and 2.0 (83.42%).

- *Polarization Methods.* A polarization element can be used to add two orthogonally polarized beams with the same geometrical parameters. Like the spectral case, the total power is $\Phi_{tot} = \Phi_1 + \Phi_2$ with the original beam parameters.
- *Coherence Methods.* Interference can be used to provide a higher flux than delineated by conservation of étendue. This is done through constructive and destructive interference.
- *Mixing Methods.* The arguments presented herein assumed that each ray was seeing the same optical system as the other rays. However, you can have an optical system where some rays see m_1 elements while another set of rays see m_2 elements. An illustrative example is a source coupled to a retroreflector over half its emission space. The light that does not interact with the retroreflector propagates out with the expected étendue, while the retroreflected light matches this étendue (assuming the source emission is symmetric in the forward and backward directions). You have doubled your flux while the étendue is half the expected value. Realistically, there is source geometry obstructing perfect performance. For example, a filament lamp will essentially have the entire retroreflected radiation incident on the coil. Some of this radiation will be absorbed, but some will scatter off the source geometry and add to the flux in the forward direction at no expense to the étendue.

Finally, for the last bullet point, see the next chapter on étendue squeezing. It shows how one can use dilution and mixing to achieve results not expected due to first-order consideration of conservation of étendue. For the first three points, it has been proposed that one could consider the physical étendue [27], such that

$$\int d\Xi = \sum_{P_i} \iiint \iint L(P_i, \lambda, x, y, p, q) d\lambda dx dy dp dq, \qquad (2.112)$$

where Ξ is the physical étendue, L is the radiance, P_i is the ith polarization state, and λ is the wavelength. Essentially, to track physical étendue, we must return to the fundamental definition of source emission—the radiance function. Due to conservation of energy, the radiance function can decrease through an optical system, but never increase.* The term of physical étendue is perfectly conserved through a lossless system.

REFERENCES

1. W.H. Steel, Luminosity, throughput, or etendue? *Appl. Opt.* **13**, 704–705 (1974).
2. C.W. McCutchen, Optical extent, *Appl. Opt.* **13**, 1537 (1974).
3. G.G. Shepherd, How about radiance response? *Appl. Opt.* **13**, 1734 (1974).
4. W.H. Steel, Luminosity, throughput, or etendue? Further comments, *Appl. Opt.* **14**, 252 (1975).
5. Minutes of the 9th Session of ICO, section 10(c), *J. Opt. Soc. Am.* **63**, 906 (1973).
6. ARTFL Project, The University of Chicago, Department of Romance Languages and Literature, http://machaut.uchicago.edu/ (last accessed 9 September 2012).

* This statement assumes there is no gain in the system. If there is gain, then additional terms must be included in Equation (2.112).

7. R.W. Boyd, *Radiometry and the Detection of Optical Radiation*, John Wiley and Sons, New York (1983).
8. W. Cassarly, Taming light, *OE Magazine* (Dec. 2002), pp. 16–18.
9. R. Winston, J.C. Miñano, and P. Benítez, *Nonimaging Optics*, Elsevier Academic Press, Burlington, MA (2005).
10. J. C. Miñano and P. Benítez, Fermat's principle and conservation of 2D étendue, *Nonimaging Optics and Efficient Illumination Systems, Proc. of SPIE*, **5529**, 87–95 (2004). SPIE, Bellingham, WA.
11. M. Born and W. Wolf, *Principles of Optics*, 7th (expanded) ed., Cambridge University Press, Cambridge, UK (1999).
12. R.C. James, *Advanced Calculus*, Wadsworth Publishing Company, Belmont, CA (1966).
13. P. Drude, *The Theory of Optics*, Part III, Dover, New York (1959).
14. R.R. Shannon, *The Art and Science of Optical Design*, Cambridge University Press, Cambridge, UK (1997).
15. P. Mouroulis and J. Macdonald, *Geometrical Optics and Optical Design*, Oxford University Press, New York (2005).
16. M. Born and E. Wolf, *Principles of Optics*, 7th (expanded) ed., Cambridge University Press, Cambridge, UK (1999).
17. P. Moon, *The Scientific Basis of Illuminating Engineering*, McGraw-Hill, New York (1936).
18. E.M. Sparrow and R.D. Cess, *Radiation Heat Transfer*, Brooks/Cole, Pacific Grove, CA (1970).
19. W.R. McCluney, *Introduction to Radiometry and Photometry*, Artech House, Boston (1994).
20. R. Siegel and J.R. Howell, *Thermal Radiation Heat Transfer*, 4th ed., Taylor and Francis, New York (2001).
21. J. R. Howell, A catalog of radiation heat transfer configuration factors, http://www.me.utexas.edu/~Howell/tablecon.html, Univ. of Texas at Austin (last accessed 9 September 2012).
22. H. Rehn, Optical properties of elliptical reflectors, *Opt. Eng.* **43**, 1480–1488 (2004).
23. M.S. Brennesholtz, Light collection efficiency for light valve projection systems, *Projection Displays II, Proc. Of SPIE*, **2650**, 71–79 (1996). SPIE, Bellingham, WA.
24. M.S. Kaminski, K.J. Garcia, M.A. Stevenson, M. Frate, and R.J. Koshel, Advanced topics in source modeling, *Modeling and Characterization of Light Sources, Proc. Of SPIE*, **4775**, 46–57 (2002). SPIE, Bellingham, WA.
25. R. Winston, J. Miñano, and P. Benítez, *Nonimaging Optics*, Elsevier Academic Press, Burlington, MA (2005).
26. J. Chaves, *Introduction to Nonimaging Optics*, CRC Press, Boca Raton, FL (2008).
27. S. A. Lerner and B. Dahlgrenn, Etendue and optical system design, *Nonimaging Optics and Efficient Illumination Systems III, Proc. of SPIE*, **6338**, 633801 (2006). SPIE, Bellingham, WA.

CHAPTER 3

SQUEEZING THE ÉTENDUE

Pablo Benítez, Juan C. Miñano, and José Blen

3.1 INTRODUCTION

When an optical design problem is stated, calculating the fundamental limitations that frame the solution is important to establish if the performance goal is achievable. However, it is not unusual that designers consider restrictions inherited from tradition, which implies limitations that are not fundamental. One example of such nonfundamental restriction is the symmetry of the optics, due to the tradition of optical systems being rotational symmetric.

In this chapter, we will think "outside of the box" to discuss a general concept, called étendue squeezing [1] or étendue remapping [2], which allows one to find devices [3] that solve design problems without solution if only plane symmetric optics are used, that is, those that contend with the limits of skew invariance as presented in the previous chapter. This concept gives us more degrees of freedom for the design, and allows improving the control of the light beam of an optical system, particularly where tight light control is needed due to efficiency and geometrical constraints. This demand is especially pertinent for applications such as automotive lighting and video projectors.

3.2 ÉTENDUE SQUEEZERS VERSUS ÉTENDUE ROTATORS

It is usual to define the ray bundle M as it passes through a *reference surface* Σ_R of the three-dimensional space such that Σ_R intersects only once the trajectories of the rays of M. It is well known that if M_{4D} is a tetra-dimensional ray bundle (i.e., $\dim(M_{4D}) = 4$), its étendue is defined in Cartesian coordinates as the value of the integral [4],

$$\xi(M_{4D}) = \int_{M_{4D}(\Sigma_R)} dxdydpdq + dxdzdpdr + dydzdqdr, \tag{3.1}$$

where $\mathbf{v} = (p, q, r)$ is the vector tangent to the ray at point $\mathbf{r} = (x, y, z)$, and $|\mathbf{v}| = n(\mathbf{r})$ where $n(\mathbf{r})$ is the refractive index of the medium. If the reference surface is a plane

Illumination Engineering: Design with Nonimaging Optics, First Edition. R. John Koshel.
© 2013 the Institute of Electrical and Electronics Engineers. Published 2013 by John Wiley & Sons, Inc.

$z = z_a$, only the first addend of the integrand in Equation (3.1) is non-null, and the étendue coincides with the volume defined by M in the phase space x–y–p–q.

The theorem of conservation of the étendue states that this is an invariant of the ray bundle when it is propagated through an optical system, that is, it is independent of the reference surface on which it is calculated. The "étendue" is one of the invariants of Poincaré, and this theorem is equivalent to the Liouville theorem in three dimensions [4].

There exists another less-known étendue that is defined for biparametric ray bundles (not necessarily coplanar). If M_{2D} is a biparametric ray bundle, its étendue is given by

$$\xi(M_{2D}) = \int_{M_{2D}(\Sigma_R)} dxdp + dydq + dzdr. \tag{3.2}$$

This second invariant is another invariant of Poincaré [1, 5], and it is equivalent to Lagrange's invariant [1, 6]. When the rays of the bundle are coplanar, this invariant is also equivalent to Liouville's theorem in two dimensions. Since the "étendue" must be conserved for any ray bundle, the differential forms—the integrands of Equations (3.1) and (3.2)—are also conserved.

Let us consider an optical system that transmits rays crossing the reference plane $z = z_a$, at which the refractive index is $n = 1$ toward increasing z-values (i.e., with $r > 0$) until the second reference plane $z = z_b$, with local Cartesian transversal coordinates x'–y'. The values of x'–y'–p'–q' of the propagated rays at $z = z_b$ are given by the mapping:

$$\begin{aligned} x' &= x'(x, y, p, q) \\ y' &= y'(x, y, p, q) \\ p' &= p'(x, y, p, q) \\ q' &= q'(x, y, p, q). \end{aligned} \tag{3.3}$$

The functions in this mapping are not independent of one another (this mapping belongs to a class of transformations called canonical), since it is caused by the propagation of rays that follow the laws of geometrical optics.

In order to illustrate some limitations of plane symmetric optics, let us consider the biparametric ray bundle M_{2D} defined at the reference plane $z = z_a$ by:

$$\begin{aligned} &x \in \{-x_0, x_0\} \\ &y = 0 \\ &p \in \{-\sin\alpha, \sin\alpha\} \\ &q = 0, \end{aligned} \tag{3.4}$$

That is, the rays at $z = z_a$ are contained in the plane $y = 0$ and are tangent per Equation (3.4) to it. Therefore, x and p are two parameters defining the rays of the bundle, and its étendue is

$$\xi(M_{2D}) = \int_{M_{2D}(z=z_0)} dxdp = (2\sin\alpha)(2x_0). \tag{3.5}$$

Étendue conservation guarantees that after the propagation, we can compute the étendue as

$$\int_{M_{2D}(z=z_b)} dx'dp' + dy'dq' = (2\sin\alpha)(2x_0). \tag{3.6}$$

If the mapping of Equation (3.3) is symmetric with respect to the plane $y = 0$, then it is fulfilled that

$$\begin{aligned} x'(x,y,p,q) &= x'(x,-y,p,-q) \\ -y'(x,y,p,q) &= y'(x,-y,p,-q) \\ p'(x,y,p,q) &= p'(x,-y,p,-q) \\ -q'(x,y,p,q) &= q'(x,-y,p,-q). \end{aligned} \tag{3.7}$$

Assuming the mapping is continuous, from the second and fourth expressions in Equation (3.7), we deduce that the rays of M_{2D} have

$$\begin{aligned} y'(x,0,p,0) &= 0 \\ q'(x,0,p,0) &= 0. \end{aligned} \tag{3.8}$$

Since this result is true for any z_b-value, Equation (3.8) is just saying that in the optical system, with the continuous symmetric mapping of Equation (3.7), the trajectories of the rays of M_{2D} are contained in the plane $y = 0$. Therefore, Equation (3.6) computes the étendue at $z = z_b$ and can be reduced to the usual étendue formula for planar geometry

$$\int_{M_{2D}(z=z_b)} dx'dp' = (2\sin\alpha)(2x_0). \tag{3.9}$$

Thus, the value for the integral in Equation (3.9) is fixed by the symmetry with respect to $y = 0$ of the continuous mapping. However, by breaking the symmetry, there exist optical systems that produce the strict inequality

$$\int_{M_{2D}(z=z_b)} dx'dp' < (2\sin\alpha)(2x_0). \tag{3.10}$$

This outcome has the result of finding efficient optical systems with reduced average angular span $\langle\Delta p'\rangle$ of the rays of M_{2D} at the plane $z = z_b$. We can identify two types of such systems:

- *Étendue Rotators.* In which the integral of term $dy'dq'$ of the integrand in Equation (3.6) is positive, not null, so the inequality of Equation (3.9) is achieved.
- *Étendue Squeezers.* In which the rays of M_{2D} at the plane $z = z_b$, when projected on to the plane x'–p', is multifolded (the number folds is the order of the squeezer). In this case, the integral of term $dy'dq'$ of the integrand in Equation (3.6) can be null.

3.2.1 Étendue Rotating Mappings

Let us illustrate first the étendue rotation principle by a simple canonical example that transforms M_{2D} in such a way that not only the integral of term $dy'dq'$ is not null, but also $dx'dp' = 0$. Such a result is obtained if the mapping of Equation (3.3) produced by the optical system is:

$$\begin{aligned} x' &= y + \sigma p \\ y' &= x + \sigma q \\ p' &= q \\ q' &= p, \end{aligned} \tag{3.11}$$

where σ is an arbitrary constant (it may be $\sigma = 0$). Since for M_{2D} we have $y = 0$ and $q = 0$, we obtain for M_{2D}

$$\begin{aligned} x' &= \sigma p \\ y' &= x \\ p' &= 0 \\ q' &= p. \end{aligned} \tag{3.12}$$

And we can then deduce

$$\begin{aligned} dx'dp' &= \begin{vmatrix} \dfrac{\partial x'}{\partial x} & \dfrac{\partial x'}{\partial p} \\ \dfrac{\partial p'}{\partial x} & \dfrac{\partial p'}{\partial p} \end{vmatrix} dxdp = \begin{vmatrix} 0 & \sigma \\ 0 & 0 \end{vmatrix} dxdp = 0 \\ dy'dq' &= \begin{vmatrix} \dfrac{\partial y'}{\partial x} & \dfrac{\partial y'}{\partial p} \\ \dfrac{\partial q'}{\partial x} & \dfrac{\partial q'}{\partial p} \end{vmatrix} dxdp = \begin{vmatrix} 1 & 0 \\ 0 & 1 \end{vmatrix} dxdp = dxdp. \end{aligned} \tag{3.13}$$

Thus, for this étendue rotator the integral of Equation (3.10) is evaluated as

$$\int_{M_{2D}(z=z_b)} dx'dp' = 0, \tag{3.14}$$

and the étendue of the bundle is visible in the y'–q' plane

$$\int_{M_{2D}(z=z_b)} dy'dq' = (2\sin\alpha)(2x_0). \tag{3.15}$$

Figure 3.1 shows the projection of the ray bundle M_{2D} on the phase-space planes x–p and y–q at $z = z_a$ (before entering the étendue rotator) and on the phase-space planes x'–p' and y'–q' at $z = z_b$ (after the étendue rotator). The area of the two rectangles equals the étendue of the bundle, $(2 \sin \alpha)(2x_0)$.

Designing an optical system that produces the mapping of Equation (3.11) exactly is a very difficult problem; however, in the paraxial domain, it is easier since the mapping of Equation (3.3) is approximated to be linear. As can be checked, this

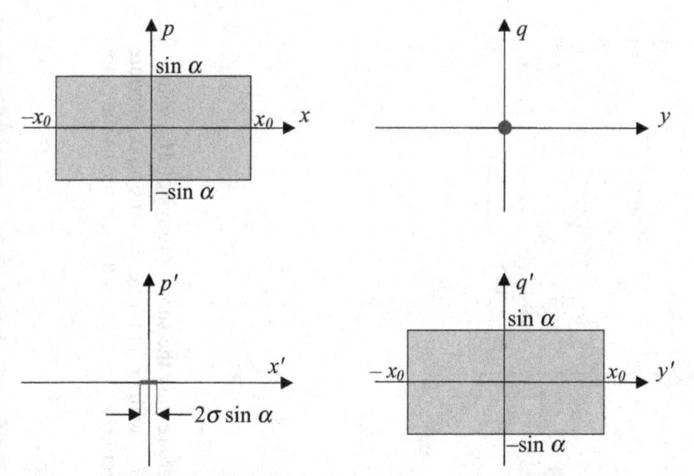

Figure 3.1 Projection of the ray bundle M_{2D} on the phase-space planes x–p and y–q at $z = z_a$ (before entering the étendue rotator) and on the phase-space planes x'–p' and y'–q' at $z = z_b$ (after the étendue rotator).

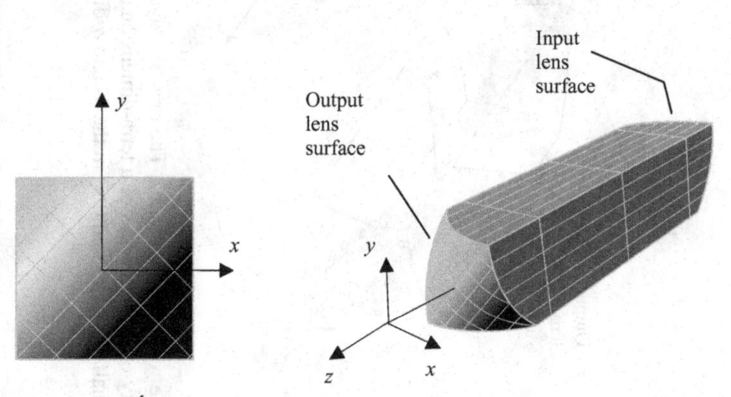

Figure 3.2 Étendue rotator made of a cylindrical lens whose symmetry direction is parallel to the straight line $x = y$ and $z = 0$, and whose cross section is a Keplerian afocal system with unit magnification.

paraxial mapping is produced by a cylindrical lens whose symmetry direction is parallel to the straight line $x = y$ and $z = 0$, and whose cross section is a Keplerian afocal system with unit magnification (see Appendix 3.B). Figure 3.2 shows the Keplerian system made with the two surfaces of a single lens. For this lens, the constant σ in the mapping of Equation (3.11) is given by $\sigma = d/n$, where d is the lens thickness (i.e., distance between the vertices or intersections of the input and output surfaces with the z-axis) and n is the refractive index of the lens.

Figure 3.3 shows ray tracing results on the étendue rotator of Figure 3.2. The rays displayed on the center are a subset of rays of M_{2D} defined by $q = 0$, $y = 0$, and $p = 0$. The plot on the right shows the intensity of the input rays of a tetra-dimensional ray bundle M_{4D} with (x, y) filling the lens input aperture and with a Gaussian intensity

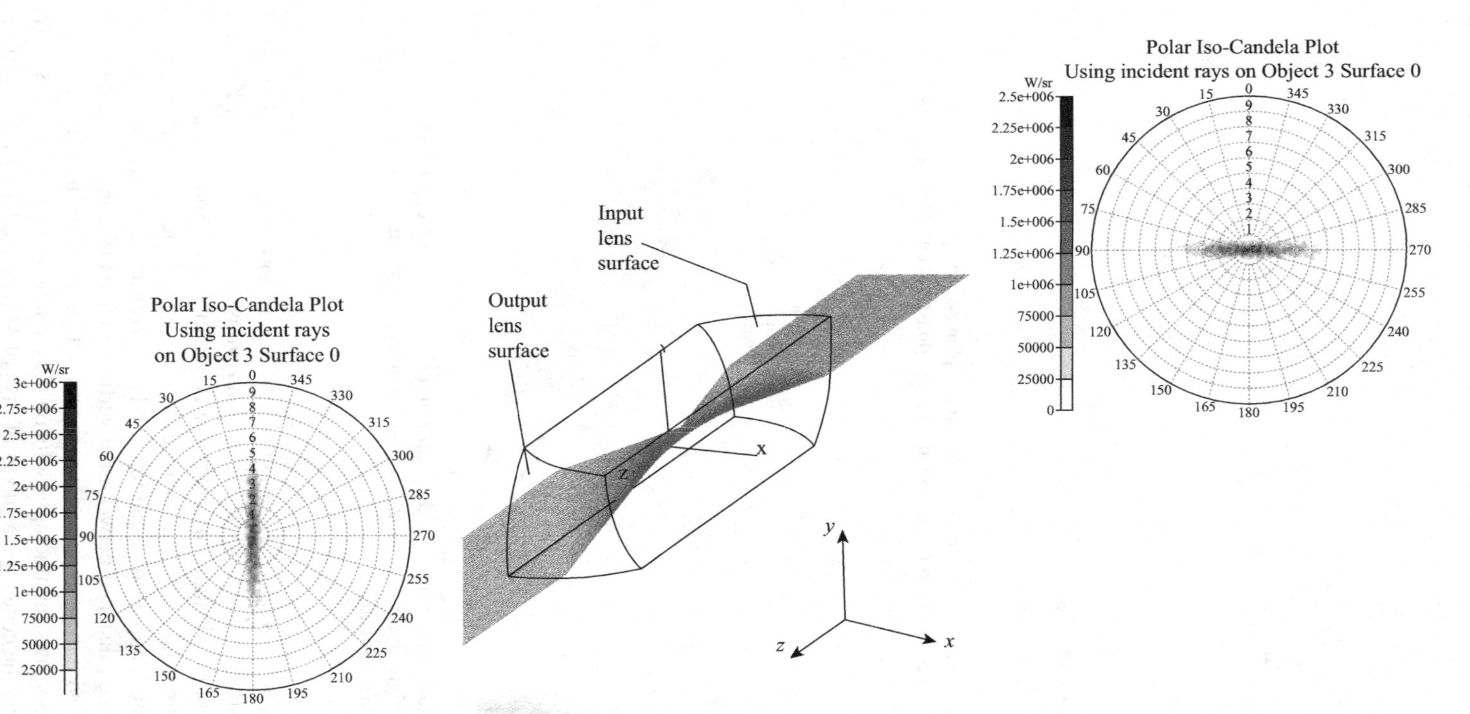

Figure 3.3 Ray tracing results on the étendue rotator of Figure 3.2. The rays displayed on the center are only the subset of rays of M_{2D} with $p = 0$. The plot on the right shows the intensity of the input rays of a tetra-dimensional ray bundle M_{4D}, with (x, y) filling the lens input aperture and with a Gaussian intensity distribution of half-angles 5° and 0.5° in the p- and q-directions, respectively. The plot on the left shows the output intensity, which is rotated as expected.

distribution of half-angles 5°and 0.5° in the p- and q-directions, respectively. The plot on the left shows the output intensity, which is rotated as expected. (Note that for the mapping per Eq. 3.11, the p and q variables are exchanged independently of x and y.)

3.2.2 Étendue Squeezing Mappings

Let us now illustrate the étendue squeezing concept. Selecting the order of the étendue squeezing as two (it will be referred to as a 2:1 squeezer), we choose the following mapping:

$$\begin{aligned} x' &= \begin{cases} 2x - x_0 + \sigma p & x > 0 \\ 2x + x_0 + \sigma p & x < 0 \end{cases} \\ y' &= \begin{cases} \frac{1}{2} y + \frac{1}{2} y_0 + \sigma q & x > 0 \\ \frac{1}{2} y - \frac{1}{2} y_0 + \sigma q & x < 0 \end{cases} \\ p' &= \frac{p}{2} \\ q' &= 2q, \end{aligned} \tag{3.16}$$

where again σ is an arbitrary constant (σ may even be 0). Since for M_{2D}, we have $y = 0$ and $q = 0$, we obtain for M_{2D}

$$\begin{aligned} x' &= \begin{cases} 2x - x_0 + \sigma p & x > 0 \\ 2x + x_0 + \sigma p & x < 0 \end{cases} \\ y' &= \begin{cases} \frac{1}{2} y_0 & x > 0 \\ -\frac{1}{2} y_0 & x < 0 \end{cases} \\ p' &= \frac{p}{2} \\ q' &= 0. \end{aligned} \tag{3.17}$$

We then deduce that

$$\begin{aligned} dx'dp' &= \begin{vmatrix} \frac{\partial x'}{\partial x} & \frac{\partial x'}{\partial p} \\ \frac{\partial p'}{\partial x} & \frac{\partial p'}{\partial p} \end{vmatrix} dxdp = \begin{vmatrix} 2 & \sigma \\ 0 & \frac{1}{2} \end{vmatrix} dxdp = dxdp \\ dy'dq' &= \begin{vmatrix} \frac{\partial y'}{\partial x} & \frac{\partial y'}{\partial p} \\ \frac{\partial q'}{\partial x} & \frac{\partial q'}{\partial p} \end{vmatrix} dxdp = \begin{vmatrix} 0 & 0 \\ 0 & 0 \end{vmatrix} dxdp = 0. \end{aligned} \tag{3.18}$$

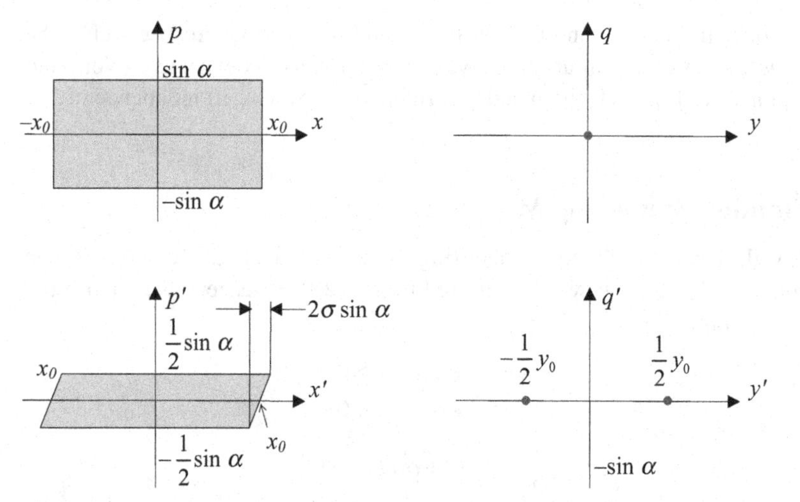

Figure 3.4 Projection of the ray bundle M_{2D} on the phase-space planes x–p and y–q at $z = z_a$ (before entering the étendue squeezer) and on the phase-space planes x'–p' and y'–q' at $z = z_b$ (after the étendue squeezer).

Before evaluating the étendue integral of Equation (3.10), let us visualize the projection of the ray bundle M_{2D} on the phase-space planes x–p and y–q at $z = z_a$ (i.e., before entering the étendue squeezer) and on the phase-space planes x'–p' and y'–q' at $z = z_b$ (i.e., after the étendue squeezer). As shown in Figure 3.4, the mapping Equation (3.16) causes the projection of the bundle on the plane x'–p' at the output $z = z_b$ to occupy half the area of the projection of the bundle in the plane x–p at the input $z = z_a$. The reason is that the bundle is folded in phase space in such a way that the two halves ($x > 0$ and $x < 0$) of the projection of the bundle in the plane x–p transform onto the same overlapped rectangle when projected on the plane x'–p'. When represented on the plane y'–q', the projection of the two halves is clearly distinguished (the $x > 0$ half projects onto the point with positive y', while the $x < 0$ half projects onto the point with negative y'). Since the étendue is preserved, the area of the two overlapped rectangles equals the étendue of the bundle, $(2 \sin\alpha)$ $(2x_0)$.

Thus, due to the overlapping when projecting the bundle on the plane x'–p', we obtain that for this 2 : 1 étendue squeezer, the integral of Equation (3.10) is evaluated as

$$\int_{M_{2D}(z=z_b)} dx'dp' = \frac{1}{2}(2\sin\alpha)(2x_0), \tag{3.19}$$

which is one-half of the value in Equation (3.9) obtained by the continuous symmetrical mapping. On the other hand

$$\int_{M_{2D}(z=z_b)} dy'dq' = 0. \tag{3.20}$$

Thus, the order of squeezing, as presented at the beginning of this section, is the factor by which étendue is reduced in one phase space plane and expanded in the other plane. The remainder of this chapter is devoted to the design of étendue squeezers. More complex examples of étendue rotators designed with the simultaneous multiple surfaces (SMS) 3D design method are given in Chapter 4 [4]. In the rest of this chapter, we will only discuss étendue squeezers.

3.3 INTRODUCTORY EXAMPLE OF ÉTENDUE SQUEEZER

Consider a simple collimating lens collecting light from a square source radiating into a square solid angle (see Fig. 3.5). This lens can be used to illuminate a distant square target having a square field of view (FOV) from its square exit aperture.

The angular extension of the lens FOV can be easily calculated from the conservation of 2D étendue. In this case, both the vertical (*V*) and the horizontal (*H*) angular spreads are the same and equal to AS_1 (angular spread) in Figure 3.5. The surfaces of such a collimating lens could simply be rotationally symmetric aspheres (with contours truncated to match the input and exit apertures).

Suppose for instance that the goal is to illuminate a rectangular (i.e., non-square) target utilizing the same lens contour, source shape, and source emitted FOV as in Figure 3.5. In this case, squeezing (or transforming) the beam FOV_1 (field of view) will be required to preserve coupling efficiency.

One way to implement étendue squeezing optics is by dividing the optical system into two or more lenticular pairs. Each lenticular pair will intercept part of the flux coming from the source and then transfer it to the target. The angular spread or compression produced by each lenticular pair will be determined by the tessellation of the pairs.

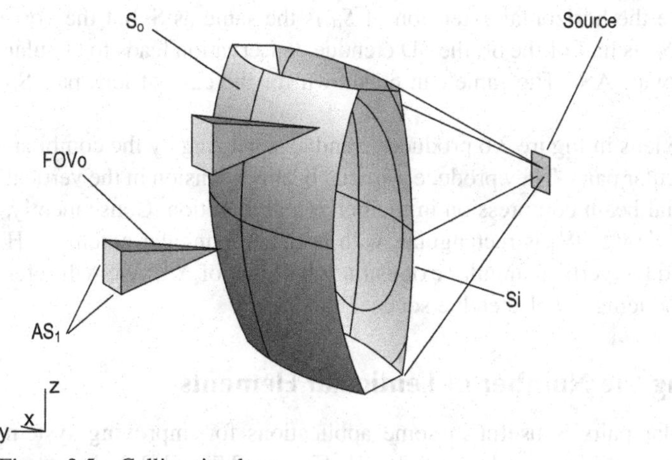

Figure 3.5 Collimating lens.

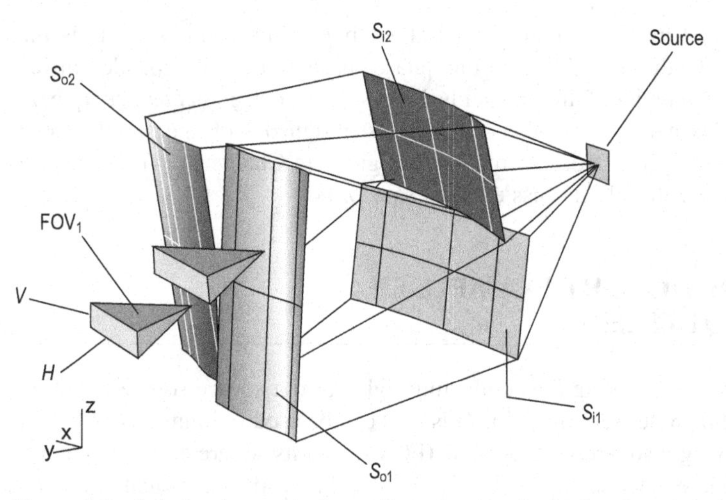

Figure 3.6 2:1 étendue squeezing optics using two lenticular element pairs.

Figure 3.6 shows how an étendue-squeezing, freeform lens collects the light from the same square source with the same square input and exit apertures as in the Figure 3.5. The input side of the lens has two lenticular elements: S_{i1} on the negative (lower) part of the as shown z-axis and S_{i2} on the positive (upper) part of z-axis, while the output side has two lenticular elements: S_{o1} on the negative part of the as shown x-axis and S_{o2} on the positive part of x-axis.

To calculate the emitted FOV of the lens of Figure 3.6, consider first the lens pair S_{i1} and S_{o1}. This pair can be designed as a conventional off-axis lens (or a nonimaging lens using, for instance, the SMS 3D design method) [4] to provide the required collimation. Again, the emission angles H and V will be given by 2D étendue conservation. On one hand, since the vertical extension of S_{i1} is half that of S_i in Figure 3.5, but the vertical extensions of S_{o1} and S_o are the same, the 2D étendue conservation leads to angular extension V being half that of AS_1 in Figure 3.5. On the other hand, since the horizontal extension of S_{i1} is the same as S_i but the horizontal extension of S_{o1} is half of the S_o, the 2D étendue conservation leads to angular extension H being twice AS_1. The same can be shown for the case of lens pair S_{i2} and S_{o2}.

Therefore, the lens in Figure 3.6 produces étendue squeezing by the combination of the two lenticular pairs. They produce a spatial beam expansion in the vertical direction and a spatial beam compression in the horizontal direction. Consequently, the emitted field of view FOV_1 is rectangular, with horizontal angular extension H twice that of AS_1 and the vertical angular extension V half that of AS_1. We will refer to this device as producing a 2:1 étendue squeeze factor.

3.3.1 Increasing the Number of Lenticular Elements

Using more lenticular pairs is useful in some applications for improving system performance, manufacturability, and light control. Figures 3.7 and 3.8 show two further examples of 2:1 étendue squeezing optics, using three and five lenticular element pairs, respectively. In these examples the positions of the lenticular arrangements in Figure 3.6 have been modified. In the example of Figure 3.7, the modifications introduce more symmetry with respect to vertical planes, while in the example of the Figure 3.8, the modifications provide higher symmetry as well, with respect to horizontal planes.

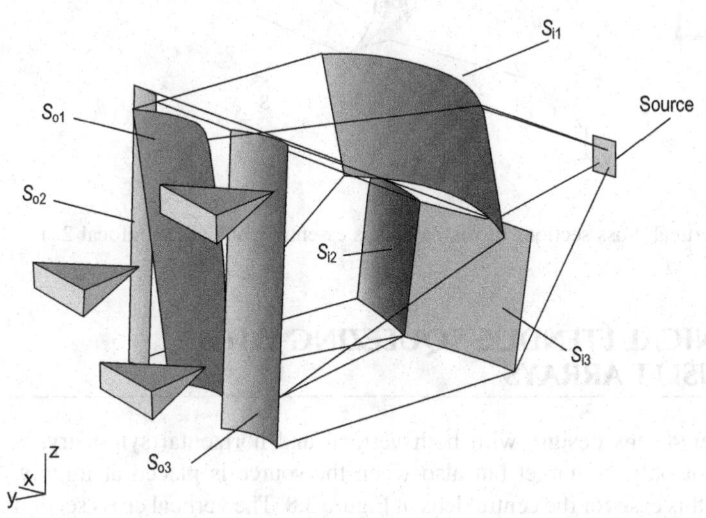

Figure 3.7 2:1 étendue squeezing optics using three lenticular element pairs.

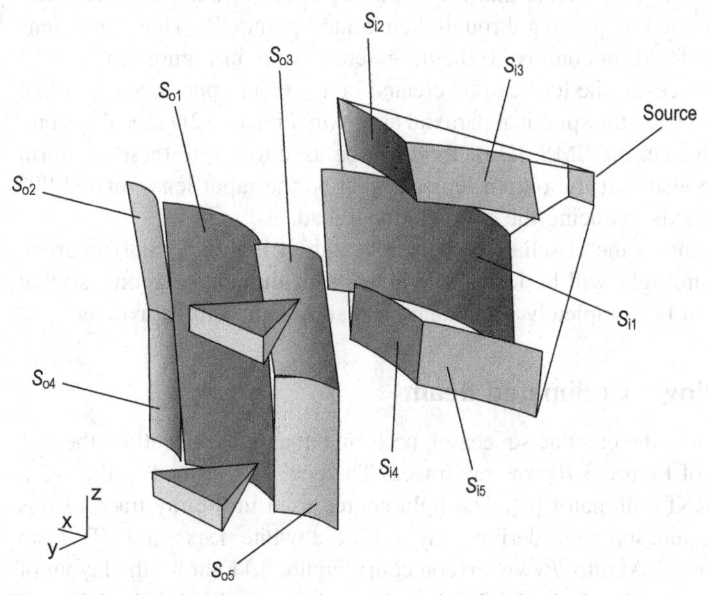

Figure 3.8 2:1 étendue squeezing optics using five lenticular element pairs.

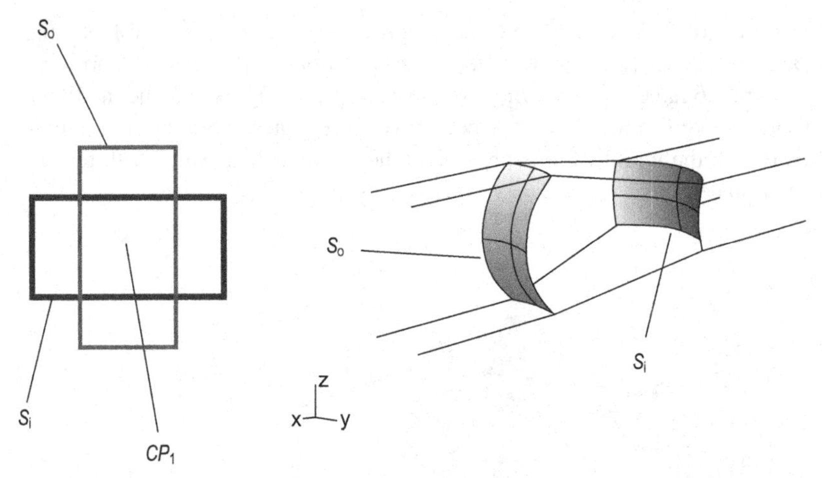

Figure 3.9 The vertical cross sections of such a lens is essentially a Galilean afocal 2:1 beam expander.

3.4 CANONICAL ÉTENDUE-SQUEEZING WITH AFOCAL LENSLET ARRAYS

A perfectly centered lens design, with both vertical and horizontal symmetry, is achieved when not only the target but also when the source is placed at infinity. Figure 3.9 shows this case for the central lens in Figure 3.8. The vertical cross section of such a lens is essentially a Galilean [5] afocal 2:1 beam expander, while the horizontal cross section is the same afocal beam expander but reversed, that is, to be used as beam compressor. Note that the symmetry enforces that the optical surfaces be normal to the line passing through their central point CP_1. Thus, these lens pairs are called centered, in contrast to the decentered lenses in Figure 3.6.

Both 3D surfaces of the lens can be created in a simpler approach as toroidal surfaces, calculated from the spherical paraxial approximation to a 2D afocal system. For better performance, the SMS 3D method can be used to create these freeform surfaces [4]. Note also that the output lens can just be the input lens, rotated 90° around the optical axis, reducing the computational load.

Figure 3.10 shows the tessellation of the lens unit of Figure 3.9 into an array. In the ideal case, no light will be lost in the junction of the lenticulations, so that the exit surface will be completely illuminated and stray light will be avoided.

3.4.1 Squeezing a Collimated Beam

In order to show how the étendue squeezing performs in a real application, the 2:1 étendue-squeezer of Figure 3.10 was ray traced. The goal was to modify the beam generated by an RXI collimator [4]. The light source used in the ray trace of this collimator was a measurement-derived ray set of a white Luxeon LED from LumiLeds (San Jose, CA; http://www.luxeon.com). Figure 3.11 shows the layout of the optical system.

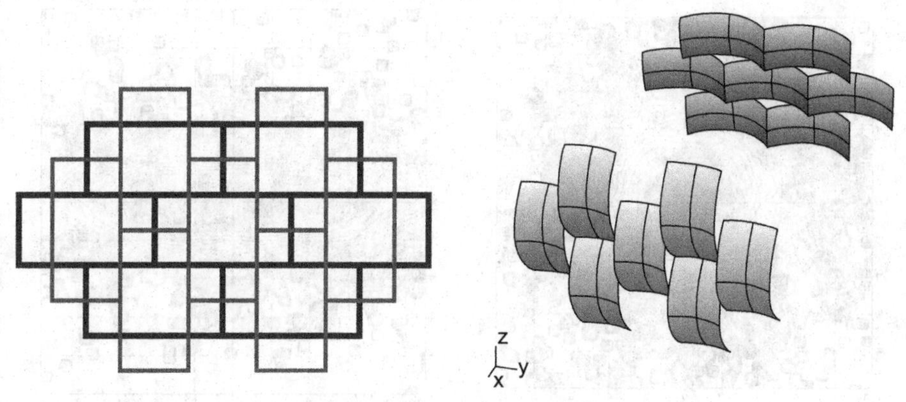

Figure 3.10 Tessellation of the lens unit of Figure 3.9 into an array.

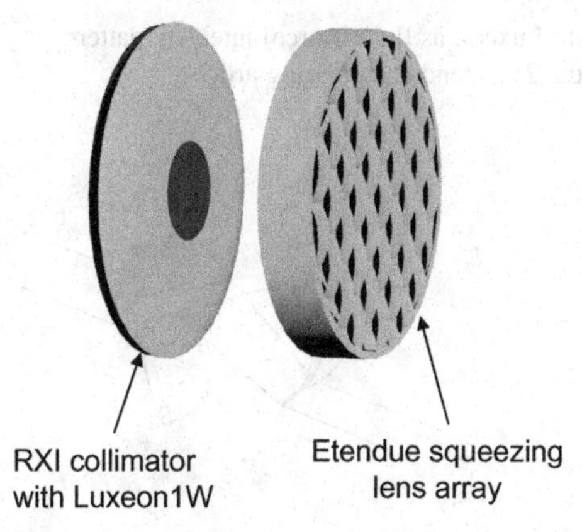

Figure 3.11 Collimator and étendue squeezing lens array layout.

Figure 3.12a shows the intensity pattern generated by only the RXI collimator. Figure 3.12b right shows how the 2 : 1 étendue squeezing array compresses the pattern in the *V* dimension and expands it in the H dimension.

An interesting feature of these afocal arrays is the capability for image formation. This means that if the 2 : 1 étendue squeezer array of Figure 3.10 is placed in front of a camera with a telephoto objective, it will anamorphically modify the FOV of the camera, making the image on the photo stretch in one dimension by a factor of two and compress the same in the other. This feature, along with the corresponding image processing, can be of interest in applications in which a low-aspect ratio (height/width) FOV is desired, but is not available in standard image sensors.

3.4.2 Other Afocal Designs

There are a further three afocal designs (see Fig. 3.13) beyond that of Figure 3.9. In Figure 3.9, the two cross sections of the lenticulations are 2D Galilean afocal designs (vertical expander and horizontal compressor), as such do not invert the far-field 2D

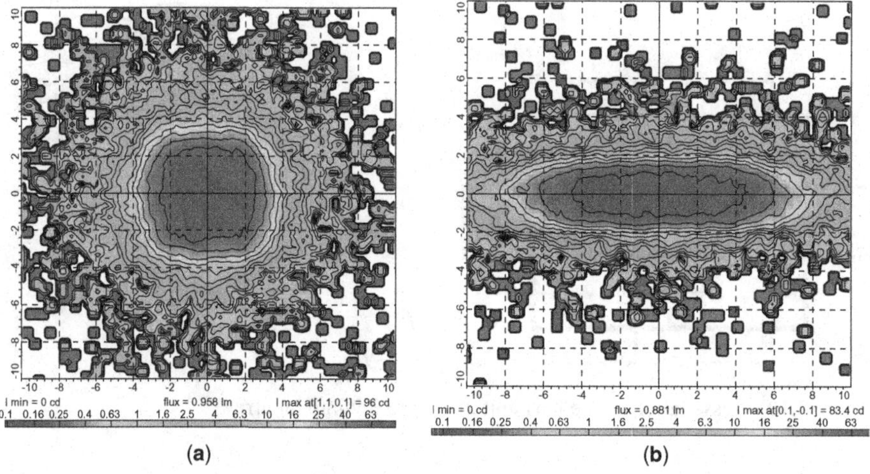

Figure 3.12 (a) RXI collimator (using a white Luxeon as light source) intensity pattern and (b) Right, collimator patter squeezed by the 2 : 1 étendue squeezing array.

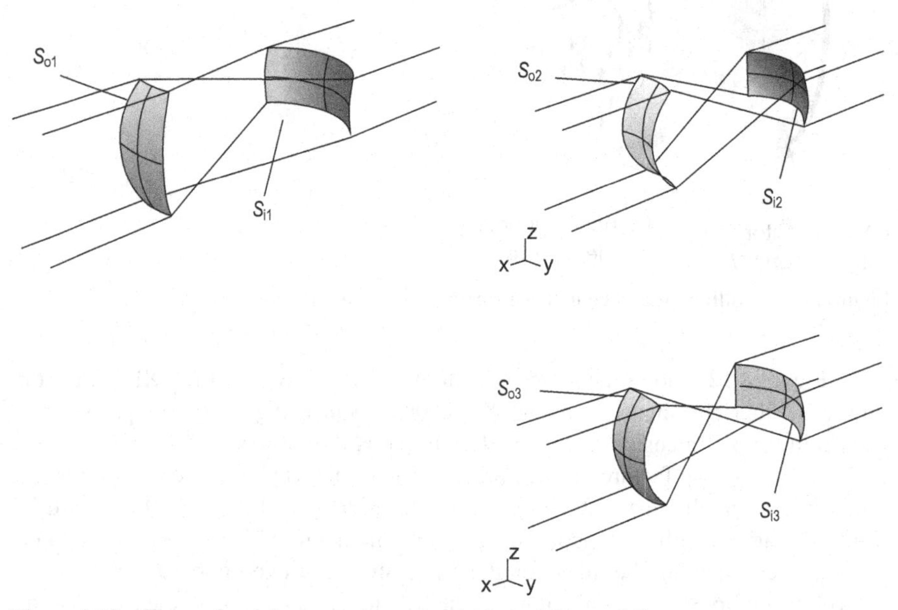

Figure 3.13 Three afocal designs, alternative to that of Figure 3.9.

image. Keplerian [5] (inverting) designs are also possible, as well as combinations of the two. Thus, in the case of lens pair S_{i1} and S_{o1}, the horizontal cross section is inverted but not the vertical one. In the case of lens pair S_{i2} and S_{o2}, the vertical cross section is inverted but not the horizontal one, while lens pair S_{i3} and S_{o3} inverts in both planes.

3.4.3 Étendue-Squeezing Lenslet Arrays with Other Squeeze-Factors

Further 2 : 1 tessellations can be obtained by scaling in the vertical dimension the contours of the tessellation diagram of Figure 3.10 (the squeeze factor can be greater or smaller than one). See, for example, the method presented in Figure 3.14. In Figure 3.14, the squeeze factor is two, so that the aspect ratio of input lens unit S_i is 2 : 2 (i.e., 1 : 1), while the aspect ratio of the output lens unit S_o is 4 : 1. Note that the lens profiles do not scale, but now the horizontal and vertical designs show different *f*-numbers. This scaling factor can also be applied to the étendue squeezers explained in Figure 3.15.

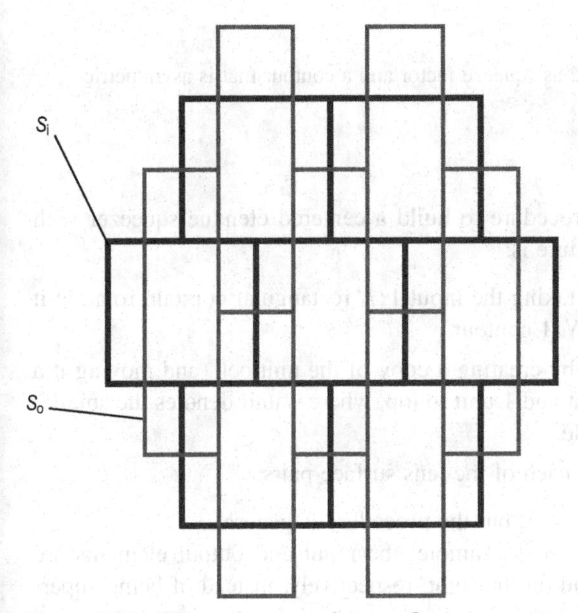

Figure 3.14 Effect of squeeze factor = 2.

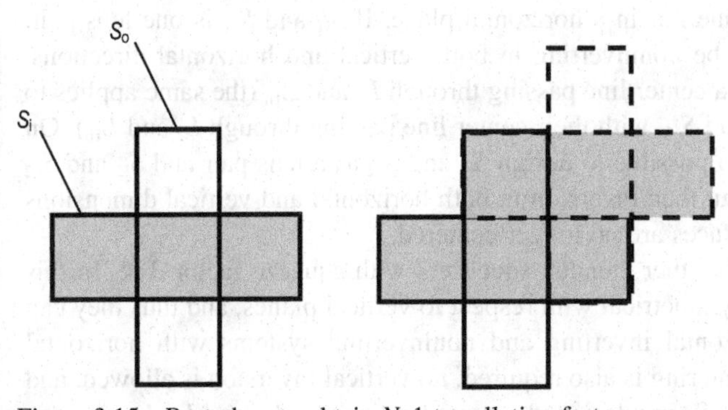

Figure 3.15 Procedure to obtain *N* : 1 tessellation factors.

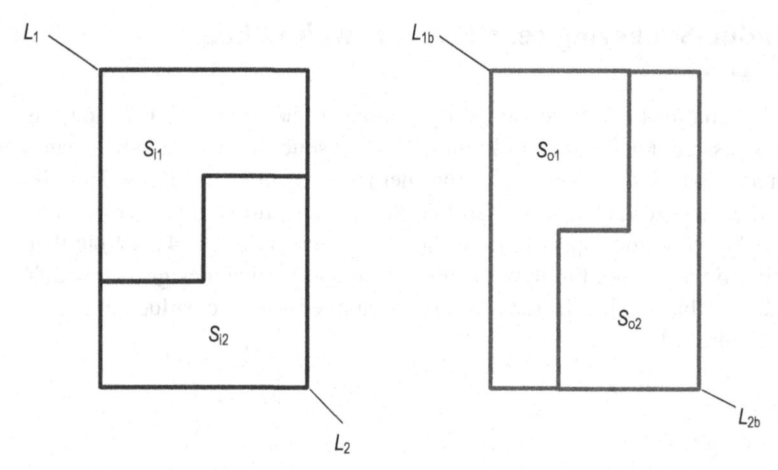

Figure 3.16 Tessellation with 3 : 2 as squeeze factor and a contour that is asymmetric about the vertical plane.

Figure 3.15 shows the procedure to build a centered étendue squeezer with squeeze factor N: 1. The procedure is:

1. Generate the unit cell by taking the input 1 : N rectangular contour, rotating it 90° to obtain the output N: 1 contour.
2. Generate the tessellation by creating a copy of the unit cell, and moving it a distance 1 unit to the right and 1 unit to top, where 1 unit denotes the smaller dimension of the rectangle.
3. Continue this process for each of the lens surface pairs.

The figure shows the case for $N = 3$, but the procedure is general

For greater clarity in the next example, the input and output elements are shown separately on the left and on the right, respectively, instead of being superimposed as used in previous Figures.

Figure 3.16 shows an étendue squeezer with squeeze-factor 3 : 2. In contrast to the previous squeezers, the contour of these lenticular elements is symmetric neither in a vertical plane nor in a horizontal plane. If S_{i1} and S_{o1} is one lens pair, the afocal design must be noninverting in both vertical and horizontal directions, and can be centered on a center line passing through L_1 and L_{1b} (the same applies to the other lens pair S_{i2} and S_{o2}, with their center line passing through L_2 and L_{2b}). On the other hand, it is also possible to design S_{i1} and S_{o2} as a lens pair and S_{i2} and S_{o2} as the other lens pair, but then inversion in both horizontal and vertical dimensions is required, and the surfaces are no longer centered.

Figure 3.17 shows other étendue squeezers with squeeze factor 3 : 2. In this case, the contours are symmetrical with respect to vertical planes, and thus they can be used for both horizontal inverting and noninverting systems with horizontal centering. If vertical centering is also required, no vertical inversion is allowed, and lens pairs S_{i1}–S_{o1} and S_{i2}–S_{o2} are designed centered at lines L_1–L_{1b} and L_2–L_{2b}, respec-

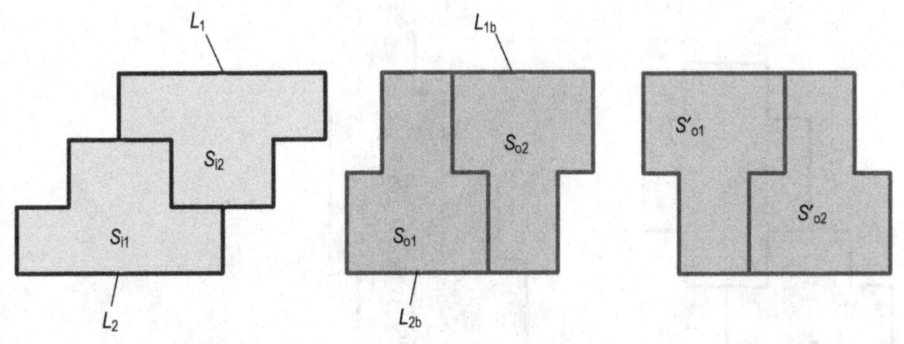

Figure 3.17 Tessellation with 3 : 2 as squeeze factor and a contour that is symmetric about the vertical plane, when vertical inversion is applied (right) and here it is not (center).

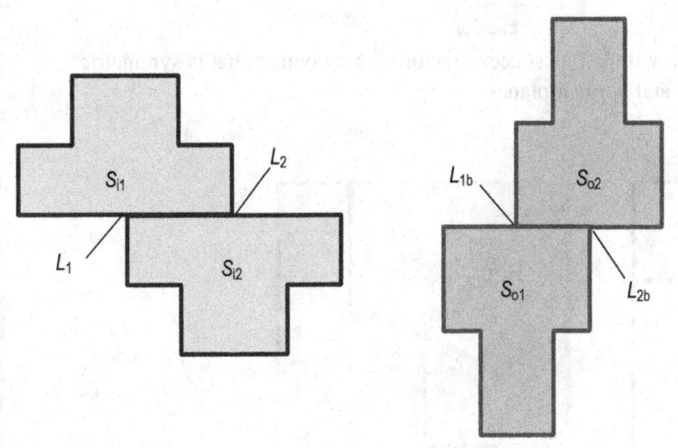

Figure 3.18 Tessellation with 3 : 2 as squeeze factor and a contour that is symmetric about the vertical plane with centered optics.

tively. In case vertical inversion is preferred, such that lens pairs S_{i1}–S'_{o1} and S_{i2}–S'_{o2} will not be centered vertically.

Figure 3.18 shows another vertically centered, inverting étendue squeezer with squeeze factor 3 : 2, in which lens pairs S_{i1}–S_{o1} and S_{i2}–S_{o2} are designed centered at lines L_1–L_{1b} and L_2–L_{2b}, respectively.

Figure 3.19 shows another étendue squeezer with squeeze factor 3 : 2 and lens pairs S_{i1}–S_{o1} and S_{i2}–S_{o2}. In contrast to the previous 3 : 2 squeezers, this is symmetric both horizontally and vertically, and thus can be designed to be centered independently of the inversion or noninversion in the horizontal and vertical directions.

Figure 3.20 shows a generalization of the previous 3 : 2 squeezers: a centered étendue squeezer similar to that of Figure 3.16 but with squeeze factor $(N + 2) : (N + 1)$, with N indicating the number of steps in the contour (e.g., $N = 1$ in the example of Fig. 3.16 and $N = 2$ in the example of Fig. 3.19).

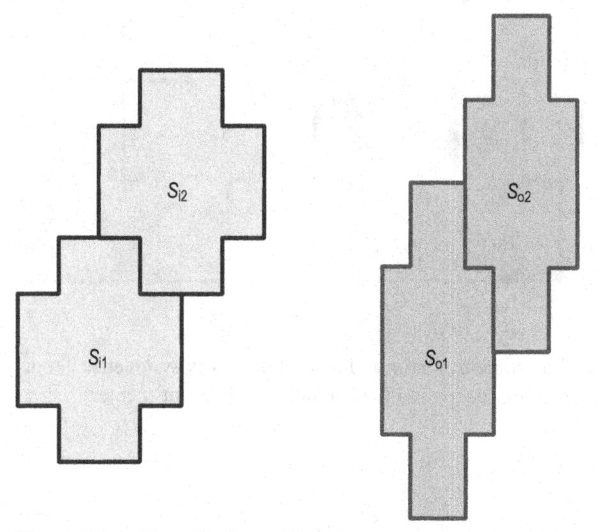

Figure 3.19 Tessellation with 3 : 2 as squeeze factor with a contour that is symmetric about both the horizontal and vertical planes.

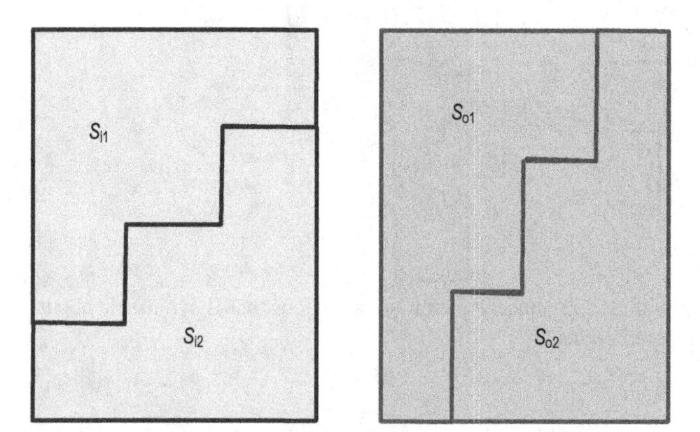

Figure 3.20 Generalization of the previous 3 : 2 squeezers.

3.5 APPLICATION TO A TWO FREEFORM MIRROR CONDENSER

The main function of the condenser optics in a projection system is to collect as much light as possible from the source and transmit it to the microdisplay, which will spatially modulate the light toward the projection optics.

Most conventional condensers use elliptic or parabolic mirrors (see Fig. 3.21). They perform far from the theoretical limits (calculated using the étendue invariance of nonimaging optics) for sources such as arc lamps or halogen bulbs. Typical small displays in the 5–15 mm^2-sr étendue range have geometrical efficiencies about 40–50% for the best condensers, although theory allows about 100%.

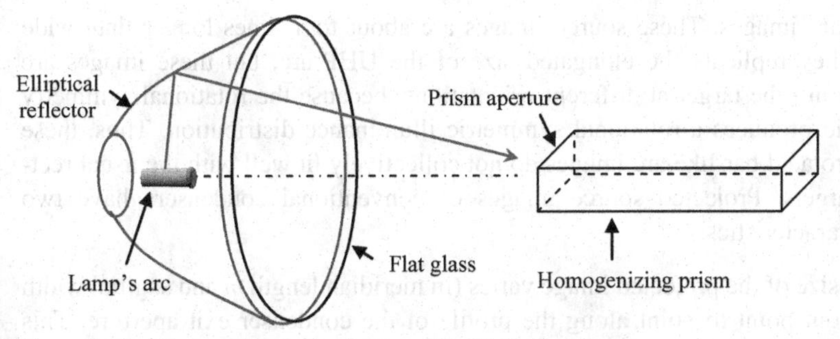

Figure 3.21 Conventional condenser with elliptical reflector.

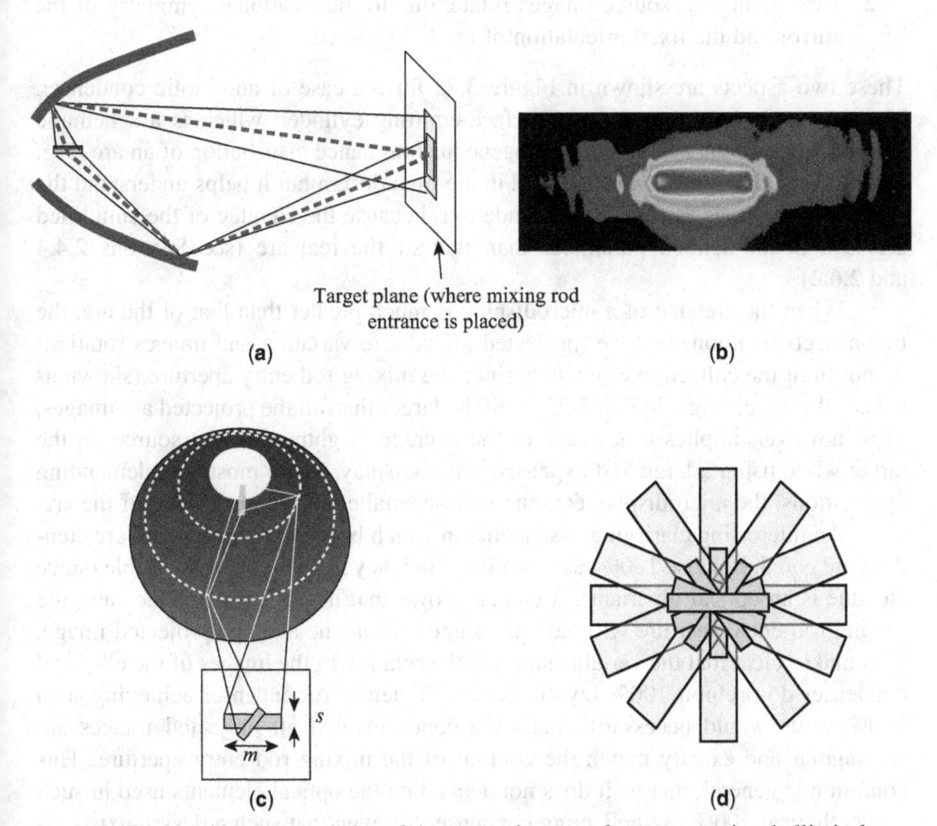

Figure 3.22 Notable characteristics of the source images of any conventional elliptical condenser: (a) variable lengths, (b) as they replicate the elongated shape of the UHP arc, they are about four times longer than wide, (c) they rotate at the target due to the condenser's rotational symmetry, and (d) multiple source images, superimposed.

To understand the limitations of conventional elliptic or parabolic condensers, it is useful to consider the concept of the projection of source images. A pinhole at the exit optical surface of the condenser will let pass a bundle of rays that bear the local image of the source (see Fig. 3.22). The limited collection efficiency for small étendues of a conventional elliptical condenser can be understood by looking at its projected arc images. These source images are about four times longer than wide (because they replicate the elongated size of the UHP arc, but these images are projected onto the target at different orientations, because the rotational symmetry of the optic produces a rotational symmetric illuminance distribution. Thus, these variously rotated bar-like arc images do not collectively fit well with the usual rectangular targets. Projected source images of conventional condensers have two notable characteristics:

1. The size of the projected image varies (in meridian length m and sagittal width s) from point to point along the profile of the condenser exit aperture. This phenomenon is due to the comatic aberration of the elliptical reflector.
2. The rectangular source images rotate, due to the rotational symmetry of the mirror and the fixed orientation of the light source.

These two aspects are shown in Figure 3.22 for the case of an elliptic condenser. The drawings show the arc as a surface-emitting cylinder, which is a schematic simplification of the actual non-homogeneous luminance distribution of an arc (Fig. 3.22b). This simplification is not used in the final design but it helps understand the inherent limitations of the elliptic condenser, because the étendue of the simplified arc is a better defined parameter than that of the real arc (see Sections 2.4.4 and 2.6.2).

When the étendue of a microdisplay is much greater than that of the arc, the two aspects mentioned above (projected image size variation and images rotation) do not limit the collection efficiency, since the mixing rod entry aperture (shown as a dashed-line rectangle in Fig. 3.22d) will be larger than all the projected arc images. This, however, implies a decrease of the average brightness of the source on the target while using a large and expensive microdisplay. In the most cost-demanding applications, the microdisplay étendue will be smaller and closer to that of the arc.

An interesting clarifying case is that in which both microdisplay and arc étendues are equal, because 100% ray coupling efficiency is theoretically possible (since étendue is an optical invariant). It can be shown that in this equi-étendue case, the mixing rod entry aperture will have the same area as the average projected image. This makes clear that the variable size and the rotation of the images of the elliptical condenser do prohibit 100% ray collection efficiency. A condenser achieving such 100% value would necessarily meet the condition that all projected images are rectangular and exactly match the contour of the mixing rod entry aperture. This condition is general, that is, it does not depend on the optical elements used in such a hypothetical 100% ray-collecting condenser, or even that such a design exists.

A recent trend in improving collection efficiency for small étendues has been to reduce the arc étendue by reducing the gap between the electrodes [6]. Furthermore, some optical designs have been developed to improve the efficiency by

1. Reducing the arc étendue via a hemispherical mirror concentric to the arc, which reflects half of the emitted light back to the arc (which is partially absorbent) and through it to increase its luminance. The light from the higher-luminance half-étendue arc is then collected by a conventional elliptical reflector [6, 7].

2. Creating a side-by-side image of the arc with a decentered hemispherical mirror, and creating a composite 1 : 1 image of that via a dual parabolic reflector [8, 9].
3. The equalization of the meridian length of the projected arc images by correcting the elliptical mirror coma, using an aspheric reflector profile and an aspheric lens, but both surfaces still being rotationally symmetric [10, 11].

These optical approaches, apart from their complexity and technological challenges, have limited gain capability because their optics are still restricted to being rotationally symmetric.

Other approaches to improve system efficiency include color recapture [12] and color scrolling [13] (which try to recover the 2/3rd losses produced by the color filtering in single microdisplay projectors), or polarization-recovery techniques (which try to recover the 50% losses produced by the need of polarized light in LCD and LCoS systems) [14]. In these approaches, however, the resulting lamp étendue is accordingly increased (doubled and tripled in polarization and color recovery systems, respectively), further limiting the performance of small microdisplays that use classical condensers.

Regarding the state of the art of manufacturing condenser optics, all present systems are based exclusively on rotational-symmetric surfaces. These are manufactured mainly from glass (due to its low cost) or by glass-ceramic (for higher thermal stability). The accuracy of both techniques is limited, so the manufactured profiles can differ substantially from those intended.

A conventional rotational symmetric condenser produces a superposition of arc images as the one shown in Figure 3.23a where both the length and the width of the arc images changes. A rotational symmetric system can control the length but not the width of the arc images as shown in Figure 3.23b. An étendue squeezing device can make the source appear, for instance, half as long (but twice as wide); see Figure 3.23. The size and rotation variation of source image projected can be also controlled if desired.

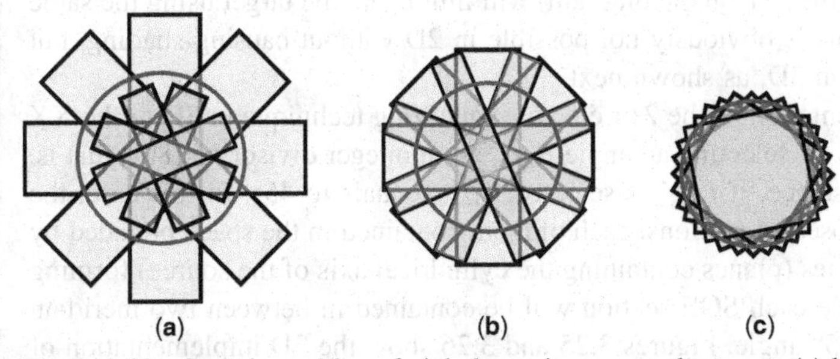

Figure 3.23 Superimposition of pinhole arc images on the target: (a) by a conventional rotational condenser; (b) by a rotational system correcting the coma, that is, equalizing the meridian length of the images; and (c) by an étendue-squeezing (ES) condenser. The ES condenser equalizes the length and the width of the projected source images.

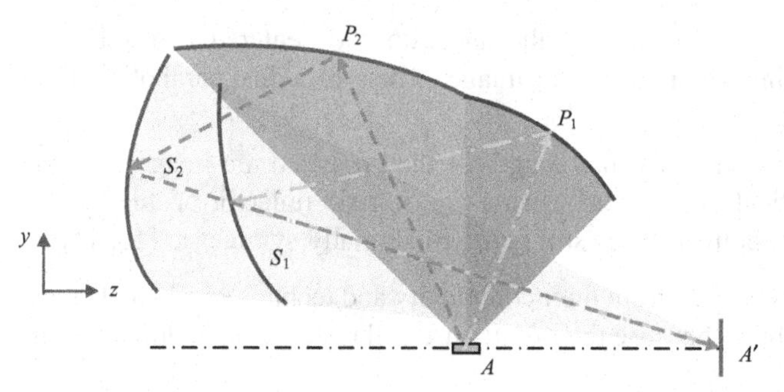

Figure 3.24 2D diagram of a 2 : 1 étendue-squeezing XX condenser.

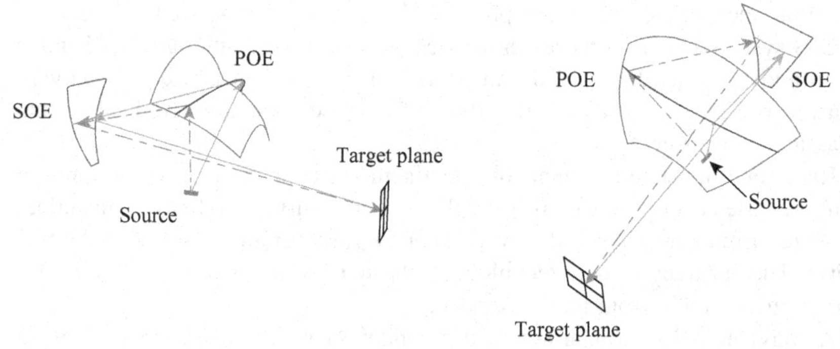

Figure 3.25 3D views of the elements P_1 and S_1 of the XX condenser with 2 : 1 étendue squeezing and $\theta_{border} = 45°$. Seed rib and mirror contours are shown.

Figure 3.24 shows the diagrammatic representation of how a XX* condenser with 2 : 1 étendue squeezing ratio performs. The primary optical element (POE) is formed by two sections: P_1, which collects the light emitted by the source towards positive z, and P_2, which collects the light emitted by the source towards negative z. Each POE section illuminates a different secondary optical element (SOE) section (S_1 and S_2) that (in this schematic diagram) will illuminate the target using the same ray trajectories. This is obviously not possible in 2D without causing shading, but it can be achieved in 3D, as shown next.

The implementation of the 2 : 1 étendue squeezing technique in 3D to the XX configuration starts by selecting an angle θ_{border} as an integer divisor of 180°, that is, $\theta_{border} = \pi/j$. For instance, if $j = 4$ is selected, θ_{border} equals to 45°. In this case, the POE will be composed of sections, each of them contained in the space bounded by two meridional planes (planes containing the cylindrical axis of the source) forming $2 \times 45° = 90°$, while each SOE section will be contained in between two meridian planes forming a 45° angle. Figures 3.25 and 3.26 show the 3D implementation of the 2D diagram in Figure 3.24.

* XX indicates a dual-reflective optic. See the next chapter and Reference [4].

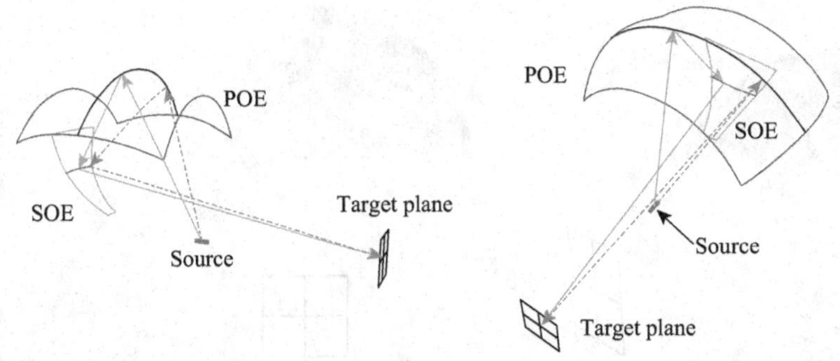

Figure 3.26 3D views of the elements P_2 and S_2 of the XX condenser with 2 : 1 étendue squeezing and $\theta_{border} = 45°$. Seed rib and mirror contours are shown.

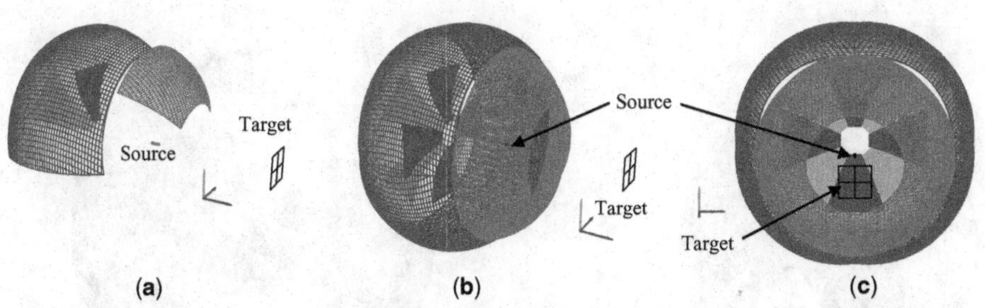

Figure 3.27 Tessellation of the adjacent elements $P_1 - P_2$ and $S_1 - S_2$ of the XX condenser with 2 : 1 étendue squeeze factor. (a) Elementary cell, (b) side view, and (c) front view of the complete device.

Each XX section is formed by a POE and its corresponding SOE. In this design, the XX section belongs to the family $N < 0$, $M < 0$ and adjacent POE and SOE halves. Seed rib and mirror boundaries are also shown. The SMS 3D calculation of the surfaces is done by the same SMS design procedure (see the next chapter). Combining the POE and SOE sections requires the tessellation initiated in Figure 3.27a and completed in Figure 3.27. Other étendue squeezing factors different from 2 : 1 can also be applied to this condenser concept. For instance, T-shaped elements introduced in Figure 3.13, which provided a 3 : 2 étendue squeezing factor, have been also applied to the XX condenser design, as shown in Figures 3.28 and 3.29.

Other étendue squeezing factors different from 2 : 1 can also be applied to this condenser concept. As an example, Figure 3.29 shows a 3 : 2 condenser. Selecting the best étendue squeezing factor depend on the source geometry (length to diameter ratio) and the target étendue. For a target of circular contour and circular FOV, and a source length $L = 1.2$ mm and diameter $D = 0.4$ mm, the 2 : 1 étendue squeeze XX design gets up to 40% gain over an elliptical condenser at 7 mm^2-sr of target étendue. As shown in Figure 3.30, the 3 : 2 design achieves a maximum gain over the elliptical condenser of 1.32 at the target étendue of 11 mm^2, and it is superior to the 2 : 1 design for target étendues greater than 10 mm^2.

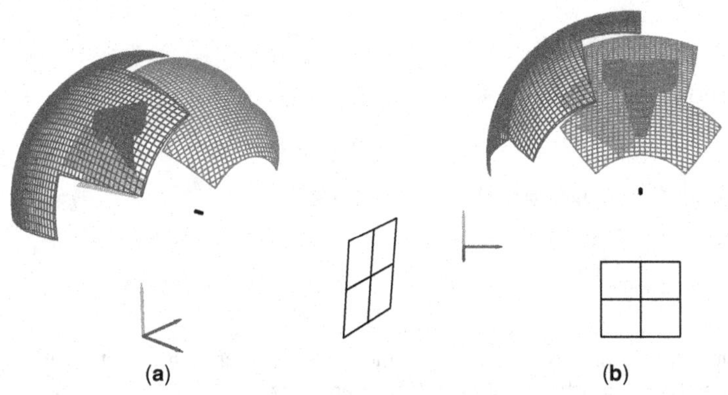

Figure 3.28 Tessellation of the adjacent elements of the XX condenser with 3 : 2 étendue.

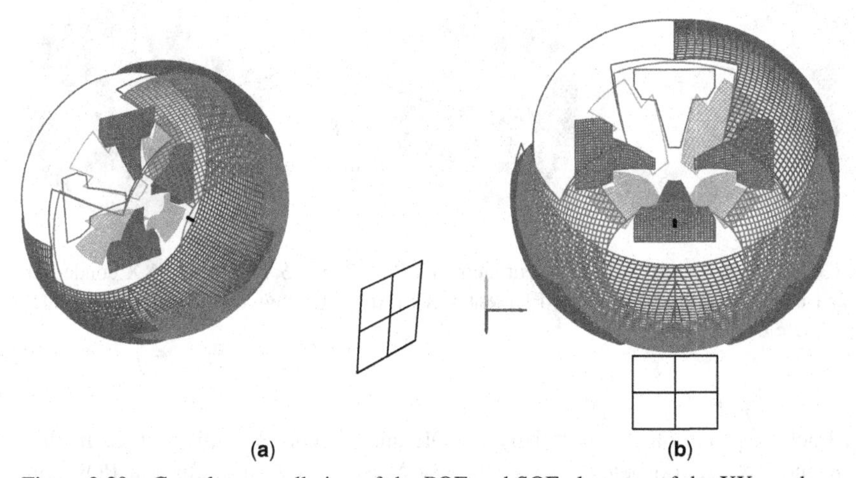

Figure 3.29 Complete tessellation of the POE and SOE elements of the XX condenser with 3 : 2 étendue squeezing.

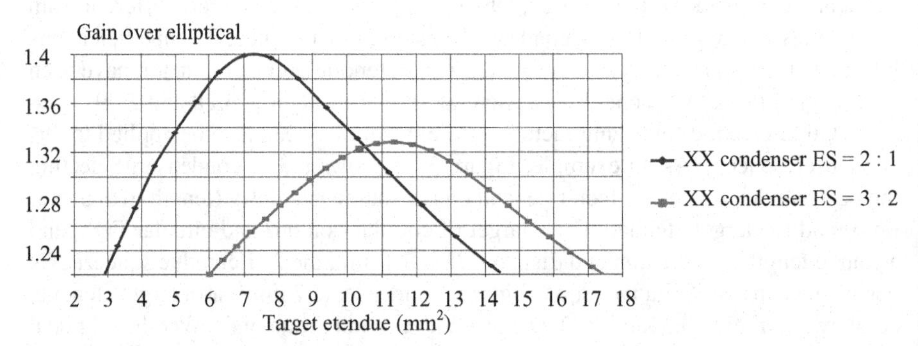

Figure 3.30 Gain of the power transmitted to the target for two cases of XX condenser with étendue squeezing.

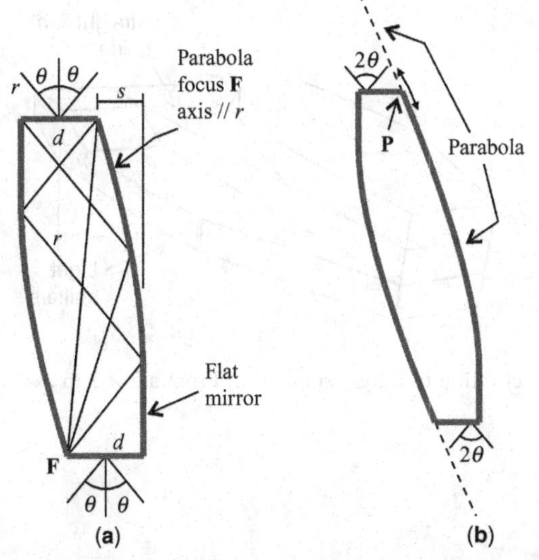

Figure 3.31 (a) Light shifters move the radiation sideways by a distance s, while maintaining the angular aperture and area of the light. (b) Moving the edge of the aperture along the parabola gives different amounts of lateral shift.

3.6 ÉTENDUE SQUEEZING IN OPTICAL MANIFOLDS

It is also possible to squeeze the étendue with light guiding manifolds, as discussed in U.S. Patent 7,520,641. The building blocks for those designs are the light shifters. These "move" light sideways while maintaining its area and angular aperture. Figure 3.31a shows an example of one of these devices. Each side is made of a flat mirror and a parabolic arc. It shifts light laterally by a distance s while maintaining its area d and angular aperture θ. In this simple device, moving the edge **P** of the aperture along the parabola makes it possible to adjust the lateral shift, as in Figure 3.31b.

As another example, consider the case in which we want to transform a source with aspect ratio 2×2 onto another source of aspect ratio 1×4. One of these devices is shown in Figure 3.32a and is made of two separate parts, each comprising two light shifters. It divides the square source in two halves (1×2 aspect ratio) and then places them side by side.

Figure 3.32b shows another example of a device that changes the source aspect ratio. It goes from a 3×3 aspect ratio to a 1×9. The optic is made of three parts. Two of them are composed of two light shifters and the middle one is a straight lightguide.

3.7 CONCLUSIONS

The étendue squeezing technique is a novel method to control the beam shape. The optical systems can be generated with conventional tools or their calculation can be

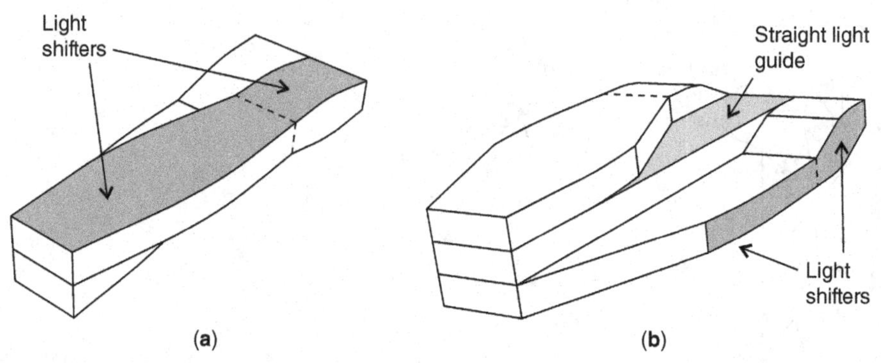

Figure 3.32 Optics that change the aspect ration of a light source: (a) From a 2 × 2 to a 1 × 4. (b) From a 3 × 3 to a 1 × 9.

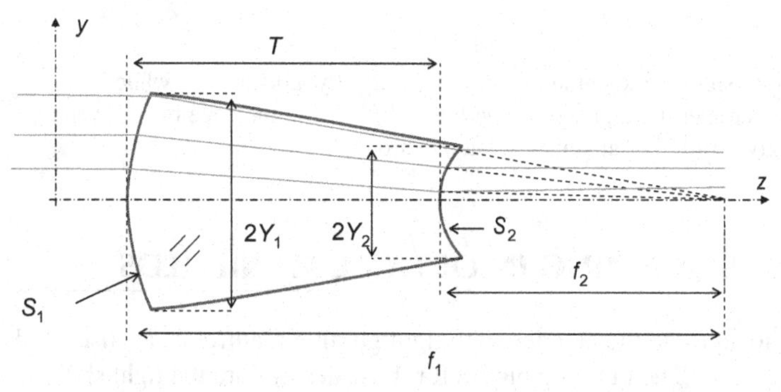

Figure 3.A.1 Galilean refractive telescope performs a beam spatial compression when light propagates from left to right and a beam spatial expansion when light propagates from right to left.

integrated into the SMS 3D design method (see the Chapter 4 and Reference [4]), in order to obtain a wider control over the light beams. With étendue squeezing designs, the efficiency of some illumination applications will be potentially improved, especially in automotive headlamp designs and in condenser designs for video projection applications. An example of a free-form XX condenser has been detailed. A comparison of power transmitted versus target étendue (done under equal conditions) shows that this XX condenser can send to a circular target up to 1.4 times the power sent by the elliptical condenser. Continued work in the field of étendue squeezing is expected to lead to new geometries and applications.

APPENDIX 3.A GALILEAN AFOCAL SYSTEM

Consider in 2D geometry the Galilean refractive telescope shown in Figure 3.A.1, comprised of a positive lens and a negative lens with coincident focal planes, lenses that bound a transparent sheet of refractive index medium $n > 1$. When light propagates from left to right, it acts as a spatial compressor of the beam, transforming the impinging parallel rays that extend over the height Y_1 onto the exiting, also parallel, rays that extend over the height $Y_2 < Y_1$.

Consider now the extended bundle comprised of the rays forming an angle $\theta_1 \leq \alpha_1$ with respect to the z-axis. After the two refractions, this extended bundle is approximately transformed into the bundle limited by the angle α_2 with respect to the z-axis, thus fulfilling:

$$Y_1 \sin\alpha_1 = Y_2 \sin\alpha_2. \tag{3.A.1}$$

Equation (3.A.1) shows the well-known fact that the (lossless) spatial compression leads to a proportional angular expansion. This equation is not only valid in the paraxial approximation, but is the well-known Lagrange invariant (also called étendue conservation theorem in 2D geometry or the Abbe Sine condition of Eq. 2.41). It is general: if the two extended bundles are perfectly coupled one to another (i.e., every ray at Y_1 within angle α_1 transforms into a ray at Y_2 at an angle less than α_2, and vice versa, every ray at Y_2 within angle α_2 comes from a ray at Y_1 at an angle less than α_1), then Equation (3.A.1) is fulfilled exactly.

The ratio Y_1/Y_2 is both the system angular magnification and the system spatial compression. If the refracting surfaces are spherical, the bundle-coupling is almost perfect if and only if:

$$Y_1 \ll T, \quad Y_2 \ll T, \quad \frac{\sin\alpha_1}{n} \ll \frac{Y_2}{T}, \quad \frac{\sin\alpha_2}{n} \ll \frac{Y_1}{T}, \tag{3.A.2}$$

where T is the system thickness. When these conditions are not fulfilled, the aberrations of the spherical refractive surfaces cause the exit rays to not fit completely inside the $\pm\alpha_2$ edge rays. If imaging aspheric surfaces are used, the aberrations can be corrected.

In terms of transmission efficiency, part of the rays at the rim of S_1 will not intersect the exit surface S_2. Reciprocally, the illuminance on the exit surface S_2 will be dimmer at the rim because exit rays close to $\pm\alpha_2$ will not come from the input surface S_1 (i.e., the actual exit rays will have an angular spread smaller than $\pm\alpha_2$ near the S_2 rim). This darker rim is usually referred to as underfilling of the exit surface. Both effects are related in this case: the lower illuminance close to the rim of S_2 is caused by the rays that should be there have been lost.

In order to avoid these efficiency losses, the usual tactic is to oversize the exit lens. Any oversizing of Y_2 will necessary imply that the minimum angle $\alpha_{2\min}$ exiting the (lossless) system, which is given by Equation (3.A.1), is smaller than the actual α_2. Therefore, the oversizing of Y_2 implies an oversizing of the angle α_2 with respect to the theoretical minimum. This is referred to as the system not being étendue limited, and is crucial in present high-demand condenser optics.

In imaging lenses, this effect of the extended nature of the transmitted bundle is minimized when

$$\frac{Y_1}{\sin\alpha_2} = \frac{Y_2}{\sin\alpha_1} \gg \frac{T}{n}, \tag{3.A.3}$$

where the equality comes from Equation (3.A.1). Therefore, the f-number of the imaging lenses must be low enough to make these extended bundle effects negligible. For example, if we consider $1 \ll 10$, for $\alpha_1 = \pm 3$ degrees, $n = 1.5$, and $Y_1/Y_2 = 2$, we obtain $f_1/Y_1 = f_2/Y_2 < 2.8$. Therefore, when light propagates right to left, the Galilean telescope performs as a spatial beam expander and an angular bundle compressor.

APPENDIX 3.B KEPLERIAN AFOCAL SYSTEM

Consider in 2D geometry the Keplerian refractive telescope shown in Figure 3.B.1, which is comprised of two positive lenses with different focal lengths but coincident focal planes. As with the Galilean case, when light propagates from left to right, the device acts as a beam spatial compressor and angular bundle expander. It produces the reverse effect when light propagates right to left. The same strategies and conclusions about oversizing the exit aperture are applicable here (including the ideality of an SMS2D design).

Let us note the well-known differences between the Galilean and Keplerian configurations: (i) the higher surface-curvatures needed in the Keplerian design for the same thickness T and (ii) the inversion of the images when used as a telescope. Therefore:

$$\frac{\sin \alpha_2}{\sin \alpha_1} \approx \pm \frac{Y_1}{Y_2}, \tag{3.B.1}$$

where the + is for the Galilean and the − is for the Keplerian system.

Any ray impinging upwards/downwards will exit the system pointing upwards/downwards in the Galilean case and downwards/upwards in the Keplerian case. This can be seen also in terms of the caustic of the rays: in the Galilean, the caustic of the rays between the lenses is virtual, while in the Keplerian, the caustic is real.

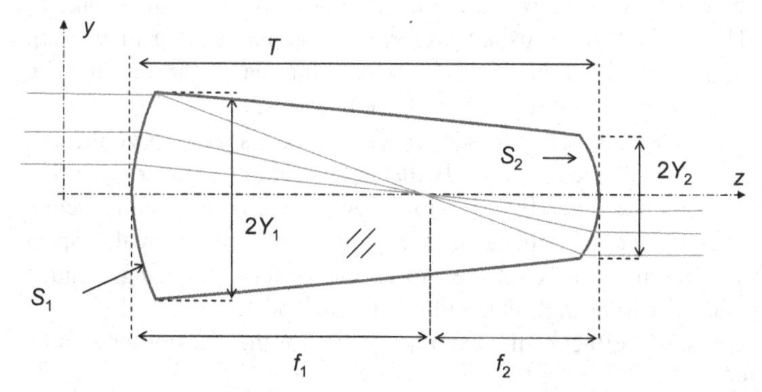

Figure 3.B.1 A Galilean refractive telescope performing a beam spatial compression when light propagates from left to right and a spatial expansion when light propagates right to left.

REFERENCES

1. J. Blen, Design of multiple free-form optical surfaces in three dimensions, PhD dissertation, Universidad Politécnica de Madrid (2007).
2. Name suggested by William J. Cassarly (pers. comm., 2009).
3. Devices shown in this chapter are protected by the following US and international patents and patents pending: U.S. 6,639,733; PCT/US2006/029464; US 6,639,733; U.S. 6,896,381; U.S. 7,152,985; U.S .7,181,378; U.S. 7,460,985; PCT/US2006/029464, July 28 (2006).
4. P. Benítez, J.C. Miñano, J. Blen, R. Mohedano, J. Chaves, O. Dross, M. Hernández, J.L. Álvarez, and W. Falicoff, SMS Optical Design Method in three dimensions, *Opt. Eng.* **43**, 1489 (2004).
5. M. Born and E. Wolf, *Principles of Optics*, Pergamon, Oxford (1975).
6. H. Moench and A. Ritz, Higher output more compact UHP lamp systems, *SID Int. Symp. Digest Tech. Papers* **33**, 1160 (2002).
7. K. Strobl, U.S. Patent 6,356,700 (2002).
8. K. Li and S. Inatsugu, Dual paraboloid reflector system configurations for increasing brightness of an arc lamp, *Proc. SPIE*, R. Winston and R.J. Koshel, eds. **5942**, 594201 (2005).
9. K.K. Li, S. Sillyman, and S. Inatsugu, Optimization of dual paraboloidal reflector and polarization system for displays using a ray-tracing model, *Opt. Eng.* **43**, 1545 (2004).
10. N. Tadaaki, Japan Patent &,174,974 (1995).
11. J.A. Shimizu, U.S. Patent 5,966,250 (1999).
12. D.S. Dewald, S.M. Penn, and M. Davis, Sequential color recapture and dynamic filtering: a method of scrolling color, *SID Int. Symp. Digest Tech. Papers* **32**, 1076 (2001).
13. J.A. Shimizu, Scrolling color LCOS for HDTV rear projection, *SID Int. Symp. Digest Tech. Papers* **32**, 1072 (2001).
14. M. Duelli and A.T. Taylor, Novel polarization conversion and integration system for projection displays, *SID Int. Symp. Digest Tech. Papers* **34**, 766 (2003).

CHAPTER 4

SMS 3D DESIGN METHOD

Juan Carlos Miñano, Pablo Benítez, Aleksandra Cvetkovic, and Rubén Mohedano

4.1 INTRODUCTION

The simultaneous multiple surfaces (SMS) 3D optical design method is a nontrivial extension of the SMS 2D [1–7] design method. The first attempt for a 3D extension was done in 1999 [8]. The SMS 3D method allows controlling several (up to three as of this writing) orthonormal bundles of rays. This means that the bundles can be prescribed at the input and the output of the optical system. The SMS 3D design will transform (by refractions and/or reflections) the input bundles into the output ones. The method has many applications in nonimaging optics because it enables better control of the light emitted by an extended light source than any other 3D design method (which are based in single freeform surface design). The SMS 3D design method is being expanded into new areas of imaging and nonimaging optics. The same features of SMS 2D devices: compactness, efficiency, and few pieces, are in general achieved by SMS 3D devices.

4.2 STATE OF THE ART OF FREEFORM OPTICAL DESIGN METHODS

Rotationally symmetric optics cannot satisfactorily solve some nonsymmetric illumination or concentration requirements. Typical examples of these cases are low-beam head lamps, condensers, and some solar concentrators. In these cases, sources and targets have asymmetric design requirements. A rotational symmetric optical device can partially solve the problem particularly if we relax conditions on efficiency of light transfer and the number of elements forming the optical system.

Freeform surfaces are optical surfaces without linear or rotational symmetry. Their freeform nature provides additional degrees of freedom, which may be used to solve asymmetric problems with advantages over symmetric designs (for instance with higher efficiency or less number of elements) at the expense of tooling freeform surfaces. Such tooling is more difficult than rotational or linear symmetric ones. The advantages given by freeform surfaces are particularly interesting for nonimaging optics applications, which has been the first type of optics to benefit from advances in freeform design and tooling. Another point of interest for nonimaging optics is

Illumination Engineering: Design with Nonimaging Optics, First Edition. R. John Koshel.
© 2013 the Institute of Electrical and Electronics Engineers. Published 2013 by John Wiley & Sons, Inc.

that its less tight requirements (surface accuracy and finishing) when compared with imaging optics have allowed illumination optics to follow closely the evolution of freeform tooling and also have allowed mass production of nonimaging devices. This last result is especially appealing because their fabrication is done under much tighter cost requirements, and in general, replication costs in mass production techniques, such as plastic molding, are essentially independent of the shape of the optical surfaces, so rotational or linear symmetry does not give a great cost reduction.

Design procedures for freeform optical devices have not evolved as fast as the tooling [9, 10]. Essentially, there are two strategies for optical design methods: numerical optimization and direct methods. In numerical optimization, for which there is a wealth of study being done, a merit function is defined with several variables chosen (typically less than 20, but some optimizations can have hundreds of variables, see Chapter 7). The merit function is typically "wild," with many local optima. There are many methods being used, such as the steepest descent, damped least squares, multistart, simulated annealing, simplex, genetic algorithms, global synthesis, escape function, and neural networks [11, 12]. These methods also work well for nonimaging devices when the system has rotational symmetry [13].

A direct method is a mathematical procedure that delivers without iterations the optical surface equations when the optical prescription is given. The direct methods start with a source and its representative emission profile and the target where the irradiance, intensity, and/or radiance are prescribed. These prescriptions are translated to conditions on wave fronts at the input and at the output of the optical system to design. Simply said, we build the transfer optics that take the light from the source distribution to the desired target distribution.

Direct design methods are much more complex to describe and to implement than optimization schemes. Their advantage is that they can reach "new territories" where optimization techniques can get lost by the infinitude of local minima or by the large number of parameters to optimize. However, a scheme that uses direct design methods to initially design the system, and then optimization schemes to marginally improve performance, investigate trade parameters, or include more complex physical phenomena (e.g., thermal effects, source tolerances, and fabrication surface quality) are attractive. At this time, there are basically three direct design methods for freeform surfaces: (i) generalized Cartesian ovals, (ii) SMS in its 3D version (SMS 3D), and (iii) Monge–Ampere type equations:

(a) A generalized Cartesian oval is an optical surface obtained by requiring the optical path length between two prescribed wave fronts to be constant. The optical prescription is given by these two wave fronts. Descartes was the first to design with such a procedure, but he restricted his attention to spherical wave fronts. The resulting surfaces, called Cartesian ovals (this name comes after him), are not freeform (strictly speaking) but aspherics [14, 15]. When the optical surface is a reflector, the Cartesian ovals obtained from spherical wave fronts are quadrics. Levi-Civitta generalized the problem to nonspherical wave fronts. In this case, the resulting optical surfaces are, in general, freeform [16].

(b) The SMS 3D design method can be seen as one step ahead of the Cartesian oval problem. To design an optical system that perfectly couples two input wave fronts into two output wave fronts, it turns out that two freeform surfaces are in general sufficient for the optical system that solves the problem. If we increase the number of wave fronts to couple, then we need to design an analogous number of additional optical surfaces. The optical prescription is given by the sets of input and output wave fronts. In general, the optical surfaces have no analytical expression and must be calculated simultaneously, point by point. This method, in its 3D version, is a powerful direct design method for illumination and concentration devices managing extended sources. It gives full control of two pairs of wave fronts at the input and at the output, and a partial control of a third pair of wave fronts. This means that the resulting device perfectly couples two pairs of prescribed input and output wave fronts and only partially the third pair.

(c) In Monge–Ampere equation problems, the input and the output intensity patterns are prescribed. The method provides the surface (refractive or reflective) transforming the input intensity pattern into the output one [17–19]. It can only be applied to point sources. Consequently, real (extended) sources must be placed far away from the optical surfaces so they look like point sources. A Monge–Ampere equation also appears when not only the input and output intensity patterns are prescribed but also the input and output wave fronts. In this case, two optical surfaces are needed. The solution is again also valid for point sources.

Recently, Fournier, Cassarly, and Rolland [20] have successfully applied method (c) to design single freeform reflectors for extended sources and prescribed intensity patterns. The idea is to approximate the source by a point, apply Oliker's method [17, 18], and then to calculate the difference between the actual intensity pattern (obtained with the actual extended source) with respect to the prescribed one. This difference is used to prescribe a new intensity pattern that takes into account these errors. Oliker's method is then applied again to this new intensity prescription in an iterative way until the desired intensity pattern is obtained. The success of this method depends on the size of the source relative to the mirror separation, the "smoothness" of the intensity pattern, and the desired approximation to the prescribed pattern. Such "iterative" direct methods are predicted to assist with problems that cannot be explicitly solved effectively with single-run direct methods, and often benefit from secondary optimization methods.

4.3. SMS 3D STATEMENT OF THE OPTICAL PROBLEM

A normal ray congruence (also called an orthotomic system of rays or orthonormal bundle) is a set of ray trajectories that are perpendicular to a one-parameter family of surfaces, namely the wave fronts. For instance, the rays emitted from a point form a normal congruence. If the medium is homogeneous, then these rays are straight lines and the wave fronts are spheres centered on the emission point. The

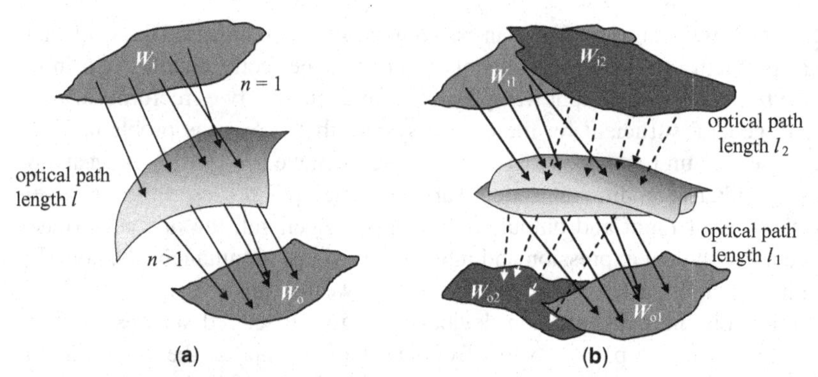

Figure 4.1 (a) Cartesian oval transforms one input congruence into an output congruence. (b) The simplest version of the simultaneous multiple surfaces (SMS) 3D method provides two surfaces that transform two input congruences into two output ones.

normal ray congruence is fully characterized by giving just one wave front. When the refractive index distribution is known, the ray trajectories of the congruence can be calculated. We can use W_j either to denote the normal congruence or one wave front surface of it. The rays of the congruence W_j are those perpendiculars to wave front W_j.

As said before, a generalized Cartesian oval is an optical surface (refractive or reflective) such that a given normal congruence, W_i is transformed into another given normal congruence W_o (see Fig. 4.1a). The SMS 3D method in its simplest form is a procedure for designing two optical surfaces whereby two given normal congruences W_{i1} and W_{i2} are transformed by a combination of refractions and/or reflections at these surfaces into another two given normal congruences, W_{o1} and W_{o2} (see Fig. 4.1b).

4.4 SMS CHAINS

Let W_{i1}, W_{i2}, W_{o1}, and W_{o2} be four normal ray congruences or wave fronts. Assume we want to design an optical system to transform the input normal congruences W_{i1} and W_{i2} into the output normal congruences W_{o1} and W_{o2}, respectively. This means that any ray of W_{i1} entering the optical system exits it as a ray of W_{o1} (the same holds for W_{i2} and W_{o2}). The wave fronts may be real or virtual. The surfaces to be designed, which will be called S_i and S_o, and they can be either reflective or refractive. For simplicity, we will assume that the rays are coming from the input wave fronts W_{i1} and W_{i2} to the output wave fronts W_{o1} and W_{o2}, respectively. Since the W_{i1}, W_{i2}, W_{o1} and W_{o2} surfaces are wave fronts, the rays are perpendicular to W_{i1} or W_{i2} before the deflection at S_i and perpendicular to W_{o1} or W_{o2} after the deflection at S_o. Only the surfaces S_i and S_o of this optical system will be calculated, and the remaining surfaces, if any, are assumed to be prescribed.

Since the optical system must couple W_{i1} and W_{o1}, the normal congruences W_{i1} and W_{o1} are the same, and any of the two wave fronts W_{i1} or W_{o1} can be used

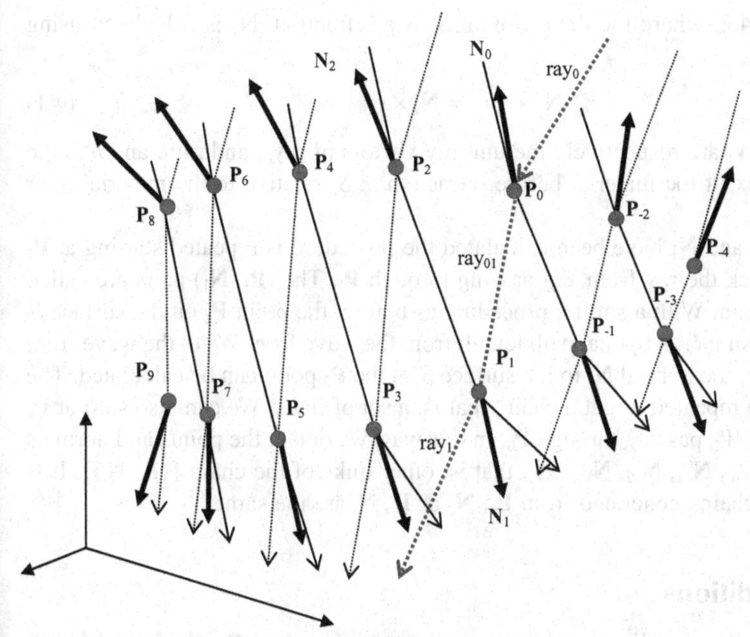

Figure 4.2 SMS chain generated from the link ($\mathbf{P}_0$, $\mathbf{N}_0$).

to characterize this normal congruence, which will be called W_1. Similarly, W_2 is the normal congruence having W_{i2} and W_{o2} as wave fronts. Let $\mathbf{P}_0$ be a point on S_i, $\mathbf{N}_0$ be the normal unit vector of the surface at $\mathbf{P}_0$, ℓ_1 be the optical path length from the wave front W_{i1} to the wave front W_{o1} (for the rays of W_1), and ℓ_2 be the optical path length from W_{i2} to W_{o2} (for the rays of W_2). Assume that $\mathbf{P}_0$, $\mathbf{N}_0$, ℓ_1, ℓ_2, W_{i1}, W_{i2}, W_{o1}, and W_{o2} are given. We will show that a set of points on the surfaces S_i and S_o and the unit vectors normal to the surfaces at these points can be calculated from this data. A set of points and unit vectors like this one will be called the SMS chain.

4.4.1 SMS Chain Generation

Let ray$_0$ be the ray of the normal congruence W_{i1} passing through $\mathbf{P}_0$ of the surface S_i (see Fig. 4.2). Since the normal $\mathbf{N}_0$ is known, we can trace ray$_{01}$ after deflection at $\mathbf{P}_0$ (refraction in Fig. 4.2). We can now calculate the point $\mathbf{P}_1$ on the surface S_o such that the optical path length from the wave front W_{i1} to the wave front W_{o1} is ℓ_1. The point $\mathbf{P}_1$ is along ray$_{01}$ after deflection at $\mathbf{P}_0$ if there are no prescribed surfaces between S_i and S_o. If this is not the case, $\mathbf{P}_1$ would be along the ray path ray$_0$ after deflection at $\mathbf{P}_0$. Such a ray path can be traced because any other surface besides S_i and S_o is prescribed.

Once $\mathbf{P}_1$ has been calculated, we calculate the normal $\mathbf{N}_1$ to the surface S_o at the point $\mathbf{P}_1$. This is done with ray$_1$ of the normal congruence W_{o1} passing through $\mathbf{P}_1$. $\mathbf{N}_1$ is the normal that allows ray$_{01}$ to be deflected into ray$_1$. In the case of the Figure 4.2, where the deflection at S_o is a refraction, $\mathbf{N}_1$ is calculated using Equation 4.1

$$\mathbf{N}_1 \times n\mathbf{v}_{01} = \mathbf{N}_1 \times \mathbf{v}_1, \tag{4.1}$$

where $\mathbf{v}_{01}$ and $\mathbf{v}_1$ are respectively the unit ray vectors of ray_{01} and ray_1, and n is the refractive index of the material before refraction at S_o relative to the material after refraction at S_o.

Once $\mathbf{P}_1$ and $\mathbf{N}_1$ have been calculated the procedure is repeated starting at $\mathbf{P}_1$ and tracing back the ray from W_{o2} passing through $\mathbf{P}_1$. The $(\mathbf{P}_1, \mathbf{N}_1)$ pairs are called links of the chain. With a similar procedure as before, the point $\mathbf{P}_2$ on the surface S_i is calculated using ℓ_2 as optical path length from the wave front W_{i2} to the wave front W_{o2}. Following, the normal $\mathbf{N}_2$ to the surface S_i at the $\mathbf{P}_2$ point can be calculated. The process can be repeated to get a chain, that is, a set of links. We can also start at $\mathbf{P}_0$ with the ray of W_{2i} passing through $\mathbf{P}_0$. In this way, we obtain the points and normals $\mathbf{P}_{-1}, \mathbf{P}_{-2}, \mathbf{P}_{-3}, \ldots, \mathbf{N}_{-1}, \mathbf{N}_{-2}, \mathbf{N}_{-3}, \ldots$, that is, other links of the chain $\{(\mathbf{P}_j, \mathbf{N}_j)\}$. It is clear that the chains generated from $\mathbf{P}_0$, $\mathbf{N}_0$ or $\mathbf{P}_j$, $\mathbf{N}_j$ are the same.

4.4.2 Conditions

Once W_{i1}, W_{i2}, W_{o1}, and W_{o2} are given, not all combinations of $\mathbf{P}_0$, $\mathbf{N}_0$, ℓ_1, and ℓ_2 will lead to an SMS chain:

1. For the generation procedure shown in Section 4.4.1, there must be one ray and only one ray of W_{i1} (and one ray and only one ray of W_{i2}) passing through the points of the chain belonging to S_i, and, similarly, that one ray and only one ray of W_{o1} (and one ray and only one ray of W_{o2}) passing through the points of the chain belonging to S_o. This condition excludes points close to any caustics of the normal congruences.
2. The ranges of values of ℓ_1 and ℓ_2 leading to a SMS chain are also limited. These ranges depend on the relative position of $\mathbf{P}_0$ and the wave fronts W_{i1}, W_{i2}, W_{o1}, and W_{o2}.

In general, it is not possible to extend the chains indefinitely in both directions, that is, there is a finite range of points (and normals). Outside this range, the application of the procedure explained in Section 4.4.1 with the two conditions above will have no solution. The range of points (and normals) of the chain will be called the length of the chain.

4.5 SMS SURFACES

This section describes the method for generating the SMS surfaces S_i and S_o such that the normal congruences, W_{i1} and W_{i2}, are respectively transformed into the normal congruences, W_{o1} and W_{o2}, with an additional condition: one of the surfaces (S_i or S_o) should contain an arbitrary curve R_0.

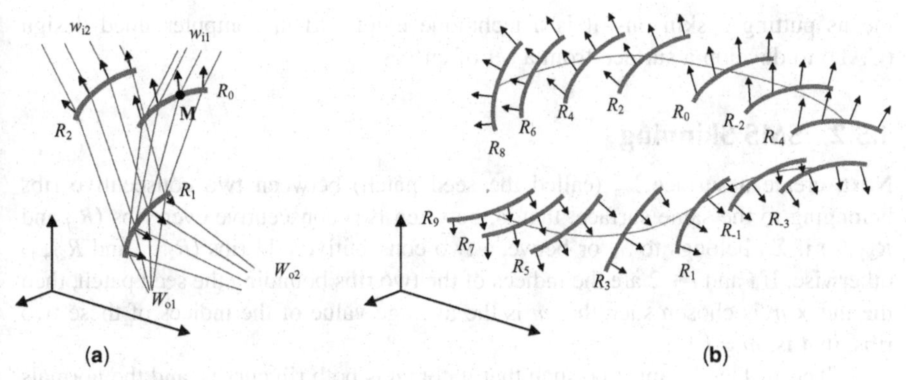

Figure 4.3 SMS ribs generated from the seed rib R_0.

4.5.1 SMS Ribs

Let R_0 be a segment of an arbitrary selected differentiable curve that will be contained, for instance, in S_i (see Fig. 4.3a). Consider that the normal vectors to S_i on R_0 are also selected to fulfill the consistency constraint that these normal vectors are perpendicular to the curve R_0. The set of surface normals $\mathbf{N}_0$ and points of the curve R_0 is called the seed rib R_0. An SMS chain can be generated from any point **M** of the curve R_0 and its corresponding normal using the procedure of Section 4.4. The set of links generated from the links of R_0, after the first step of the SMS chain generation, form what we call an SMS rib, R_1. Note that the calculation of the curve R_1 is the calculation of a curve contained in a Cartesian oval surface when a one-parameter set of rays of one of the normal congruences is known (the one-parameter set of rays is formed by the rays of W_{i1} after deflection at the curve R_0). This ensures that the normals $\mathbf{N}_1$ are perpendicular to the curve segment R_1.

Subsequent steps will produce ribs R_{2i} belonging to the surface S_i and ribs R_{2i+1} belonging to the surface S_o (see Fig. 4.3b). Assume that the equation of the seed rib curve R_0 is given in parametric form as

$$\mathbf{P} = \mathbf{R}_0(u). \tag{4.2}$$

A natural parameterization $\mathbf{P} = \mathbf{R}_i(u)$ is induced in the other rib curves generated by the SMS method. With this parameterization, points corresponding to the same u value are points that belong to the same SMS chain, that is, each value of u defines an SMS chain.

The selection of the seed rib R_0 is quite arbitrary, although not completely so. For example, assume that the curve C is the one obtained as an interpolation of the points of a chain. If we use C as a seed rib, any other rib of the same surface coincides with C at least at the points of the original chain, and in general, it is not possible to create a skin that covers these ribs (see Section 4.5.2). Although the word skinning conventionally refers to the removal of an organism's skin, in this case, it means putting a skin on—it is a technique employed in computer-aided design (CAD) to develop a surface from a set of curves.

4.5.2 SMS Skinning

Next, create a surface Σ_m (called the seed patch) between two consecutive ribs belonging to the same surface, that is, between two consecutive even ribs (R_{2j} and $R_{2(j+1)}$) if Σ_m belongs to S_i, or between two consecutive odd ribs (R_{2j-1} and R_{2j+1}) otherwise. If i and $i + 2$ are the indices of the two ribs bounding the seed patch, then the index m is chosen such that m is the average value of the indices of these two ribs, that is, $m = i + 1$.

The surface Σ_m must be such that it contains both rib curves, and the normals at its edges coincide with the surface normals of the ribs. Surface Σ_m can be calculated as a lofted surface between the ribs i and $i + 2$ (with prescribed surface normals) [21, 22]. Since the surface normal at every point of Σ_m is known, we can calculate the SMS chains generated by any point of Σ_m. In this way, we can calculate different connected patches Σ_j, which form the surfaces S_i and S_o (see Fig. 4.4). At each step of the SMS chain-generation, a new patch of the surface S_i or S_o is obtained. Note that this patch calculation is a Cartesian oval calculation. This ensures that consistent points and normal unit vectors are obtained.

The SMS method induces a parameterization on the surfaces derived from the parameterization used in the seed patch. One of the parameters is the parameter induced from the seed rib (called u previously). The curves with u = constant are called spines. In general, an arbitrary parameterization of the seed patch Σ_m leads to spines that are not C^0 curves. In order to have C^0, C^1, and so on spines, continuity conditions must be forced upon the seed patch parameterization, at the two supporting ribs R_{2j} and $R_{2(j+1)}$. Note that such a continuity level is not relevant from a

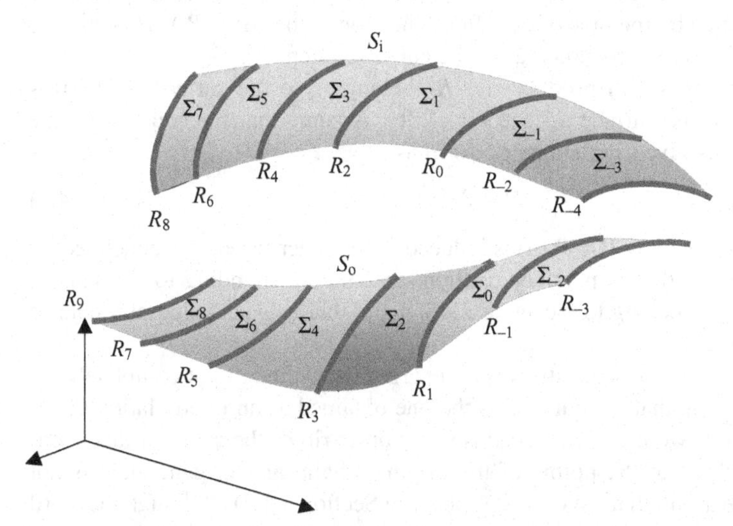

Figure 4.4 SMS skinning. The SMS surfaces are generated from a seed patch (the blue patch Σ_2) hanging on two SMS ribs (R_3 and R_1).

theoretical point of view (in any case, the surfaces remain the same, only their parameterization changes). In practice, however, the surfaces are to be modeled in the calculation process of current examples (for instance, using b-splines), and then, suitable parameterizations are required for accurate surface modeling [22].

The seed patch selection may lead, through the skinning, to surfaces that have mathematical meaning but are not physically possible (for instance, having self-intersections). This is because the caustics of the wave fronts in between the two freeform surfaces have crossed the optical surfaces. When this occurs, the design should be restarted with the selection of a different seed patch. For easier manufacturability in practical devices, it is desirable to choose the seed patch that maximizes surface smoothness.

4.5.3 Choosing the Seed Rib

The selection of the seed rib in the procedure for the SMS surface generation gives an important degree of freedom for the design. We can use this degree of freedom to attain other properties of the optical system. As an example of how to use this degree of freedom, consider a third set of input and output wave fronts W_{i3} and W_{o3}. Build up an SMS chain using the wave fronts (W_{i3}, W_{o3}) and (W_{i1}, W_{o1}) and choose an initial point $\mathbf{P}_0$ and the optical path lengths ℓ_1 and ℓ_3. Let R_0 be the interpolating curve passing through the points of this SMS chain. Choose R_0 as the seed rib in the SMS surface generation procedure for the wave fronts (W_{i1}, W_{o1}) and (W_{i2}, W_{o2}), and choose the optical lengths ℓ_1 and ℓ_2. The resulting optical system will couple the normal congruence W_{i1} with W_{o1} and W_{i2} with W_{o2}. The surface unit normal vectors at the seed rib R_0 will not coincide, in general, with the unit normal vectors calculated in the generation of the chain using the optical path lengths ℓ_1 and ℓ_3 and the wave fronts W_{i1}, W_{o1}, W_{i3} and W_{o3}, except in some particular examples. If these normal vectors coincide, then the optical system would couple not only W_{i1} with W_{o1} and W_{i2} with W_{o2}, but also it would couple the rays of W_{i3} and W_{o3} passing through the curve R_0. In practice, there are many examples in which the normal vectors at the points of R_0 obtained from W_{i1}, W_{o1}, W_{i3}, and W_{o3} (in the SMS chain generation) and the normal vectors obtained from W_{i1}, W_{o1}, W_{i2}, and W_{o2} (in the SMS surface generation) are close. In these cases, the optical system approximately couples the rays of W_{i3} and W_{o3} passing through the curve R_i. The remaining rays of W_{i3} and W_{o3} are approximately coupled, that is, we can get an optical system that perfectly couples W_{i1} with W_{o1} and W_{i2} with W_{o2} and, approximately, W_{i3} with W_{o3}. Another possibility for choosing the seed rib is given in Section 4.6.3 (and also in Reference [8])

4.6 DESIGN EXAMPLES

According to the procedure shown in the previous section, an SMS design is fully defined by giving:

1. The definition of the input and output wave fronts W_{i1} and W_{o1}, and the optical path length ℓ_1 between them

2. The definition of the input and output wave fronts W_{i2} and W_{o2} and the optical path length ℓ_2 between them
3. The nature of the optical surfaces to design, that is, the type of optical deflection of each surface
4. The refractive indices of the media on either side of both surfaces, and
5. A seed rib, R_0, and the reference to the surface on which the seed rib is to lie.

This set of parameters and curves will be called the input data. Not every possible set of input data gives a solution, that is, an SMS design. In fact, finding the wave fronts of steps 1 and 2 can be an art in itself, since there may not be a simple and direct process for selecting these wave fronts from a particular illumination prescription, especially with large extended sources compared with the desired size of the optical system.

In the examples below, we sometimes make use of *H*–*V* (i.e., horizontal–vertical) coordinates, which are the usual directional variables in numerous illumination applications. If the left–up–front directions correspond to the *x*–*y*–*z*-axes, *H* will be the angle, in degrees, between the ray and the projection of this ray on the vertical up–front plane (and *H* is positive when the ray points to the right), while *V* will be the angle, in degrees, between the ray and the projection of this ray on the horizontal right–front plane (and *V* is positive when the ray points up). Moreover, the most common nomenclature, used above, substitutes the sign by the explicit addition of the reference to the direction (R = right, L = left, U = up, and D = down). Equivalently, *H* and *V* are the angles defined by the change of variables:

$$H = -\sin^{-1}(p) \quad V = \sin^{-1}(q) \tag{4.3}$$

where $\mathbf{v} = (p, q, +(1 - p^2 - q)^{1/2})$ is the unit vector of the ray. The selection of the positive sign of the square root embodies the assumption that all the rays will travel in the positive *z* direction.

4.6.1 SMS Design with a Prescribed Seed Rib

This first example begins with this input data:

1. The specification of the input and output wave fronts W_{i1} and W_{o1} and the optical path length ℓ_1 between them: the sphere W_{o1} is centered at the point (1, 0, 0) and has a radius equal to 1. W_{i1} is a flat wave front containing the point (0, 0, 25) whose rays point toward (3.43R, *V*) (see Fig. 4.5). The optical path length ℓ_1 is 51.
2. The definition of the input and output wave fronts W_{i2} and W_{o2} and the optical path length ℓ_2 between them: the sphere W_{o2} is centered at the point (−1, 0, 0) and has a radius equal to 1. W_{i2} is a flat wave front containing the point (0, 0, 25) whose rays point toward (3.43L, *V*) (see Fig. 4.5). The optical path length $\ell_2 = \ell_1$.
3. The nature of the optical surfaces to be designed, that is, the type of optical deflection of each surface: both surfaces are reflective. This type of design is

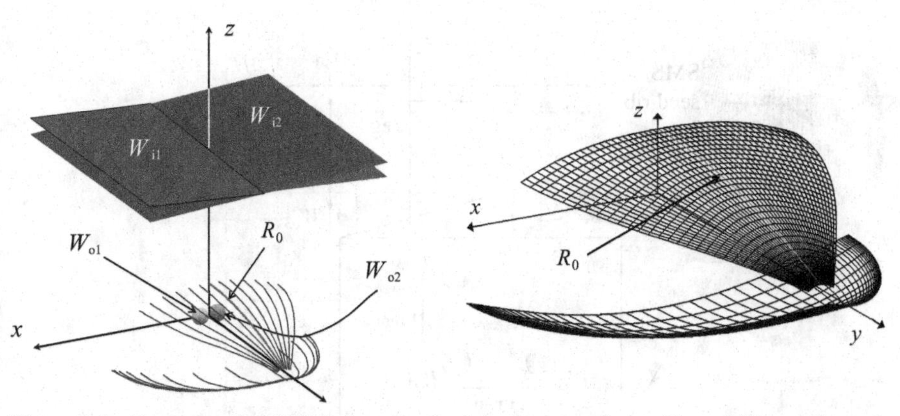

Figure 4.5 Example of an XX design. On the left, the wave fronts W_{i1} and W_{i2} are flat and the wave fronts W_{o1} and W_{o2} are spherical and centered at two points located on the x-axis. The seed rib R_0 is an arc of circumference (dotted line). The remaining lines are the ribs derived from this seed rib. On the right, the seed rib R_0 is also shown. The grid on the surfaces is formed by ribs and spines.

called an XX. Any blocking between mirrors is not considered in the design procedure. Once the design is finished, blocking losses can be evaluated as reasonable or not. The blocking of one surface by the other is a practical aspect not considered in this academic example, and it could be potentially addressed through optimization to ascertain the trade between shading and not shading.

4. The refractive indices of the media on each side of both surfaces: all media have $n = 1$.
5. A seed rib, R_0, and the reference to the surface where we want this rib to be lying on: the seed rib and its normal vectors are contained in the plane $x = 0$, as shown in Figure 4.5. This condition, along with the data in steps 1, 2, 3, and 4, imply that the solution is symmetric with respect to the $x = 0$ plane. The seed rib in this example is a circular arc.

Figure 4.5 shows the resulting ribs and spines of the XX surfaces obtained.

4.6.2 SMS Design with an SMS Spine as Seed Rib

The next figures show other examples of XX designs in which the seed rib is designed from an SMS chain, as explained in Section 4.5.3. In every case, the spherical input wave fronts W_{i1}, W_{i2}, and W_{i3} are centered at $(x, y, z) = (0.5, 0, 0)$, $(x, y, z) = (-0.5, 0, 0)$, and $(x, y, z) = (0.5, 1, 0)$, respectively.

For the design shown in Figure 4.6, the exit wave fronts W_{o1}, W_{o2}, and W_{o3} are planes whose rays are aimed in the (H, V) directions equal to (2.5R, V), (2.5L, V), and (2.5R, 5D), respectively. Since this wave front coupling could be approximately produced by a rotational imaging system (with the largest discrepancy being that of W_{o3} wave front), the surfaces of the design in Figure 4.6 are close to having such

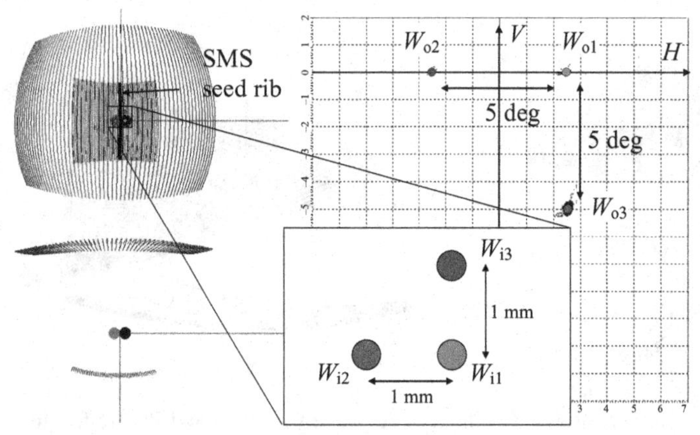

Figure 4.6 Pseudo-rotational XX design. The surfaces are nearly rotationally symmetric, although their contours are not.

symmetry (although their contours are not, because they have been fixed just by the condition that all the SMS chains have the same number of links).

As can also be seen in Figure 4.6, the ray trace from the input wave fronts through the optical systems shows how they are transformed at the exit. The transformation of W_{i1} and W_{i2} into W_{o1} and W_{o2} is nearly perfect as expected (except the noise due to the surface approximation by the CAD construct called a nonuniform rational b-spline [NURBS]). By design, only the rays of the seed rib are partially controlled with respect to W_{i3} and W_{o3}. In this pseudo-rotational case, however, the coupling of W_{i3} and W_{o3} is also good. If a rotationally symmetric SMS XX were designed, the coupling between the input and output wave fronts would be imperfect (as is that of W_3), but equally so for the three point sources (in contrast to this pseudo-rotational case).

Figure 4.7 shows two other designs obtained by modification of the exit wave fronts of the design in Figure 4.6. For the design in part a, the only modification affects W_{o3}, which has been angularly shifted down to (2.5R, 10D). Due to this change, the design maintains the width of the optics but the height is reduced by a factor of two. This reduction can be understood by the application of the étendue invariant of two-parameter bundles in 3D space to the rays linking the SMS seed rib with the adjacent rib of the other mirror. The ray trace on this device shows that the rays of W_{i3} and W_{o3} are also very well coupled, as is the case of the previous pseudo-rotational design.

In part b of Figure 4.7 is the dual case of the design on the left, in which the design modification consists of the increase from 5° to 10° of the horizontal angular separation between W_{o1} and W_{o2}. Dually, the width is halved in this case. Again, the coupling between W_{i3} and W_{o3} is good.

In all previous examples, the coupling was excellent between W_{i3} and W_{o3}, although only partially forced on the seed rib. The design of Figure 4.8, however,

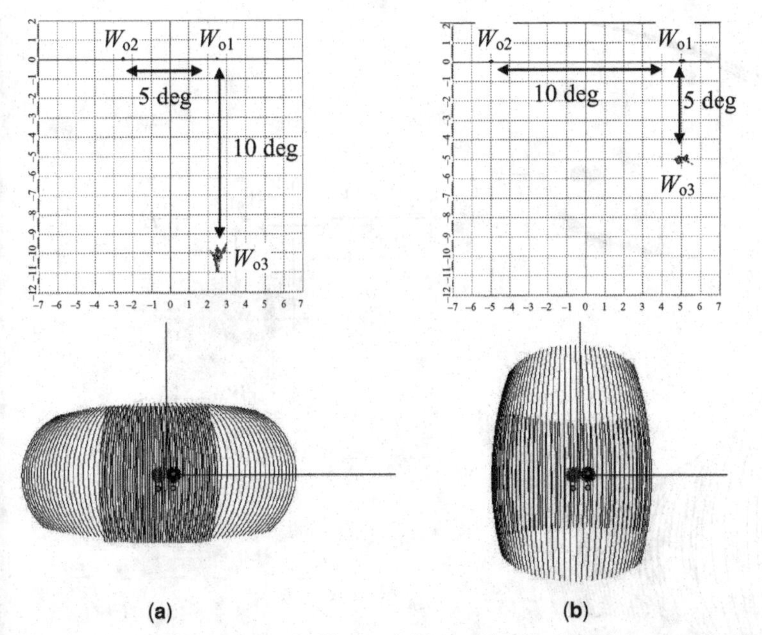

Figure 4.7 (a) Designed as in Figure 4.6 but with W_{o3} shifted down to (2.5R, 10D). (b) The dual case of the design shown in panel a, in which the modification from the design in Figure 4.6 consists of the increase from 5° to 10° of the horizontal angular separation between W_{o1} and W_{o2}.

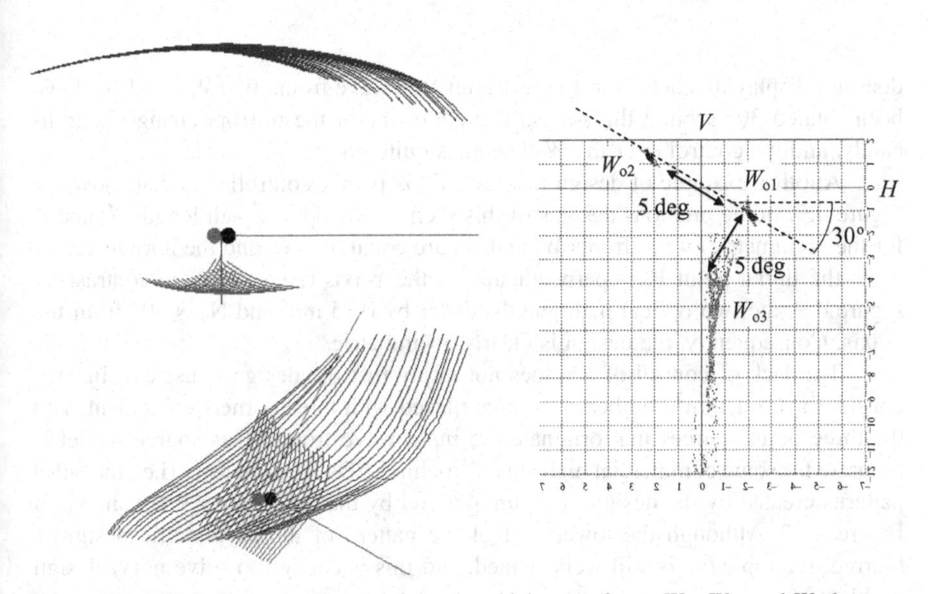

Figure 4.8 XX design as in Figure 4.6, but in which wave fronts W_{o1}, W_{o2}, and W_{o3} have been rotated 30 degrees around the z-axis.

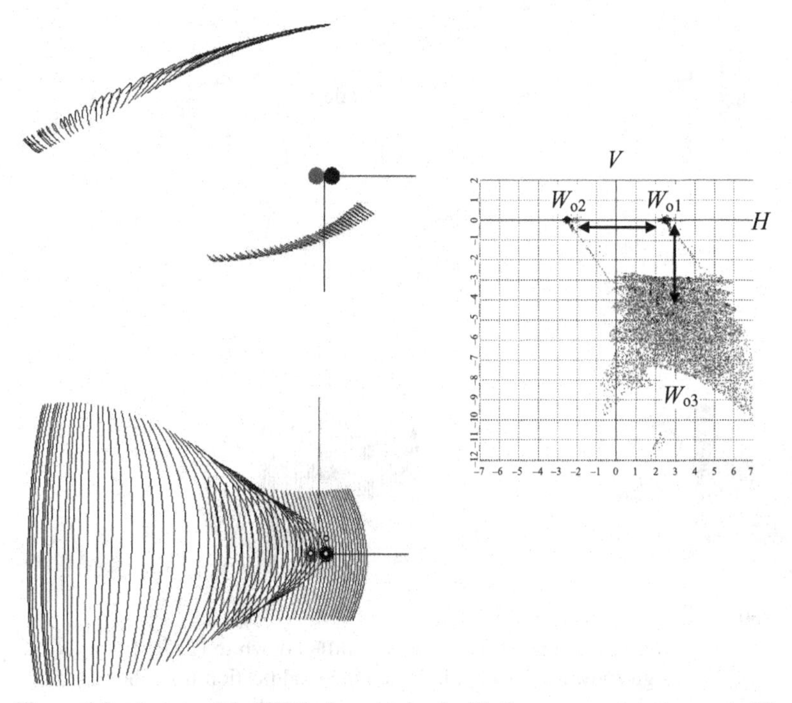

Figure 4.9 Asymmetric XX design obtained with the same parameters as in Figure 4.6 except W_1 and W_2 do not have equal path lengths, that is, $\ell_1 \neq \ell_2$.

does not display this behavior. In this design, the wave fronts W_{o1}, W_{o2}, and W_{o3} have been rotated 30° around the z-axis. The geometry of the mirrors changes significantly, and the control over the W_{i3} becomes quite poor.

Another example of design in which W_3 is poorly controlled is that shown in Figure 4.9. In the previous designs of this section, the optical path lengths ℓ_1 and ℓ_2 for the (symmetric) wave fronts W_1 and W_2 are equal, $\ell_1 = \ell_2$, and the normal vector $\mathbf{N}_0$ at the initial point $\mathbf{P}_0$ is perpendicular to the x-axis (i.e., $p = 0$). In contrast, in Figure 4.9, the two optical path lengths differ by 0.85 mm and $\mathbf{N}_0$ is 70° from the x-axis. Consequently, the design is clearly asymmetric.

The lack of control of W_3 does not imply that the design is useless. In fact, consider a 1×1 mm^2 Lambertian source placed with three corners coincident with the three point sources that originate the input wave fronts. This source model is accurate for some commercial high-flux, flip-chip LEDs. The intensity (i.e., far-field) patterns created by the design in Figure 4.6 and by that of Figure 4.9 are shown in Figure 4.10. Although the lower part of the pattern of the asymmetric design is blurred, the top edge is still well defined, and this is enough to solve many design problems.

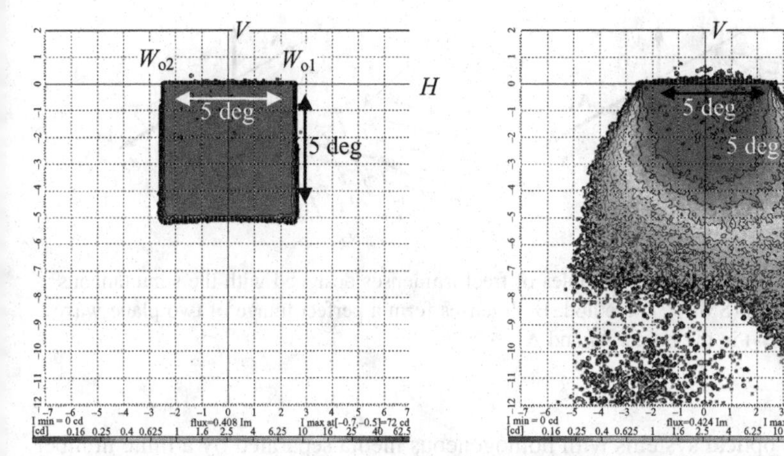

Figure 4.10 Intensity patterns produced by the designs in Figures 4.6 and 4.9 when a 1×1 mm^2 Lambertian source placed with three corners coincident with the three point sources that provide the input wave fronts.

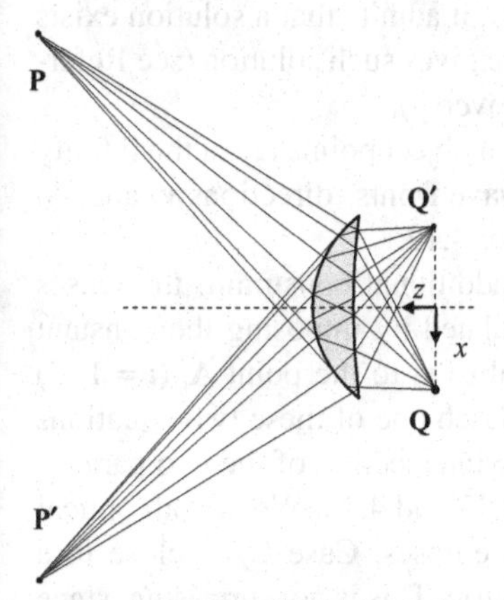

Figure 4.11 RR (SMS lens) that perfectly focuses two object points (**P** and **P′**) to two image points (**Q** and **Q′**).

4.6.3 Design of a Lens (RR) with Thin Edge

Perhaps, the simplest SMS 2D case is a lens that perfectly focuses two object points (**P** and **P′**) to two image points (**Q** and **Q′**), as the one shown in Figure 4.11.

An equivalent design can be done in 3D geometry, that is, a lens such that it focuses perfectly two object points to two image points. The impossibility of such perfect imaging of an off-axis point in three dimensions (3D) by an axially symmetric optical system is the subject of a theorem proven in the late 1970s [23]. It is

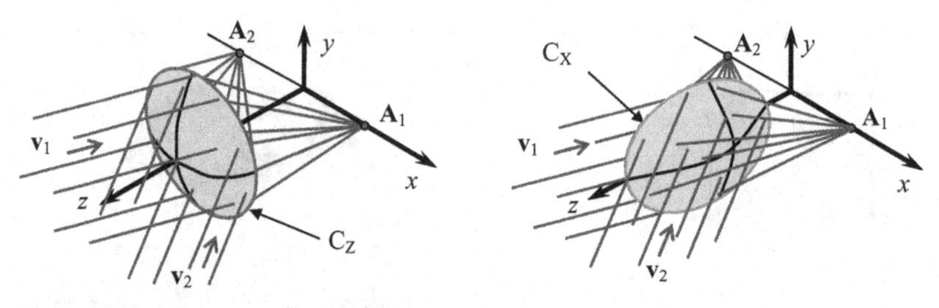

Figure 4.12 Two canonical examples of freeform lenses designed with the simultaneous multiple surfaces (SMS) 3D method. Both lenses form a perfect image of two plane wave fronts ($\mathbf{v}_1$ and $\mathbf{v}_2$) in the points $\mathbf{A}_1$ and $\mathbf{A}_2$.

restricted to optical systems with homogeneous media separated by a finite number of optical surfaces, but it provided exceptions from some trivial cases and systems with rotational symmetry (as the sphere with aplanatic points).

This theorem, reduced to two dimensions (2D), or equivalently, to the meridional rays of the axial symmetric system, caused some controversy. In Reference [24], it is stated that there is no solution (excluding also some exceptions), but in the later article [23], which corrected Reference [24], it admits that a solution exists although none is provided. The design presented here gives such solution (see Reference [5], appendix E, for more details of this controversy).

We are going to consider the case where the two object points are at the infinity and thus the lens is perfectly focusing two planar wave fronts (directions $\mathbf{v}_1$ and $\mathbf{v}_2$) in two image points ($\mathbf{A}_1$ and $\mathbf{A}_2$), as shown in Figure 4.12.

The designs shown in Figure 4.12 have an additional constraint: the lenses must have a thin edge. This thin edge can be calculated by imposing the constant optical path length from a plane wave front normal to $\mathbf{v}_i$ to the point $\mathbf{A}_i$ ($i = 1, 2$) through the rays that pass through the edge points. Each one of these two equations define a paraboloid with focus at $\mathbf{A}_i$ and axis $\mathbf{v}_i$. The intersection of the two paraboloids gives the curves C_Z and C_X shown in Figures 4.12 and 4.13. When both optical path lengths are equal, the edges (C_Z and C_X) are ellipses. Case C_Z is close to a rotationally symmetric lens, but it is slightly asymmetric. This is not surprising, since we know that a rotational symmetric lens almost forms perfect images of two off-axis points, although, as we said before, it cannot form a perfect image.

Figure 4.14 shows two cross sections of the SMS lens generated from the ellipse normal C_Z. These thin-edge lenses have been designed using the edge as the seed rib (see Section 4.5.3). In this case, it is necessary to extend approximately the surfaces of the lens in the neighborhood around the seed rib using the surface normal vectors. These normal vectors can be calculated once the seed rib curve is known by imposing the double refraction of the rays of both design bundles. Note that the unitary normal vectors $\mathbf{N}_1$ and $\mathbf{N}_2$ associated with each refractive surface are given by the unique solution of the system of equations that is derived from applying Snell's law to the four incidences (two per ray) at every point of the edge.

The tangent plane approximation is used to extend one refractive surface in the neighborhood of the points of the edge. This extension allows one to calculate

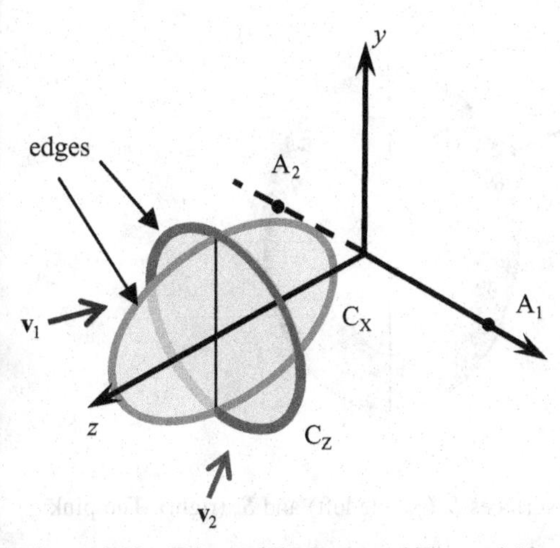

Figure 4.13 Edges of the lenses of Figure 4.12.

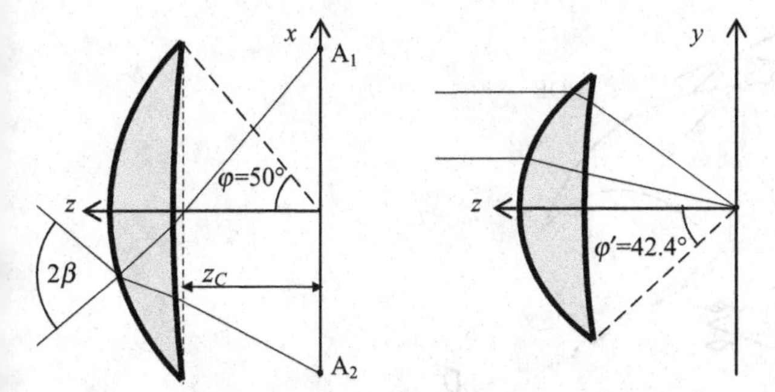

Figure 4.14 Cross sections ($y = 0$ plane and $x = 0$ plane) of the SMS lens generated from the ellipse normal to the z-axis. The design parameters are $n = 1.5$, $\beta = \pm 40°$, and $z_C = 0.8801$.

another curve on the refractive surface close to the edge but not quite at it. This step is necessary because all the SMS ribs calculated from the edge by the application of the SMS method are the same edge, so no additional ribs can be calculated from it. Once we have a curve not coincident with the edge but as close as desired to it, it is possible to generate SMS ribs on both refractive surfaces.

Figure 4.15 shows a plane view of the SMS ribs from a curve R_0 close to the lens edge. By choosing different curves R_0 close to the edge but not coincident among them, it is possible to calculate as many points as desired on both refractive surfaces. This eliminates the need of surface skinning (Section 4.5.2).

4.6.4 Design of an XX Condenser for a Cylindrical Source

As said before, the simplest application of the SMS 3D design method provides full control of the coupling of two wave fronts and the partial control of a third coupling

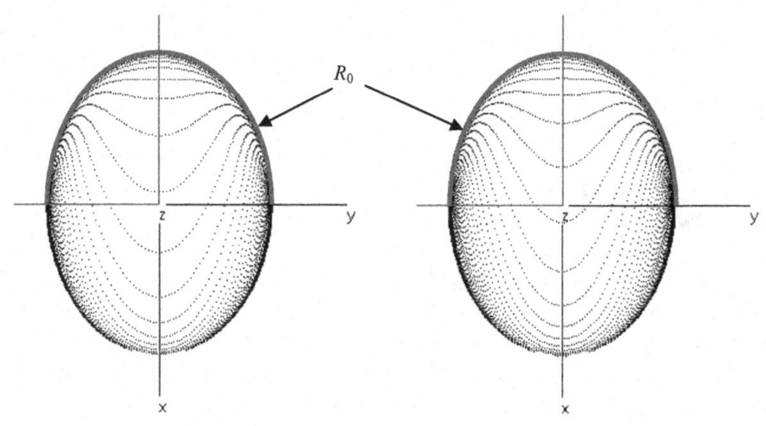

Figure 4.15 Plan view of the ribs on the surfaces S_i (on the left) and S_o (right). The pink curve is the seed rib R_0.

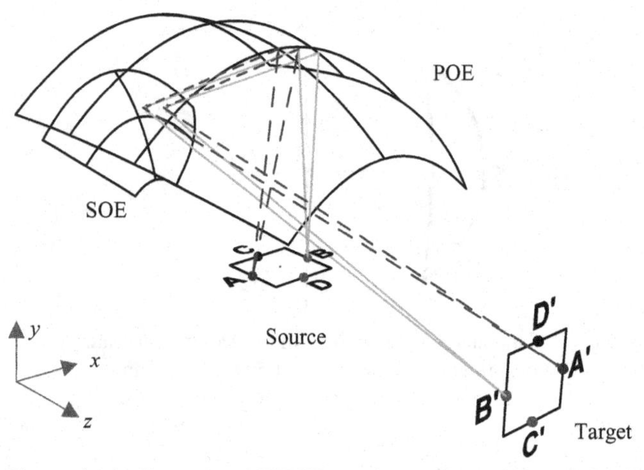

Figure 4.16 Example of XX 3D condenser, formed by two freeform mirrors, the primary optical element (POE) and a secondary optical element (SOE).

with just two optical surfaces (mirrors or dioptrics). Thus, the size, position, and orientation of the projected source images of an illumination optical system can be controlled to an unprecedented degree for a two-surface design. SMS-designed, dual-reflector devices are termed XX (each X refers to a reflection as rays propagate from the source to the target). In order to clarify the various XX condenser families, this section considers the design of an XX that must collect the light emitted by a rectangular flat source and transfer it to a rectangular flat target, in the geometric configuration shown in Figure 4.16. The light source is placed at the $y = 0$ plane and emits in the $y > 0$ hemisphere of directions. The target is placed at a $z =$ constant plane and will receive the light from the $z < 0$ hemisphere of directions.

Assume for a while that the source size is small enough compared with the condenser size, and that the XX is able to image the source on the target so the following linear mapping holds (note that the SMS can guarantee a sharp image of 2 points, for instance **A** and **B** into **A′** and **B′**, and a partial image of the remaining ones **C** and **D**):

$$\begin{pmatrix} x' \\ y' \end{pmatrix} = \begin{pmatrix} N & 0 \\ 0 & M \end{pmatrix} \begin{pmatrix} x \\ z \end{pmatrix} + \begin{pmatrix} c_1 \\ c_2 \end{pmatrix}, \tag{4.4}$$

where (x', y') is a point on the target, (x, z) is a point on the source (the same global coordinate system x–y–z for source and target is being used), and c_1 and c_2 are constants that define the mapping of the center of the source to the center of the target. This mapping implies that to first-order approximation (valid for a small source), the SMS method provides an image-forming design, whereby the light source is placed on the object plane and the target on the image plane. In Figure 4.16, points **A**, **B**, **C**, and **D** are object points, while **A′**, **B′**, **C′**, and **D′** are their corresponding image points. The diagonal of the matrix in the previous equation defines the magnifications of the optical system. Constants M and N are defined as:

- Magnification M: ratio between the segment of the target **C′D′**, and the segment **CD** of the source.
- Magnification N: ratio between the segment of the target **A′B′**, and the segment **AB** of the source.

Parameters M and N can be either positive or negative, so four families of XX can be considered.

For the calculation of the SMS seed rib (see Section 4.5.3), two point sources (**A** and **B** in Figure 4.16) are placed such that the line joining them is parallel to the x-axis of the coordinate system. The line that joins the target points **A′** and **B′** will be parallel to the x-axis of the coordinate system as well, see Figure 4.17.

For the SMS-ribs calculation, two point sources (C and D in Figure 4.16) are placed such that the line joining them is parallel to the z-axis of the coordinate

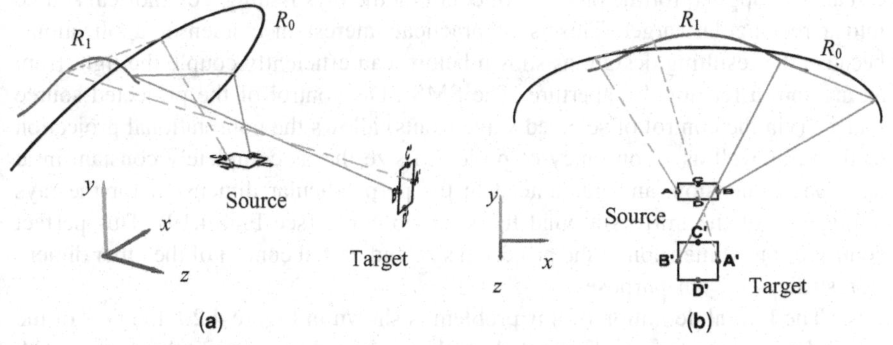

Figure 4.17 Initial curve R_0 (seed rib) calculation, which is done with source spherical wave fronts emitted from points **A** and **B** and with target spherical wave fronts centered on **A′** and **B′**. This two-point mapping defines magnification N (negative in this figure).

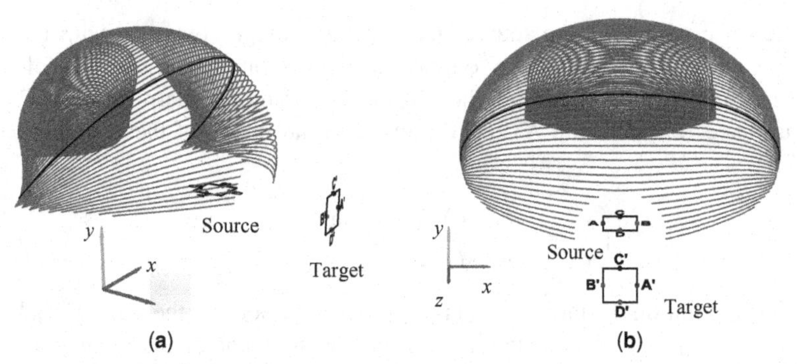

Figure 4.18 SMS-ribs calculation, which is done with source spherical wave fronts emitted from points **C** and **D** and with target spherical wave fronts centered on **C′** and **D′**. This two-point mapping defines magnification M (negative in this figure).

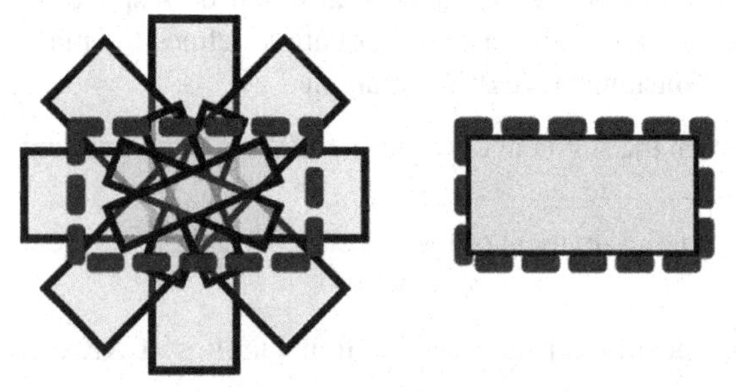

Figure 4.19 The XX 3D condenser, in contrast to conventional condensers (left), can be designed to produce no rotation of projected arc images so that they all can fit rectangular apertures.

system. The line that joins the target points **C′** and **D′** will be parallel to the y-axis of the coordinate system. Figure 4.18 shows some ribs on an XX.

The XX configuration and geometry just introduced in the previous paragraphs can also be applied for the problem of coupling the rays issuing a cylindrical source into a rectangular target. This is of practical interest in condenser applications, because the resulting design, as shown below, can efficiently couple the light from an arc into a rectangular aperture. The SMS 3D's control of the projected source images (via the control of selected wave fronts) allows the nonrotational projection of them, as well as a constancy of projected size that is completely constant in at least one dimension, and guaranteed in the perpendicular dimension for the rays hitting one of the mirrors around the seed rib curve (see Fig. 4.19). This perfect control of one dimension of the projected size and partial control of the other dimension suffices for our purposes.

The formal definition of this problem is shown in Figure 4.20: the rays of the source have all the same radiance and are those emitted from a cylinder surface with directions forming an angle greater than β_{MIN} with the cylinder's axis. The rays accepted by the target are those reaching a rectangle forming an angle smaller than

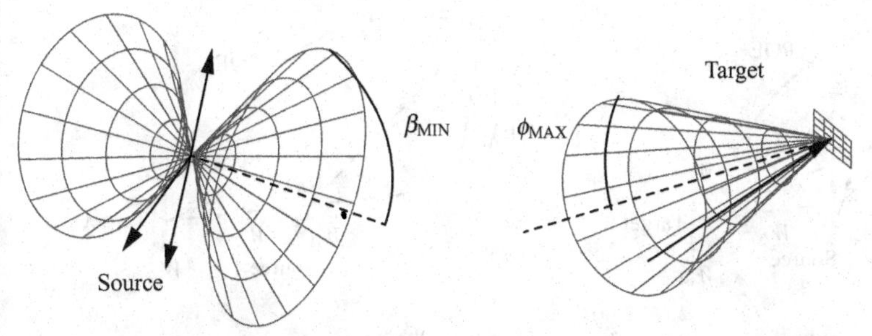

Figure 4.20 Source and target definitions: Every point of the source surface (a cylindrical surface) emits in its entire open hemisphere excepting in the directions forming an angle smaller than β_{MIN} with the cylinder axis. The target is a rectangle accepting radiation coming within a cone normal of ϕ_{MAX} to the target surface.

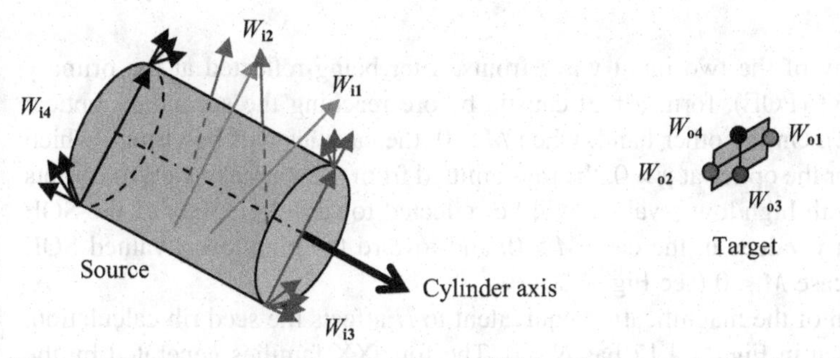

Figure 4.21 Rays of the input wave fronts (W_{ix}) and center points of the output wave fronts (W_{ox}) used for the SMS 3D design.

ϕ_{MAX} with the normal to the rectangle. The condenser has to maximize the power transferred from the source to the target.

Figure 4.21 shows the input and output wave fronts for the SMS 3D design process. All of them belong to the edge rays of the bundles defined for the source and the target. All of the output wave fronts W_{ox} are spherical and centered at the midpoints of the sides of the rectangular target. Two of the wave front couples W_{i3}–W_{o3} and W_{i4}–W_{o4} are used for the calculations of the SMS chains while the other two couples W_{i1}–W_{o1} and W_{i2}–W_{o2} are only used only for the seed rib calculation. This is why there is only a partial control of one dimension of the source images. W_{i3} and W_{i4} are orthonormal rays issuing form the cylinder edges. W_{i1} and W_{i2} are also orthonormal bundles formed by tangent rays to the cylinder.

Four families of solutions can be defined according to the signs of the magnifications. Since the input source is no longer a plane, the classical definition of magnification does not apply, but the four families still appear as a consequence of the wave front-pair assignment, and for simplification of the families' nomenclature, the terms M and N and the magnification signs will remain. To illustrate this, Figure 4.22 shows the spines contained on the plane $x = 0$ for families with the two possible signs of magnification equivalent to M. Note that in this 2D section, when $M < 0$,

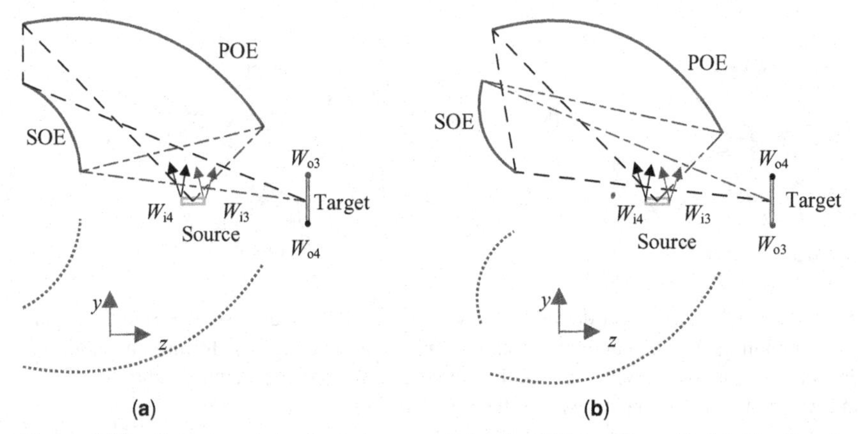

Figure 4.22 Spines contained on plane $x = 0$ for the XX families with (a) $M > 0$ and (b) $M < 0$.

the rays of any of the two input wave fronts, after being reflected at the primary optical element (POE), form a real caustic before reaching the secondary optical element (SOE). On the other hand, when $M > 0$, the caustics will be virtual, which implies that for the optics at $y > 0$, the rays emitted from the source toward the points of the POE with high/low z-values will be reflected toward the points of the SOE with low/high y-values in the case $M > 0$, and toward the high/low y-valued SOE points in the case $M < 0$ (see Fig. 4.22).

The sign of the magnification equivalent to N affects the seed rib calculation. The case shown in Figure 4.17 has $N < 0$. The four XX families generated by the two possible signs of M and N can be equivalently described by the real or virtual nature of their two caustic surfaces in 3D.

Since the cylindrical source emits light toward both the $y > 0$ and $y < 0$ half-spaces, there is still another Boolean variable to be added to the signs of magnifications M and N, raising the four families of XX solutions to eight. This third Boolean variable arises from the additional possibility of choosing that the half of the POE mirror at $y > 0$ reflects the light toward the half of the SOE at $y > 0$ (as shown in all previous figures), or toward the half of the SOE at $y < 0$, as shown in Figure 4.23. Consider the next XX design example: $M < 0$, $N < 0$, with nonadjacent POE and SOE paired halves (as in Fig. 4.23). The distance from source center to target plane is fixed at 30 mm. The input parameters of this design are:

- Cylindrical source: length $L = 1.2$ mm; diameter $D = 0.3$ mm; and $\beta_{MIN} = 45°$.
- Rectangular flat target: aspect ratio = 4 : 1 and $\phi_{MAX} = 19°$.

Figure 4.24 shows the POE and SOE mirrors of this XX design. Figure 4.25 shows the standard top, side, and front views, indicating the dimensions.

In order to evaluate performance, the collection efficiency versus target etendue has been calculated by ray tracing (using the commercial ray tracing package LightTools® by Synopsys (Mountain View, CA) [25]). The set of optical surface points obtained from the SMS design method were used to create the freeform surface patches with Rhino3D [26]. These surfaces are then exported to LightTools

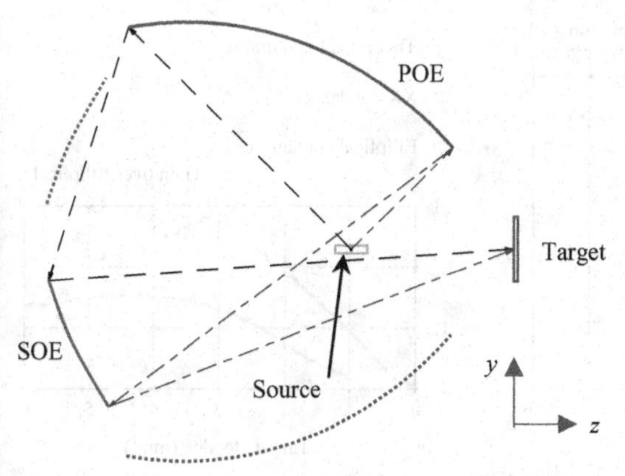

Figure 4.23 $x = 0$ section of one design variation in which the half of the POE mirror at $y > 0$ reflects the light toward the half of the SOE at $y < 0$; the paired POE and SOE halves are not adjacent.

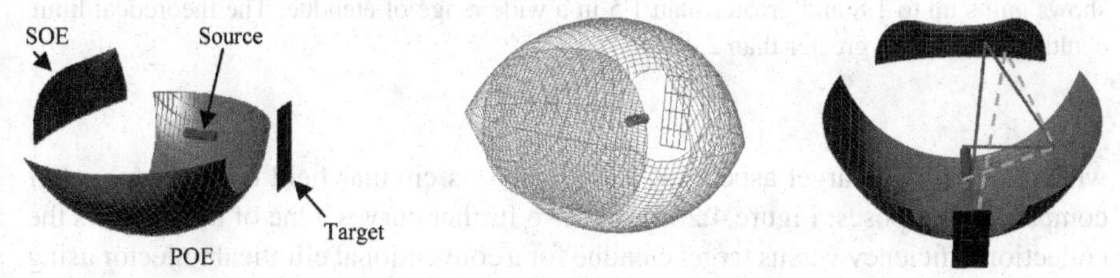

Figure 4.24 Selected XX design of the family $M < 0$, $N < 0$, and nonadjacent POE and SOE paired halves (only POE with $y < 0$ and SOE with $y > 0$ are shown at the left and at right hand sides). Source and targets are not shown to scale.

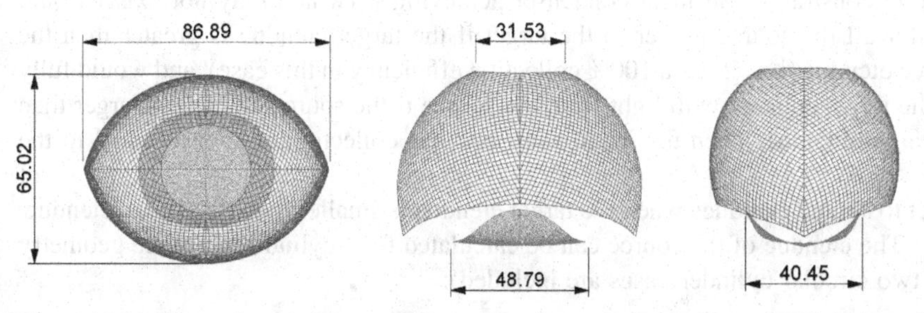

Figure 4.25 Three views with some dimensions (in mm) of the selected XX design of Figure 4.24.

in SAT format. The source and the target are created in LightTools with its own CAD capability.

Figure 4.26 shows efficiency versus target étendue. This étendue is given by $E_{\text{target}} = A_{\text{target}}\pi\sin^2(\phi_{\text{MAX}})$, where A_{target} is the target area and $\phi_{\text{MAX}} = \pm 19°$ is the target acceptance angle. The étendue of the target is varied by varying the target area A_{target}

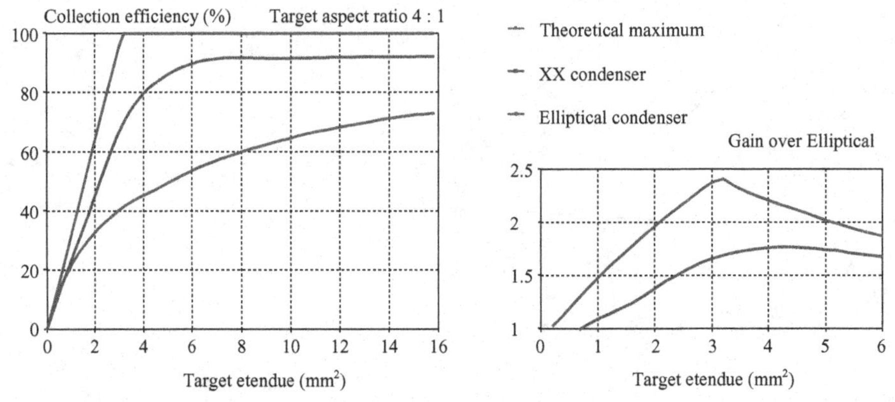

Figure 4.26 Ray tracing results: collection efficiency versus étendue of the target for the condenser in Figure 4.24. The maximum theoretical performance and that of conventional elliptical condenser are also shown for comparison. The étendue of the source is 3.13 mm^2-sr. All mirrors have 100% reflectivity. Gain defined as the ratio of the collection efficiencies of a given condenser to that of the elliptical condenser. XX in Figure 4.24 shows gains up to 1.8 and greater than 1.5 in a wide range of étendue. The theoretical limit could achieve gains greater than 2.4.

while freezing the target aspect ratio (4:1) and its circular field of view ϕ_{MAX}. For comparison purposes, Figure 4.26 shows two further curves. One of them shows the collection efficiency versus target étendue for a conventional elliptical reflector using the same source and target with the same aspect ratio (4:1). The eccentricity of the ellipsoid has been set equal to 0.8 (which is the optimized standard) and the target field of view of $\phi_{MAX} = \pm 30°$ (which is also the standard value in the market). The third curve in Figure 4.26 corresponds to the theoretical limit, which is imposed by étendue constraints: an ideal condenser achieving it (which may not exist) would transfer all the source power to the target if the target étendue is greater than the source etendue (i.e., it has a 100% collection efficiency in this case), and would fully fill the target étendue with light from the source if the source étendue is larger than the target étendue. Then the ideal condenser has collection efficiency equal to the ratio of target to source étendues when the target étendue is smaller than the source étendue.

The étendue of the source can be calculated for a cylindrical source geometry (the two circular cylinder bases are included):

$$E_{\text{source}} = \pi DL(\pi + \sin(2\beta_{\text{MIN}}) - 2\beta_{\text{MIN}}) + \frac{1}{2}\pi^2 D^2 (1 - \sin^2 \beta_{\text{MIN}}). \tag{4.5}$$

For the above input data, $E_{\text{source}} = 3.13$ mm^2-sr. Figure 4.26 shows that the XX performs better than the elliptical reflector (for all mirrors, specular reflectivity has been set equal to 1), getting close to the theoretical limit. There are three factors, however, that prevent the XX from reaching the theoretical limit:

1. When target étendue is large, some rays that reflect off the POE will miss the SOE, so that the XX curve cannot reach 100% collection efficiency.

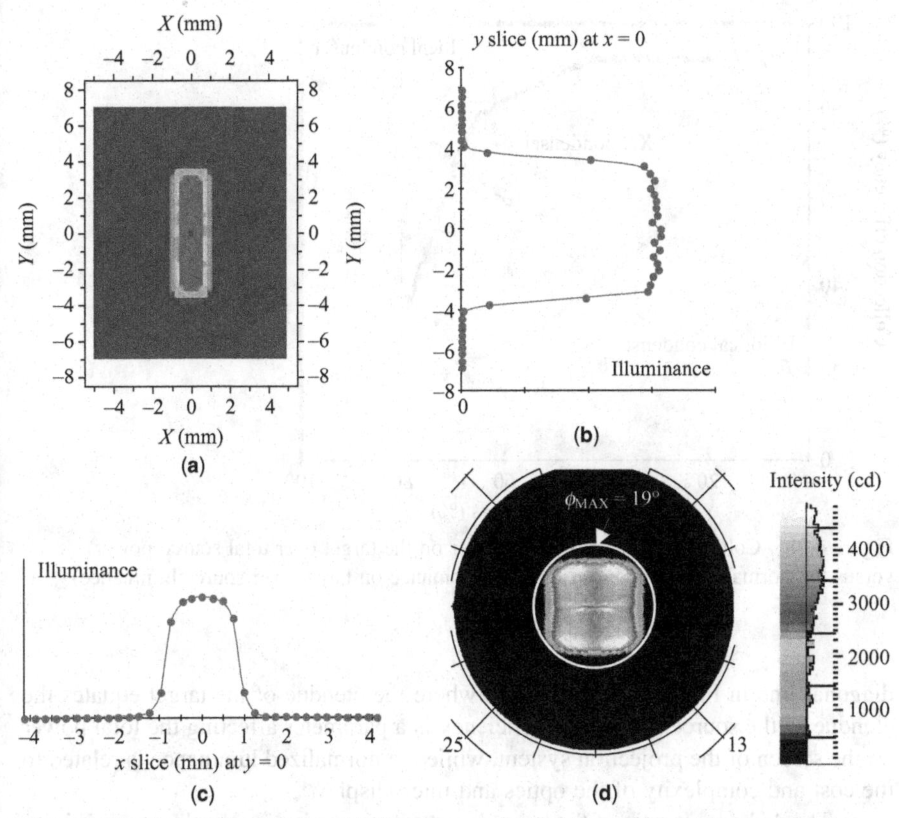

Figure 4.27 Ray tracing results: (a) illuminance distribution on the target plane for the condenser of Figure 4.24. (b) Illuminance along a y-slice, (c) illuminance along an x-slice, (d) intensity distribution at the target assuming the source has a flux of 1000 lm.

2. The "shoulder" of the XX efficiency curve is rounded (in contrast to the theoretical one, which shows a slope discontinuity). This is due to the nonstepped transitions of the illuminance distribution on the target plane and also to the rounded contour lines of that illuminance distribution (shown in Fig. 4.27).
3. When target étendue is small, the slope of the XX curve in Figure 4.26 is less than the theoretical slope because the XX does not fill completely and uniformly fill the circular field of view (see the intensity distribution in Fig. 4.27).

Figure 4.28 shows another way to represent the data of Figure 4.26, which is more interesting, at least from an academic point of view. In this figure, the collection efficiency is represented versus the normalized brightness, which is defined as the ratio of the average brightness on the target and the source brightness. The average brightness on the target is simply the ratio of the power on the target over the target's etendue. The source brightness (or luminance) is the ratio of the source power and the source étendue. This calculation assumes that the luminance of the source is the same for all the source rays (as it is in our model). The points with the same target étendue in this representation form a straight line crossing the origin. The dashed

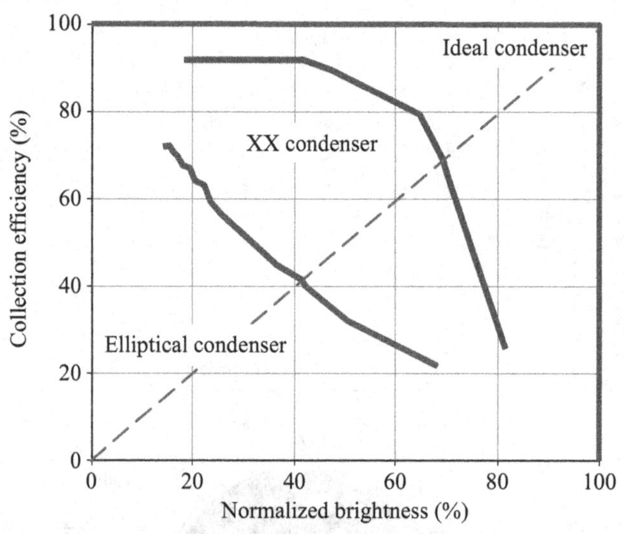

Figure 4.28 Collection efficiency (i.e., power on the target over total source power) versus the normalized luminance (average luminance on target over source luminance).

diagonal line in Figure 4.28 is the line where the étendue of the target equates the étendue of the source. Collection efficiency is a parameter affecting the total power on the screen of the projection system, while the normalized luminance is related to the cost and complexity of the optics and microdisplay.

The 4 : 1 aspect ratio of the target is not seen in projection display applications. On the contrary, the 16 : 9 format is considered the present standard. An XX condenser similar to that in Figure 4.24, designed for a cylindrical source with diameter 0.62 mm (source étendue = 6.96 mm^2-sr) and keeping the rest of parameters unchanged, produces a 16 : 9 illuminance distribution on the target plane. Figure 4.29 shows the ray tracing results. The XX with target's circular field of view of $\phi_{MAX} = 19°$ still performs better than elliptical reflector, although the gain is reduced to 1.5, due to the lower aspect ratio of the target (again, for all mirrors, specular reflectivity has been set equal to 1). The theoretical limit gain is also reduced to two.

It is interesting to note that these XX condensers perform closer to the theoretical limit if a square field of view for the target is considered. Figure 4.29 also shows the collection efficiency and target étendue using a square field of view of 28° × 28° for the XX at the target (this square field of view is almost inscribed in the preceding 19°-radius circular field of view). This result is more than just an academic consideration if we take into account that the angular acceptance field of the dielectric-filled mixing prism working with total internal reflection has a square shape. The theoretical limit stays unchanged, but the XX performs much better because its intensity distribution matches better with the square field of view. Note that the slope of the efficiency curve of the XX with square field of view now becomes very close to the theoretical limit near the origin, indicating a uniform and well-filled field of view.

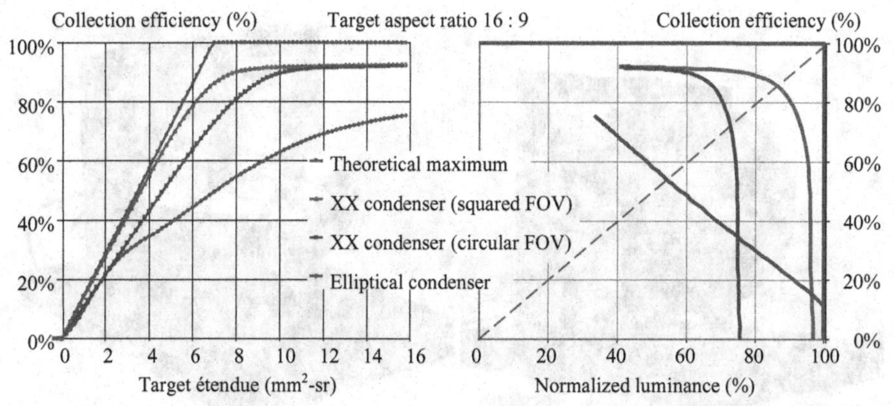

Figure 4.29 Ray tracing results: (left) collection efficiency of an XX condenser versus the 3D etendue of a 16:9 target with circular and square field of view (FOV). The maximum theoretical performance and the conventional elliptical condenser (with circular FOV) are also shown for comparison. All mirrors have 100% reflectivity. The 3D étendue of the source is 6.96 mm^2-sr. The right hand side shows the same results in a collection-efficiency versus normalized brightness plane (normalized brightness = average brightness on target/ source brightness).

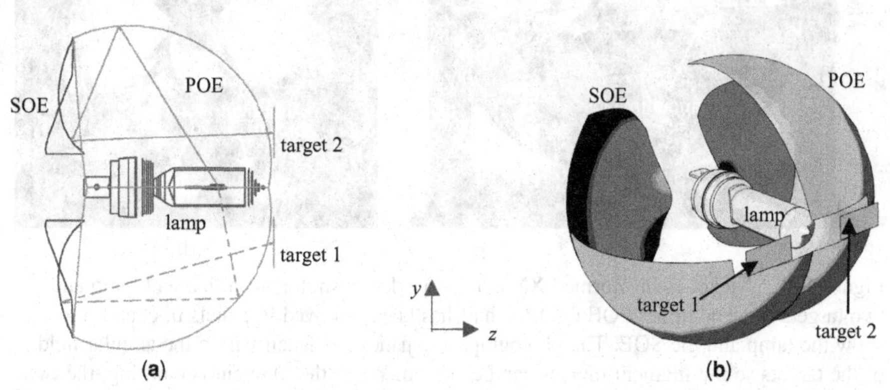

Figure 4.30 Demonstrator prototype diagrams; (a) cross section; (b) cutaway perspective view.

An XX demonstrator prototype was been fabricated to validate these design principles. For easier handling, instead of a projector arc lamp (whose high flux cannot be easily dimmed), an automotive H7 halogen lamp is selected. The H7 filament is a spiral enclosed by a cylinder of length $L = 4.3$ mm and diameter $D = 1.55$ mm. The lamp geometry constrained the choice to the XX configuration, so as to avoid shading and mirror to lamp interferences.

The design selected is an XX of the family $N < 0$ and $M < 0$ with adjacent POE and SOE halves. Moreover, each half has a separate rectangular target, as shown in Figure 4.30. The 3D drawing in Figure 4.30b shows the device rotated 90° around the cylindrical source axis, in contrast to the device in Figure 4.24.

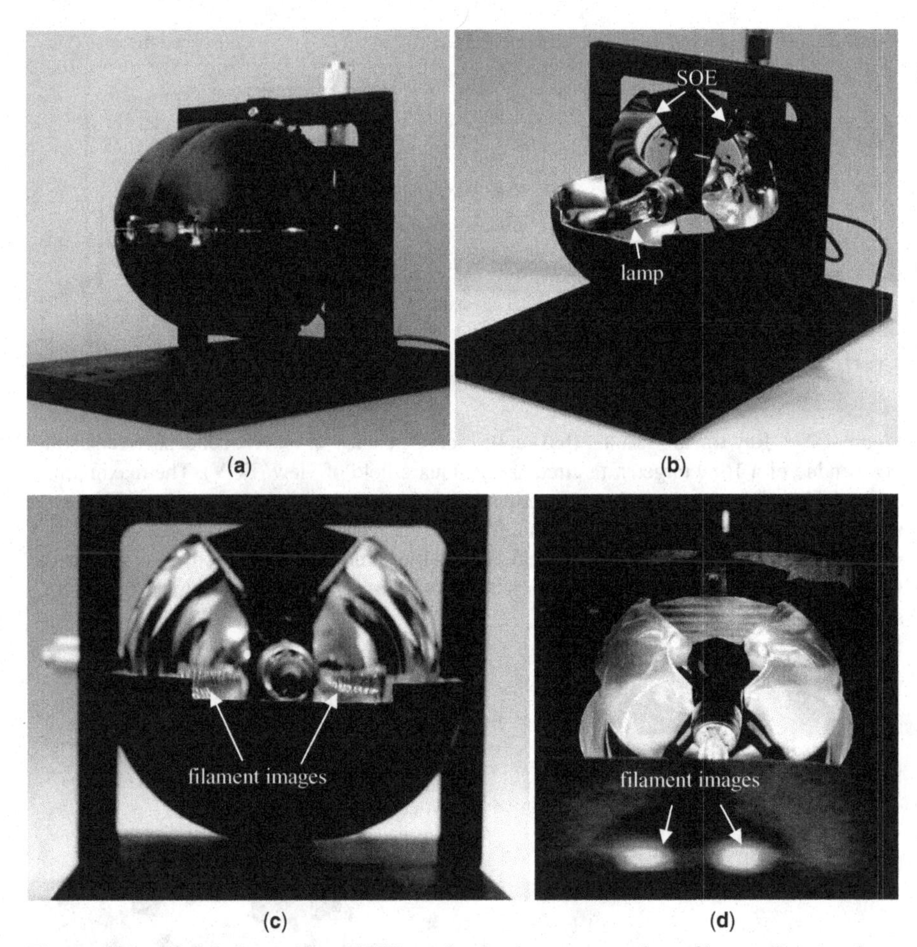

Figure 4.31 Nickel electroformed XX condenser demonstrator for a filament lamp and two targets (a). The upper POE reflector half has been removed in panels b, c, and d to show the lamp and the SOE. The photography in panel c is taken within the angular field of the targets so the filament images can been seen. When the filament is emitting, the two images are formed on a paper placed at the target plane (d).

The prototype, manufactured by LPI [27] using nickel electroforming for the reflectors, is shown in Figure 4.31. The mirror coating is made of evaporated aluminum. The $y > 0$ and $y < 0$ halves of the SOE mirror are made as identical replicas of the mold. Similarly, the POE is made of two halves, but in this case, the POE is split into $x > 0$ and $x < 0$ halves, for easier mold release. Figure 4.31a shows the entire condenser. The upper POE has bee taken out in Figure 4.31b showing the lamp and the SOE. The photography of Figure 4.31c was taken within the angular field of the target. It shows the two images of the filament formed at the exit plane. Figure 4.31d shows the two spots formed by the condenser on a paper placed at the condenser target plane. For a clearer understanding of this arrangement, the upper POE has been taken out from Figure 4.31b–d.

Although no optical tolerance analysis has been done, some qualitative considerations can be established. Regarding the effect of tolerances on the manufacturing of the XX condenser, we distinguish between the source tolerances (which include the variability of the arc electrode, bulb shape, time-varying arc position, etc.) and the optical surface tolerances. The source tolerances can be seen as a variation of the apparent source position and size, which will cause a corresponding modification of pinhole projected images. The source tolerances affect the XX in a different way compared with a conventional elliptical reflector, as the pinhole-projected images of both devices differ significantly. Because the projected images of the XX have all similar sizes, we foresee a more relaxed tolerance for it than for the elliptic reflector. Nevertheless, regarding optic surface tolerance, we expect the XX to be less tolerant simply because the rays suffer two reflections instead of the single reflection of the conventional elliptic reflector. Therefore, variation due to the first surface reflection not having the design input wave front is enhanced upon the second reflection. The novelty of the freeform surface shapes means that no effective standard of specification and testing is available yet for the optic tolerances.

4.6.5 Freeform XR for Photovoltaics Applications

The design of photovoltaic concentrators (CPV) introduces a very specific optical design problem, with specific features. We are going to focus our attention on high concentration photovoltaics (HCPV), that is, when the range of concentration is above, say 400. This means that the aperture of the photovoltaic concentrator is at least 400 times the active area of the cell. This is the type of application that is usually found when the photovoltaic (PV) cells are the most efficient ones. At present, the PV solar to electric conversion efficiency record is 43.5% [28]. It has been achieved by the company Solar Junction with a multijunction (MJ) cell. The relatively high cost of these high efficiency cells calls for a solution like CPV in which the light received from the sun is concentrated in the small area where the cell is placed. The aim of such a device is to reduce the cost of electricity generated by PV conversion. For this purpose, it is necessary that the optical system (concentrator) collecting and concentrating the light be very efficient. It must also be suitable for low cost mass production, capable of concentrating the light on a small region (high concentration), insensitive to manufacturing and mounting inaccuracies, and able to provide uniform illumination of the cell (for efficiency and reliability purposes, since MJ cells tend to perform poorly without uniformity). These conditions are translated to specific conditions in the design process: an efficient and low-cost optical system implies, among other things, a small number of optical components. Insensitivity to manufacturing and mounting inaccuracies and capability of high concentration imply a concentrator acceptance angle close to the thermodynamic limit. All conditions together imply that each optical component must perform several optical functions.

The thermodynamic limit of concentration (see for instance Reference [5]) relates the acceptance angle α of a concentrator with its concentration C_g, in a simple way: $C_g \sin^2\alpha \leq n^2$, where n is the refractive index in the concentrated region. For

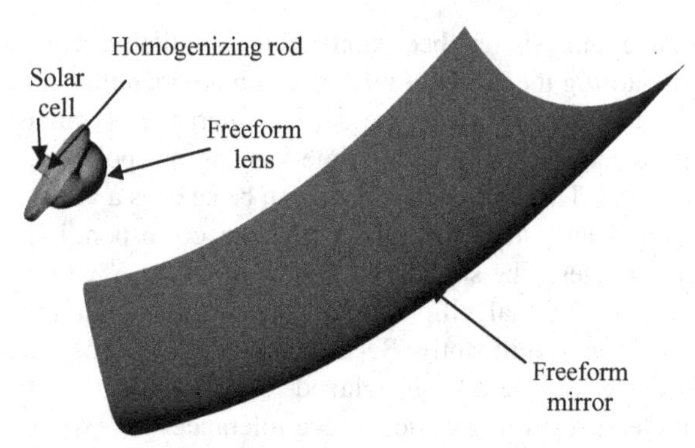

Figure 4.32 Freeform XR with homogenizing rod.

cost and reliability reasons, n is about 1.5 in the best situations. The equality in this condition is called the thermodynamic limit. The concentration acceptance product (*CAP*) gives one way to quantify how close the performance of a concentrator to the thermodynamic limit. CAP is defined as CAP = $\sqrt{C_g}\,\sin\alpha$. Then CAP $\leq n$.

A high *CAP* is equivalent to the highest possible acceptance angle (for a given C_g). This allows tolerance in mass production of all components, relaxes the module assembly and system installation, and reduces structural costs. Reducing the number of elements and achieving high acceptance angle relaxes optical and mechanical requirements, such as accuracy of the optical surfaces profiles, module assembly tolerances, precision of installation and of the supporting structure, and so on. The XR configuration, which we develop in this section (in which the rays coming from the sun hit a mirror surface [X] and are reflected toward the lens, where they are refracted [R] toward the cell), is suitable for the concentrating photovoltaic application because it can meet the previously mentioned conditions. This section describes the design procedure for this specific application and follows Reference [29].

The optical device is shown in Figure 4.32 [30]. It is formed by a mirror (X), a lens (R), and a transparent rod. The PV cell is optically adhered to the end of the short transparent rod, preferably glass, which is molded in one piece with the freeform lens. This piece is called a secondary optical element (SOE). The rod is not mirror coated. The function of the rod is to ensure a uniform irradiance on the cell. The rod is long enough to produce a desired degree of homogeneization, as will be shown later. As illustrated in the Figure 4.33, total internal reflection (TIR) at point T is possible for any ray beyond the critical angle $\theta_c = \sin^{-1}(1/n)$, where n is refractive index of the secondary lens, which limits how close to the surface normal the interior rays can hit the sides of the rode and not escape. For a rod with sides with draft angle γ (as shown in Figure 4.33) and when the angle of incidence is $\theta = \theta_c$, then the angle β (the angle that incoming rays enter to the rod) is $\beta = 90° - \beta - \gamma$ [31].

Such a prism is a well-known homogenizing device in nonimaging optics and is based on the same principle as the kaleidoscope. Homogenizing rods are commonly used in CPV system [32–34], but their length is usually much longer than the cell size (typically four to ten times), while in the XR, the rod may be much

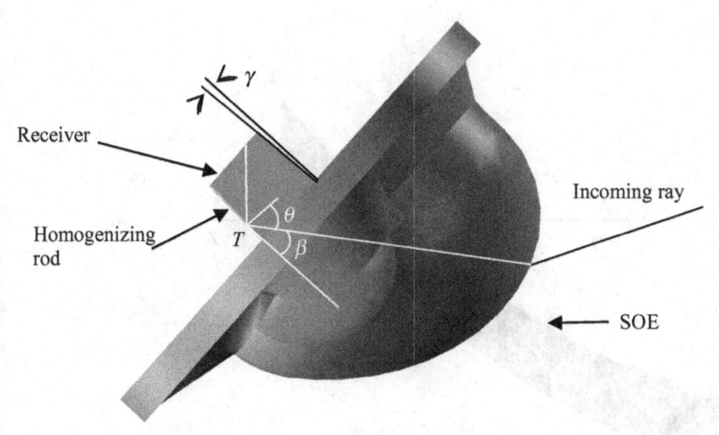

Figure 4.33 SOE formed by a freeform lens with homogenizing rod.

shorter (from 0.5 to 1 times the cell size). The sufficiently good homogenization in a short length is possible because the illumination angle of the cell in the XR is wide.

4.6.5.1 The XR Design Procedure

The freeform lens and mirror of the XR configuration are designed using the SMS 3D method. This design procedure can be divided into four stages:

(a) Defining the system parameters

(b) Defining the input SMS 3D data

(c) Designing the initial curve

(d) Designing the two SMS 3D freeform surfaces

Following the algorithm, in step (a), we define the system parameters, which are:

- the secondary lens refraction index n;
- the inclination angle θ of the receiver with respect to the x–y plane (positive as shown in Fig. 4.34);
- an initial point $\mathbf{P}_0 = (0, y_{P0}, z_{P0})$ on the mirror, and the normal vector $\mathbf{N}_0$ to the mirror at $\mathbf{P}_0$, in this example, parallel to plane $x = 0$;
- the y-coordinate Y_{S0} of a point $\mathbf{S}_0$ on the lens;
- the rod input aperture length, which is segment **PQ**;
- the acceptance angle α, which is determined by the etendue conservation law assuming that the span of the mirror on the y-dimension is approximately y_{P0}–y_{P0}, so $\alpha = \sin^{-1}(n(y_{P0}\text{–}y_{S0})\sin(\beta)/|\mathbf{PQ}|)$; and
- a factor $b < 1$, which may be useful to adjust the density of calculated points, as explained below.

Once the general parameters have been selected, (step b) the additional data required for the SMS 3D design are defined (Fig. 4.35). These additional data refer to the

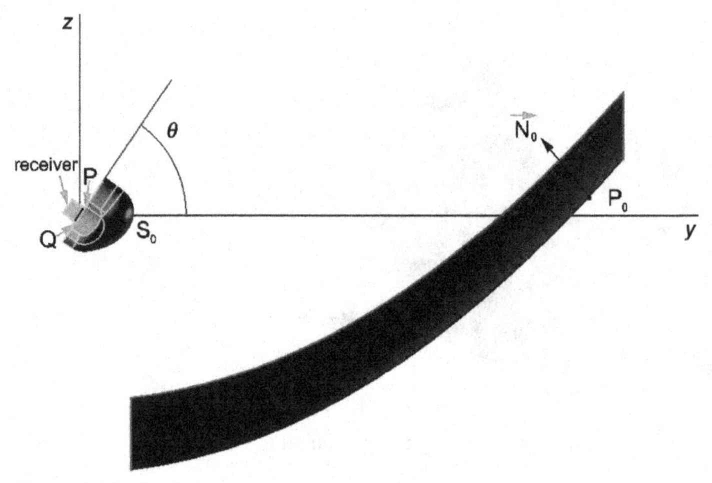

Figure 4.34 Input parameters.

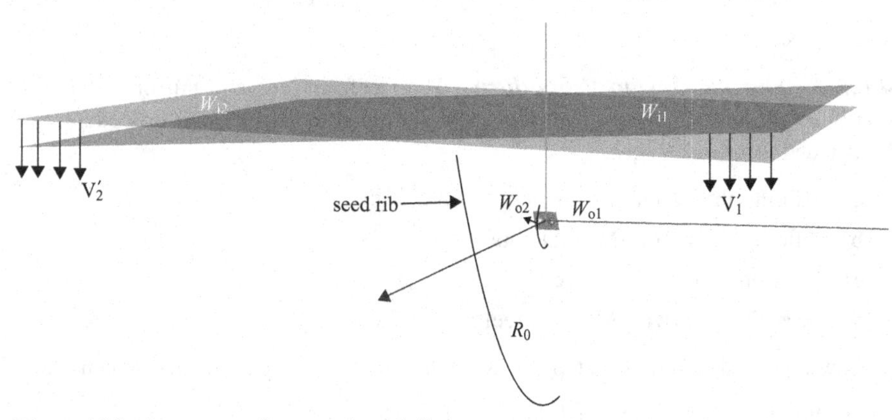

Figure 4.35 Input wave fronts and seed rib.

bundles of rays that are transformed by the XR. Each one of these (orthonormal) bundles is characterized by a wave front. There are four wave fronts to define: two of them refer to two bundles of rays located at the entry of the concentrator (input wave fronts) and the remaining two refers to the same bundles once they have crossed the concentrator (output wave fronts). The SMS 3D method gives an optical system that transforms the input into the output wave fronts. The number of surfaces to design determines how many input and output bundles can be coupled. In the present case, there are two surfaces to design: the reflective X and the refractive R one. This means that there are two wave fronts at the input that can be perfectly coupled with two wave fronts at the output. There is a third pair of wave fronts that are partially coupled (see Section 4.5.3). We now have to choose the two pairs of perfectly coupled wave fronts.

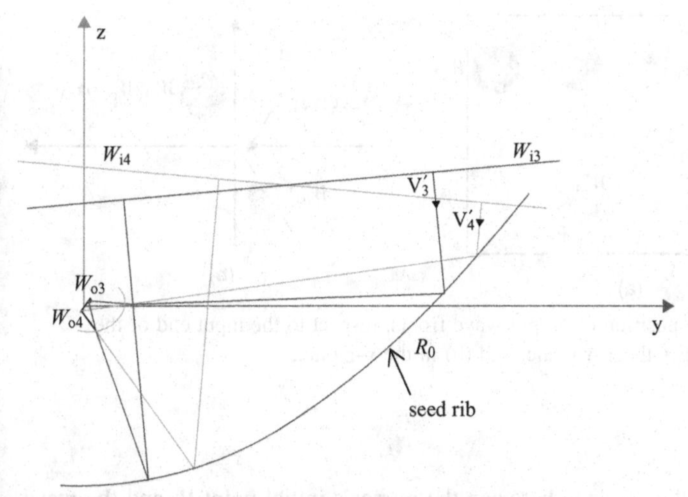

Figure 4.36 Input wave fronts for initial curve design.

To choose them, we will make use of the edge ray theorem [5, 35]. This theorem establishes that in order to concentrate the rays coming from the sun and its neighborhood into the cell, it is enough to design an optical system that couples the edge rays of the incoming bundle into the edge rays of the outgoing bundle. Since these edge rays form a three-parameter bundle of rays, it is in general not possible to couple them having only two surfaces to design. The SMS 3D method can only guarantee that two biparametric set of rays at the input are perfectly coupled with another two at the output (each wave front defines one biparametric set of rays), so we will have to choose two wave fronts formed by edge rays at the input and another pair of wave fronts of edge rays at the output. This will approximate the condition established by the edge ray theorem. The same idea can be applied to the third pair of wave fronts used in the XR design (step c).

- The input and output wave fronts W_{i1} and W_{o1} are specified. W_{i1} is a flat wave front the rays of which point in the direction $\mathbf{v}_i' = (p', q', -(1 - p'^2 - q'^2)^{1/2})$, with $(p', q') = (+b\sin(\alpha), +b\sin(\alpha))$. W_{o1} is a spherical wave front centered at the point $(x, y, z) = (b|PQ|/2, b\cos\theta|PQ|/2, b\sin\theta|PQ|/2)$ (see Figs.4.35–4.37)
- The optical path length ℓ_1 between wave fronts W_{i1} and W_{o1} is calculated using the condition that the ray of W_{i1} impinging on $\mathbf{P}_0$ of Figure 4.34 will be reflected toward S_0 and then refracted thereupon to become a ray of W_{o1}. Before calculating ℓ_1, the coordinates of point $\mathbf{S}_0$ of Figure 4.34 must be calculated by reflection of the ray of W_{i1} on $\mathbf{P}_0$ (note that the normal vector $\mathbf{N}_0$ at $\mathbf{P}_0$ was specified as an input datum), and intersecting that reflected ray with the plane $y = y_{S0}$ (since y_{S0} was also given). Typically, $x_{S0} \neq 0$. Then, ℓ_1 is calculated as

$$\ell_1 = d(\mathbf{P}_0, W_{i1}) + d(\mathbf{S}_0, \mathbf{P}_0) + d(\mathbf{S}_0, W_{o1})n, \qquad (4.6)$$

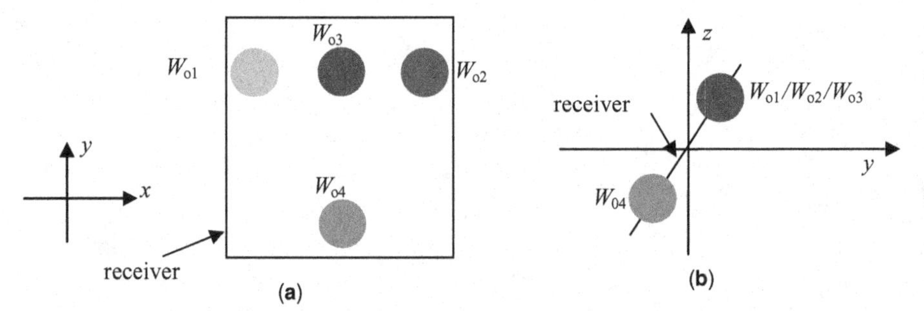

Figure 4.37 Relative position of output wave fronts, respect to the input end of the homogenizing rod: (a) in the x–y plane, and (b) in the y–z plane.

where,

d ($\mathbf{P}_0$, W_{i1}) is the distance between the mirror's initial point $\mathbf{P}_0$ and the wave front W_{i1};

d ($\mathbf{S}_0$, W_{01}) is the distance between the lens and the mirror's initial point $\mathbf{S}_0$ and the wave front W_{01}; and

d ($\mathbf{S}_0$, $\mathbf{P}_0$) is the distance between the points $\mathbf{S}_0$ and $\mathbf{P}_0$.

The absolute positions of W_{i1} and W_{o1} are not defined, but that does not affect the calculation.

- The input and output wave fronts W_{i2} and W_{o2} are selected, in this example, as symmetric to W_{i1} and W_{o1} with respect to plane $x = 0$. So, their definition has been done the same way as was for the first pair of wave fronts. Then, W_{i2} is a flat wave front whose rays point in the direction $\mathbf{v}_2' = (p', q', -(1 - p'^2 - q'^2)^{1/2})$, with $(p', q') = (-b\sin(\alpha), +b\sin(\alpha))$. W_{o2} is a spherical wave front centered at the point $(x, y, z) = (-\mathrm{b}|PQ|/2, \mathrm{b}\cos\theta\ |PQ|/2, \mathrm{b}\sin\theta\ |PQ|/2)$.The optical path length ℓ_2 fulfils $\ell_2 = \ell_1$ in this x-symmetric example.
- A seed rib, R_0 and the reference to the surface where we want the seed rib to be are calculated as follows. The seed rib R_0 can be obtained by an SMS 2D calculation in the plane $x = 0$ using two pairs of wave fronts W_{i3}, W_{o3}, and W_{i4}, W_{o4}, as shown in Figure 4.36. W_{i3}, is a flat wave front the rays of which point in the direction $\mathbf{v}_3' = (p', q', -(1 - p'^2 - q'^2)^{1/2})$, with $(p', q') = (0,+b\sin(\alpha))$, and it couples W_{o3}, which is a spherical wave front centered at the point $(x, y, z) = (0, b\cos\theta\ |PQ|/2, b\sin\theta\ |PQ|/2)$. W_{i4} is a flat wave front for which the rays point in the direction $\mathbf{v}_4' = (p', q', -(1 - p'^2 - q'^2)^{1/2})$, with $(p', q') = (0,-b\sin(\alpha))$, and it couples W_{o4}, which is spherical wave front centered at the point $(x, y, z) = (0, -b\cos\theta\ |PQ|/2, -b\sin\theta\ |PQ|/2)$. For simplicity, a single parameter b is used everywhere. The optical path lengths ℓ_3 between wave fronts W_{i3} and W_{o3} can be selected equal to ℓ_1 and ℓ_2. An optical path length ℓ_4 between wave fronts W_{i4} and W_{o4} is chosen (its value will be adjusted next).

Figure 4.37 shows the relative positions of the four exiting wave fronts W_{o1}, W_{o2}, W_{o3}, W_{o4}, relative to the input end of the homogenizing rod [31].

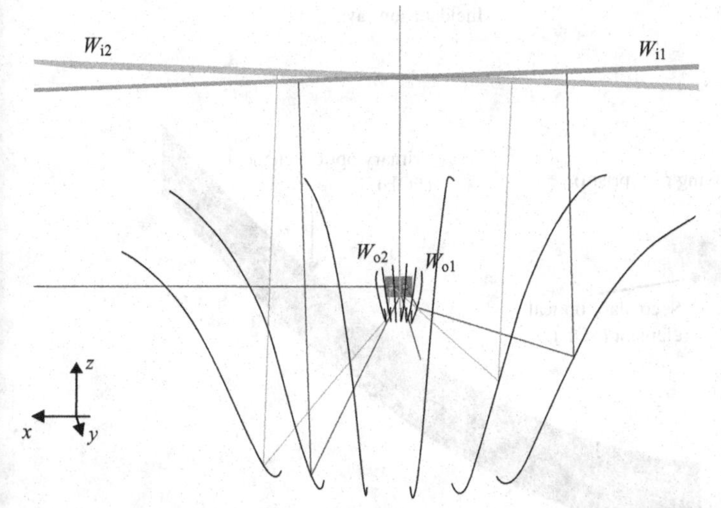

Figure 4.38 Calculated SMS ribs.

Figure 4.39 Side and top views of XR.

Once the input parameters are selected, the SMS design proceeds as described in Section 4.5. The initial curve can be seen in the Figure 4.33, and the SMS ribs in Figure 4.38.

The surfaces of the mirror and lens are calculated as an interpolated surface between the SMS ribs (consistent with the normal vectors). Such an interpolation can be easily done, for instance, using a lofted surface interpolation available in most CAD packages. The parameter $b < 1$ set at the beginning can be used to select the number of points along the seed rib and the number of ribs to be designed (the smaller b is, the higher the number of points and ribs). The resultant system can be seen in Figures 4.39 and 4.40. Figure 4.41 shows a close up of the SOE.

4.6.5.2 Results of Ray Tracing Analysis This concentrator was ray-traced to determine the angular transmission, optical efficiency, and irradiance distribution on

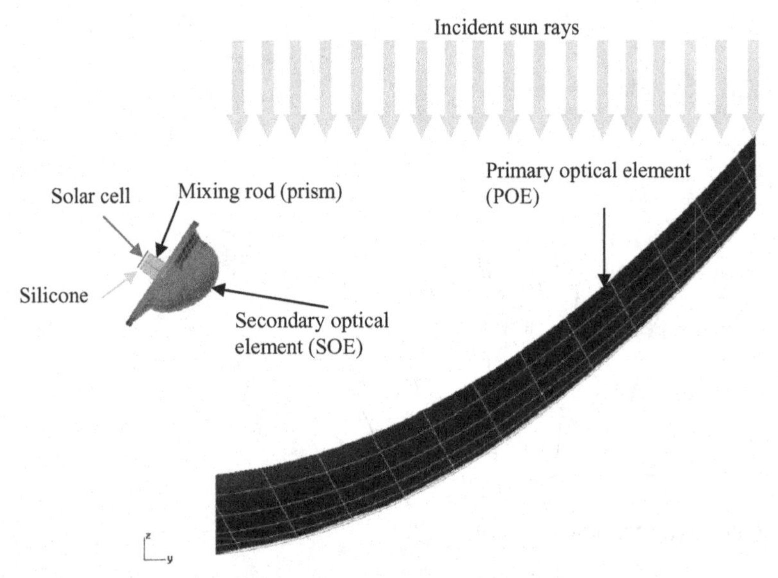

Figure 4.40 SMS3D XR concentrator (side view).

Figure 4.41 SOE of the XR designed for HCPV applications, showing the freeform lens.

the concentrator exit aperture using the commercial software package TracePro [36]. More details of the PV concentrator derived from it (including experimental results) is found in [31, 37–39]. Figure 4.42 shows an experimental prototype of an HCPV array based in this XR design [39].

One of the characteristics needed for the evaluation of the concentrator optical performance is the angular transmission of the concentrator, defined for a given

Figure 4.42 Experimental prototype of an HCPV array based in the XR design of Figure 4.39.

incident beam of parallel rays as the power reaching the cell surface over the power incident on the concentrator aperture (i.e., the optical efficiency as a function of the angle of incidence of the parallel beam). For this calculation, we assume all the rays in the parallel beam have the same radiance. In the stray light community, this is akin to a point source transmission (PST) calculation. The solar direct radiation can be modeled as a set of parallel beams (with the same radiance), with directions pointing inside the solar disk, of nominal angular radius 0.26°.

Two important merit parameters of the concentrator are derived from the angular transmission curve: the optical efficiency (η_{opt}) and the acceptance angle (α). The optical efficiency of the concentrator is the maximum value of the angular transmission usually found when the angle of incidence is zero (normal incidence). Note that the power incident the cell is not necessarily that entering the cell because its coating may reflect some of the radiation. Therefore, we use a definition of optical efficiency of the power incident on the cell divided by the power of a parallel beam reaching the entry aperture.

The acceptance angle is that between the direction at which the angular transmission peaks and the direction at which the optical efficiency lowers to 90% of the maximum. This is an arbitrary definition that has become widespread. Some authors prefer to define the acceptance angle at 95% of the maximum. Obviously, this last definition is more restrictive, but is in agreement with two standard deviations of a normal distribution. None of these definitions fit exactly with the one needed for establishing that the CAP should be smaller than n. Nevertheless, they are close to it, and the definition for the 90% is usually taken to evaluate the CAP of the concentrator.

The angular transmission is usually represented in a normalized form, known as the relative angular transmission, because it is normalized to its maximum value, so that the maximum value of the relative angular transmission must be unity. The angular transmission (not relative) can be obtained by multiplying the relative

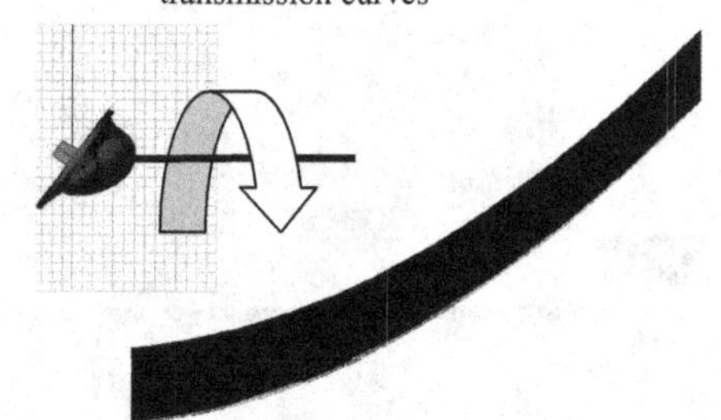

Positive direction of rotation in order to obtain the x-section for transmission curves

Figure 4.43 Definition of cross sections for transmission curves analysis.

angular transmission times the optical efficiency. Hereinafter, all transmission curves will be of relative angular transmission.

The transmission function T of any incident beam of parallel rays is the percent of its power that reaches the cell surface, assuming that all rays of a parallel beam have the same radiance. The direction of the incident parallel beam can be specified by two direction cosines (p, q), so that in general, T is the function $T(p, q)$. The transmission curve in the x-section is the function $T(0, q)$ and the transmission curve in the y-section is $T(p, 0)$. Figure 4.43 explains both sections. In practice, instead of representing $T(0, q)$ or $T(p, 0)$, we represent $T(\alpha)$, where α is either $\arccos(p)$ or $\arccos(q)$, depending on which section is considered.

The input information for the ray tracing is:

(1) Mirror reflectivity: 95%

(2) Dielectric parts: $n = 1.52$ (glass), used for refraction and Fresnel losses

(3) Active solar cell area: 10×10 mm^2

(4) Mixing rod length: 9 mm

(5) Mixing rod aperture area = 9×9 mm^2. There is a ± 0.5-mm difference between the size of the rod aperture and that of the cell active area for tolerance purposes

(6) No rounding of lateral edges of the mixing rod

(7) Silicone refractive index: 1.52

(8) Silicone layer thickness: 50 μm

(9) Direct-beam input irradiance: 900 W/m^2, and

(10) Cover transmission: 91%.

The resulting transmission curves for both sections are shown in Figure 4.44.

Sometimes, the relative angular transmission is calculated taking into account the angular size of the sun instead of using a beam of identical parallel rays. This is a straightforward way to show the concentrator angular performance under more realistic conditions. In this case the resulting relative angular transmission is a convolution of the sun's disc with the relative angular transmission calculated using a

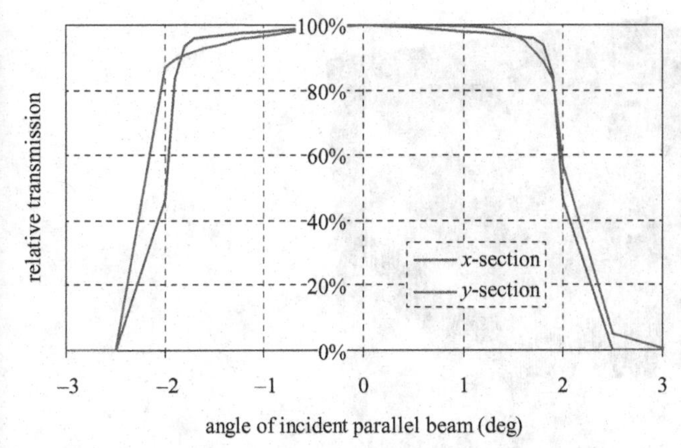

Figure 4.44 Transmission curves at nominal position.

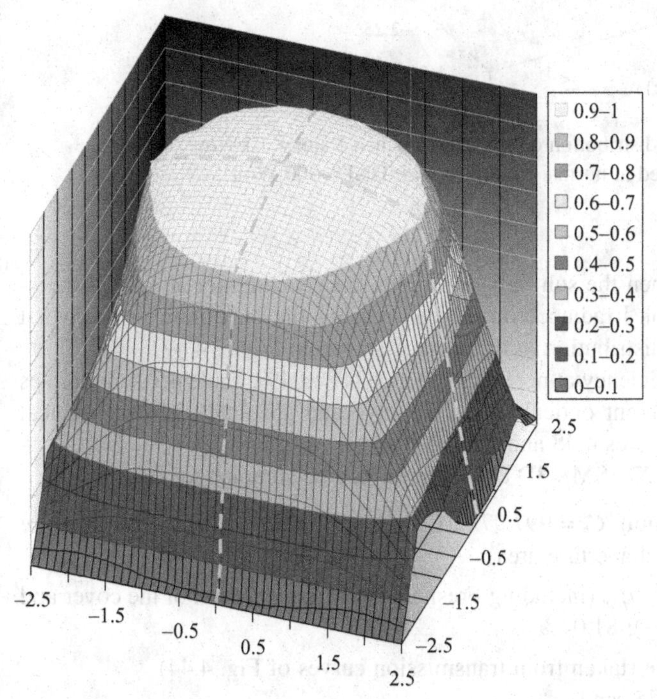

Figure 4.45 Relative angular transmission in 3D versus the angles (in degrees) of the incoming bundle of parallel rays. The cross sections (yellow and blue lines) are the *x*- and *y*-sections.

parallels ray beam. Figure 4.45 shows the relative angular transmission calculated taking into account the sun size (nominal angular radius 0.26°).

Figure 4.46 shows the resulting irradiance distribution on the cell surface when the sun is at normal incidence and irradiating the concentrator aperture at 900 W/m^2. Peak irradiance was 883 kW/m^2, which shows the homogenizing effect

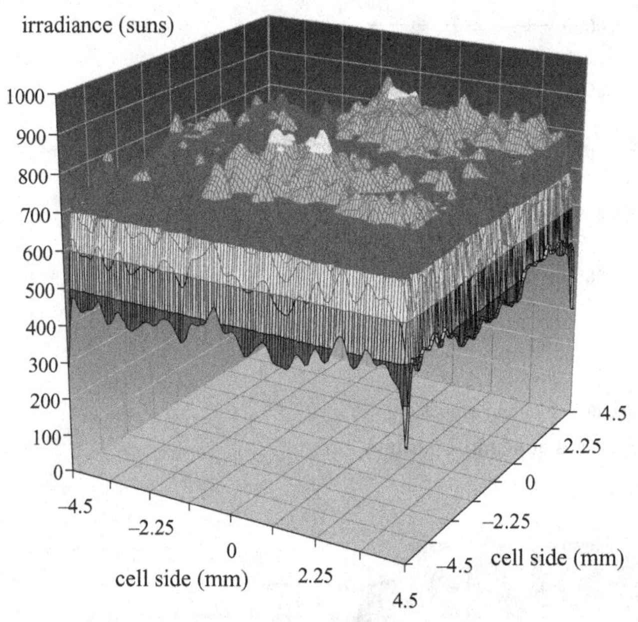

Figure 4.46 Irradiance distribution on the cell (in suns, 1 sun = 1 kW/m^2) when the concentrator is illuminated at normal incidence with DNI = 900 W/m^2.

of the mixing rod. When the sun radiation lights the concentrator at 1.5° in the *y*-direction, then the peak irradiance on the cell reaches the maximum value (worst case). The irradiance distribution in this case is shown in Figure 4.47.

The resulting efficiency at normal incidence is 81.04%. Table 4.1 summarizes the results of the different optical losses (relative and absolute). Their graphical representation is in Figures 4.48 and 4.49.

The behavior of XR SMS 3D concentrator can be summarized as:

(1) Concentration ratio $C_g = 997.97x$ ($C_g = A_{\text{entry aperture}}/A_{\text{exit aperture}}$; exit aperture area = mixing rod aperture area)

(2) Optical efficiency η_{opt} (including Fresnel losses, transmission of the cover and mirror reflectivity): 81.04%

(3) Acceptance angle (taken from transmission curves of Fig. 4.44)
✓ *x*-section ±1.75° and
✓ *y*-section ±1.77°; and

(4) Peak irradiance on the cell 883x (for no tracking error).

4.7 CONCLUSIONS

In Section 4.6, we have reviewed several SMS 3D examples. There are many additional examples that can be shown. For example, Figure 4.50 shows a freeform RXI

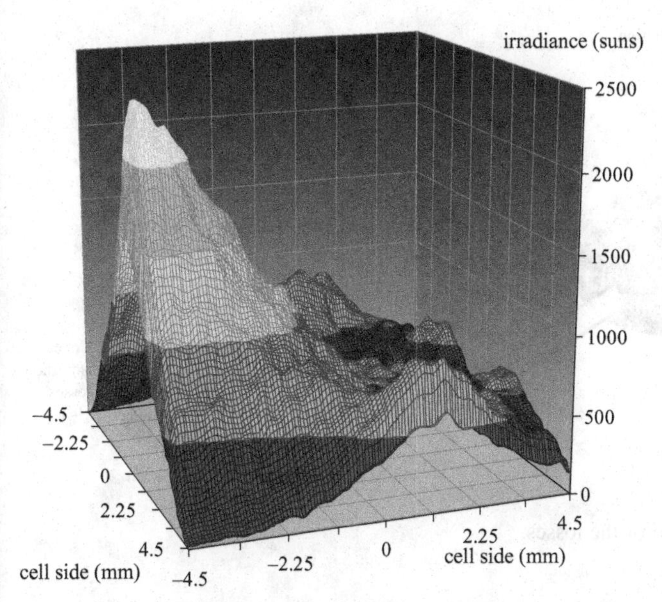

Figure 4.47 Irradiance distribution on the cell (in suns, 1 sun = 1 kW/m^2) when the concentrator is illuminated at off axis +1.5° y-direction (worst case) with DNI = 900 W/m^2. Peak irradiance on the cell is 2304 suns.

TABLE 4.1 Summary of Optical Losses in the XR System

Optical losses	Relative (%)	Absolute (%)
Transmission after SOE shading	100.0	100.0
Cover transmission (no AR coated)	91.0	91.0
Mirror fill factor	100.0	91.0
Mirror reflectivity	95.0	86.4
SOE transmission (no AR coated)	95.0	82.1
Mixing rod rounding	98.7	81.0

designed with the SMS 3D method. The RXI is a single-piece optical device that collimates (or concentrates) the light very near the thermodynamic limit and in a compact way [3]. Its design is more complex than the ones shown here because one of the optical surfaces (the upper one of Fig. 4.50) is used twice by the rays: once as a reflector by TIR and another time by refraction.

Figure 4.51 shows another RXI that is not only a collimator but also a Köhler integrator [40, 41]. This is a design more complex than the one of Figure 4.50. Nevertheless, they share the basic idea of the SMS 3D method applied to devices with two optical surfaces: it is possible to design together two freeform surfaces that control two wave fronts and partially a third one. This provides a unique technique for designing optical systems with high versatility and the ability to control the light to a degree unavailable with any other design method.

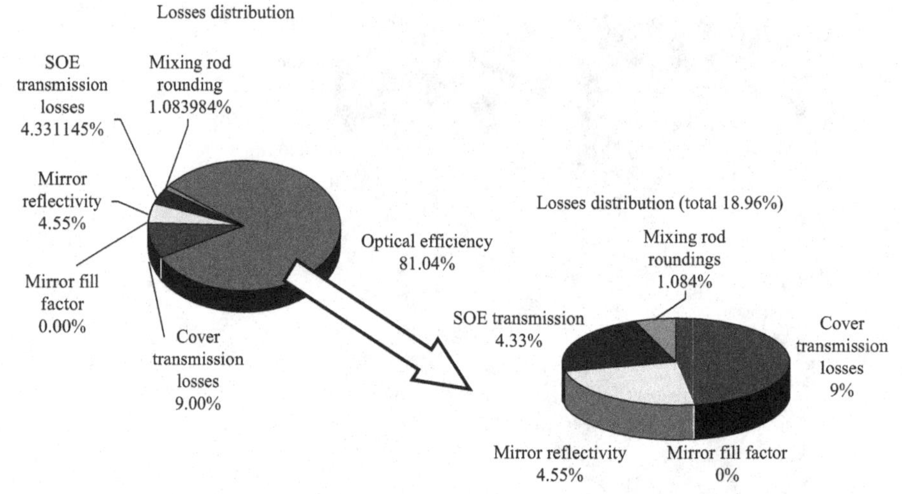

Figure 4.48 Distribution of the losses.

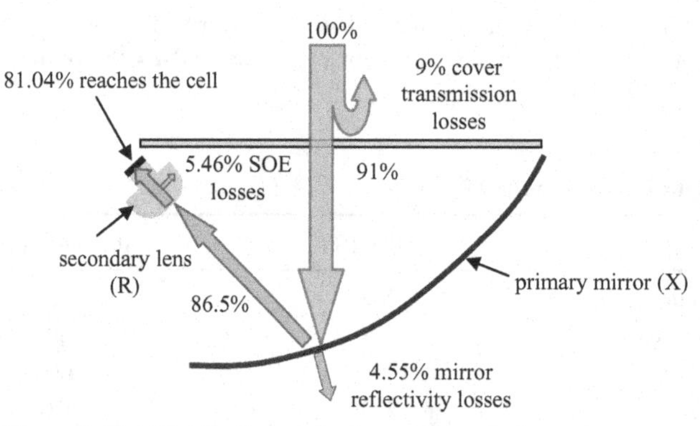

Figure 4.49 Distribution of the losses through the system.

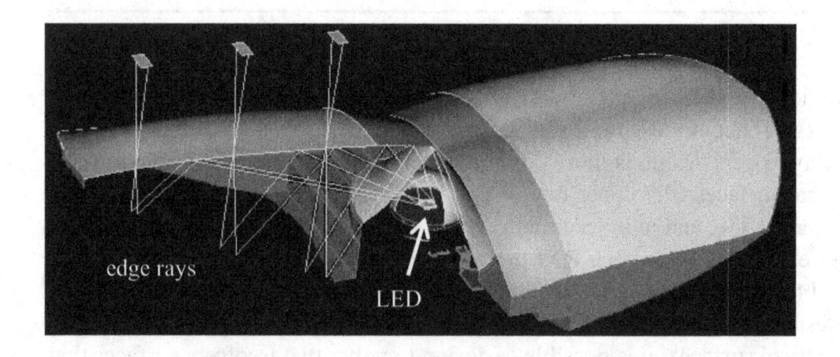

Figure 4.50 Freeform RXI for an LED showing the LED chip and some edge rays. The simultaneous multiple surfaces (SMS) 3D method controls the bundle of rays issuing from the LED chip corners.

Figure 4.51 Practical execution of a Freeform Köhler integrator RXI for illumination. The light source LED is not shown.

Apart from its theoretical interest, the SMS 3D method has excellent practical potential. The examples shown here prove this assertion. The method has also applications in imaging devices, although its potential in this field is little explored so far. Imaging applications of the SMS 2D method for two rotational surfaces and ultra-high numerical aperture gave better imaging performance than conventional two-surface aplanatic designs [23]. In fact, aplanatic designs can be viewed as a particular case of SMS 2D designs. This particular case appears when the two pairs of bundles to be coupled by the optical system converge to a single (on axis) pair. The equivalence of the aplanatic design with a particular SMS 2D case was recognized in the early stages of the SMS 2D method development [4], and has been commonly used since then [41, 42].

The SMS 2D method has been successfully used to design ultra-compact, wide-angle optical systems. For instance, increasing the compactness and illumination angles (compared with the conventional systems) of a 4:3 projector for a screen of 60" and in the design of a 16:9 external projector for a screen of 80" where the projection distance was considerably reduced [41], from several meters (in conventional systems) to 34 cm. Since the SMS 3D method introduces more degrees of freedom in the design process, it can be expected to be an improvement over SMS 2D. Progress there is twofold: (i) increasing the number of rotationally symmetric surfaces to satisfy more design constraint (such as the achromatic condition [43]) for high-quality ultra-compact wide-angle imaging designs and (ii) designing without rotational symmetry. The latter applications are expected to be useful when the image is far from rotational symmetry, such the 16:9 HDTV applications.

One of the examples presented (Section 4.6.5) is the SMS 3D design of an XR photovoltaic concentrator, which can summarize the capabilities of the method. The goal of this design is high concentration with high efficiency and high acceptance angle by a compact configuration convenient for thermal and mechanical management. The reason why the system has to be freeform is to avoid the shadowing of the heat sink system on the entry aperture that occurs in a rotational symmetric XR.

The XR concentrator in this system has been chosen for its excellent aspect ratio (around 0.6, as compared with 1–1.5 of classical systems) and for its ability to perform near the thermodynamic limit. For the geometrical concentration of 1000×, the simulation results show the acceptance angle is ±1.8°, which relaxes the manufacturing tolerances of all optical and mechanical components, especially the concentrator itself. Such a large acceptance angle is key to a cost-competitive photovoltaic generator. The optical efficiency (power on the cell surface over the power of a parallel beam reaching the entry aperture of the concentrator considering Fresnel losses, absorption losses, and mirror reflectivity [95%]) is 81% (which includes cover losses by a transmission of 91%). These features have made this design an excellent option for concentrating photovoltaics.

REFERENCES

1. J.C. Miñano and J.C. González, New method of design of nonimaging concentrators, *Appl. Opt.* **31**, 3051–3060 (1992). http://www.opticsinfobase.org/ao/abstract.cfm?URI=ao-31-16-3051
2. J.C. Miñano, P. Benítez, and J.C. González, RX: a nonimaging concentrator, *Appl. Opt.* **34**, 2226–2235 (1995). http://www.opticsinfobase.org/abstract.cfm?URI=ao-34-13-2226
3. J.C. Miñano, J.C. González, and P. Benítez, A high-gain, compact, nonimaging concentrator: RXI, *Appl. Opt.* **34**, 7850–7856 (1995). http://www.opticsinfobase.org/abstract.cfm?URI=ao-34-34-7850
4. P. Benítez and J.C. Miñano, Ultrahigh-numerical-aperture imaging concentrator, *J. Opt. Soc. Am. A* **14**, 1988–1997 (1997). http://www.opticsinfobase.org/abstract.cfm?URI=josaa-14-8-1988
5. R. Winston, J.C. Miñano, and P. Benítez, *Nonimaging Optics*, Academic Press, New York (2005).
6. P. Benítez, J.C. Miñano, J. Blen, R. Mohedano, J. Chaves, O. Dross, M. Hernández, J.L. Alvarez, and W. Falicoff, *Opt. Eng.* **43**, 1489 (2004). doi: 10.1117/1.1752918.
7. J. Chaves, *Introduction to Nonimaging Optics*, CRC Press, Boca Ratón, FL (2008).
8. P. Benítez, R. Mohedano, and J.C. Miñano, Design in 3D geometry with the simultaneous multiple surface design method of nonimaging optics, Nonimaging Optics: Maximum Efficiency Light Transfer V, Roland Winston, SPIE, Denver, EE.UU (1999).
9. J. Schaefer, Single point diamond turning: progress in precision, in International Optical Design, Technical Digest (CD), Optical Society of America (2006), Paper ThB1. http://www.opticsinfobase.org/abstract.cfm?URI=IODC-2006-ThB1
10. M. Thomas and M. Sander, Improving optical freeform production. Photonics Spectra, September 2006 http://www.photonics.com/Content/ReadArticle.aspx?ArticleID=26649
11. O. Marinescu and F. Bociort, *Appl. Opt.* **46**, 8385 (2007).
12. F. Bociort, Optical system optimization, in R.G. Driggers, ed., *Encyclopedia of Optical Engineering*, Marcel Dekker, New York (2003), pp. 1843–1850.
13. M. Nicholson, How to perform freeform optical design, ZEMAX Users' Knowledge Base, 2009 http://www.zemax.com/kb/articles/265/1/How-to-Perform-Freeform-Optical-Design/Page1.html
14. O.N. Stravoudis, *The Optics of Rays, Wave fronts and Caustics*, Academic Press, New York (1972).
15. R.K. Luneburg, *Mathematical Theory of Optics*, University of California, Berkeley (1964).
16. T. Levi-Civita, Atti della Reale Accademia dei Lincei. Rendiconti della Clase de ScienzeFisiche, *Matematiche e Naturali* **9**, 185 and 237 (1900).
17. V.I. Oliker, On the geometry of convex reflectors, in B. Opozda, U. Simon, and M. Wiehe, eds., *PDE's, Submanifolds and Affine Differential Geometry*, Banach-Center Publications, Warsaw (2002), p. 155.

18. V.I. Oliker, Mathematical aspects of design of beam shaping surfaces in geometrical optics, in M. Kirkilionis, S. Kromker, R. Rannacher, and F. Tomi, eds., *Trends in Nonlinear Analysis*, Springer-Verlag, Berlin (2003), p. 193.
19. H. Ries and J. Muschaweck, Tailored freeform optical surfaces, *J. Opt. Soc. Am. A* **19**, 590 (2002).
20. F. Fournier, W. Cassarly, and J. Rolland, Designing freeform reflectors for extended sources, *Proc. SPIE* **7423**, 742302 (2009).
21. L. Piegl and W. Tiller, *The NURBS Book*, 2nd ed.. Springer-Verlag, Berlin (1997).
22. D. Hansford and G. Farin, Curve and surface construction, in G. Farin, J. Hoschek, and M.-S. Kim, eds., *Handbook of Computer Aided Geometric Design*, North-Holland, Amsterdam (2002), pp. 165–191.
23. W.T. Welford and R. Winston, On the problem of ideal flux concentrators: addendum, *J. Opt. Soc. Am.* **69**, 367–367 (1979).http://www.opticsinfobase.org/abstract.cfm?URI=josa-69-2-367
24. W.T. Welford and R. Winston, On the problem of ideal flux concentrators, *J. Opt. Soc. Am.* **68**, 531–534 (1978).http://www.opticsinfobase.org/abstract.cfm?URI=josa-68-4-531
25. http://www.synopsys.com
26. http://www.rhino3d.com/
27. http://www.lpi-llc.com/
28. M.A. Green, K. Emery, Y. Hishikawa, W. Warta, E.D. Dunlop, Solar cell efficiency tables (version 40), *Progress in Photovoltaics*, **20**, 606–614 (2012).
29. A. Cvetkovic, Novel theoretical concepts, design, analysis and construction of nonimaging devices. Ph.D. Thesis. Universidad Politécnica de Madrid, Spain (2009). http://www-app.etsit.upm.es/tesis_etsit/documentos_biblioteca/masinformacion.php?sgt=TESIS-09-006
30. Devices shown in this paper are protected by the following U.S. Patents and U.S. and International Patents pending. 6,639,733; 2001069300; CA2402687; 2003282552; 6,896,381; 20040246606; 20050086032; 2005012951; PCT60703667; Solar Concentrator for Photovoltaics.
31. A. Cvetkovic, M. Hernandez, P. Benitez, J.C. Miñano, J. Schwartz, A. Plesniak, R. Jones, and D. Whelan, The freeform XR concentrator: a high performance SMS3D design, *High and Low Concentration for Solar Electric Applications III, Proc. SPIE* **7043**, 70430E (2008).
32. H. Ries, J.M. Gordon, and M. Laxen, High-flux photovoltaic solar concentrators with Kaleidoscope based optical designs, *Solar Energy* **60**(1), 11–16 (1997).
33. J.J. O'Ghallagher and R. Winston, Nonimaging solar concentrator with near-uniform irradiance for photovoltaic arrays, in R. Winston, ed., *Nonimaging Optics: Maximum Efficiency Light Transfer VI, Proc. SPIE* **4446**, 60–64 (2001).
34. D.G. Jenkins, High-uniformity solar concentrators for photovoltaic systems, in R. Winston, ed., *Nonimaging Optics: Maximum Efficiency Light Transfer VI, Proc. SPIE* **4446**, 52–59 (2001).
35. P. Benítez and J.C. Miñano, Offence against the edge ray theorem?, in R. Winston, and J. Koshel, eds., *Nonimaging Optics and Efficient Illumination Systems, SPIE. Proc. SPIE* **5529**, 108–119 (2004). San Diego, CA.
36. http://www.lambdares.com/
37. M. Hernandez, P. Benitez, J.C. Minano, A. Cvetkovic, R. Mohedano, O. Dross, R. Jones, D. Whelan, G.S. Kinsey, and R. Alvarez, The XR nonimaging photovoltaic concentrator, *Nonimaging Optics and Efficient Illumination Systems IV, Proc. SPIE* **6670**, 667005 (2007).
38. R. Plesniak, J. Jones, G. Schwartz, D. Martins, P. Whelan, J.C. Benítez, A. Miñano, M. Cvetkovíc, O. Hernandez, R. Dross, and R. Alvarez, paper presented at International Conference on Solar Concentrators for the Generation of Electricity ICSC—5, Palm Desert, CA (2008).
39. A. Plesniak, R. Jones, J. Schwartz, G. Martins, J. Hall, A. Narayanan, D. Whelan, P. Benítez, J.C. Miñano, A. Cvetkovic, M. Hernandez, O. Dross, and R. Alvarez, High performance concentrating photovoltaic module designs for utility scale power generation, *34th IEEE PVSC*, Philadelphia (2009).
40. J.C. Miñano, M. Hernández, P. Benítez, J. Blen, O. Dross, R. Mohedano, and A. Santamaría, Freeform integrator array optics, in R. Winston and R.J. Koshel eds, *Nonimaging Optics and Efficient Illumination Systems II, SPIE Proc.* (2005). http://dx.doi.org/10.1117/12.620240
41. F. Muñoz, Sistemas *ó*pticos avanzados de gran compatibilidad con aplicaciones en formaci*ó*n de imag*é*n y en iluminaci*ó*n. Tesis Doctoral, Departamento de Electrónica Física. E.T.S.I. de Telecomunicaciones. Universidad Politécnica de Madrid (2004).

42. J.C. Miñano, P. Benítez, W. Lin, F. Muñoz, J. Infante, and A. Santamaría, Overview of the SMS design method applied to imaging optics, *Proc. SPIE* **7429**, 74290C (2009). doi: 10.1117/12.827068.
43. J.C. Miñano, P. Benítez, and F. Muñoz, Application of the 2D etendue conservation to the design of achromatic aplanatic aspheric doublets, in R. Winston ed., *Nonimaging Optics: Maximum Efficiency Light Transfer VI, Proc. of SPIE* **4446**, 11–19 (2001).

CHAPTER 5

SOLAR CONCENTRATORS

Julio Chaves and Maikel Hernández

5.1 CONCENTRATED SOLAR RADIATION

Traditionally, solar concentrators have been designed as focusing optics, mainly as parabolic mirrors or Fresnel lenses. However, in the last decades, nonimaging optics [1–6] has emerged as the tool of choice for the design of optimized solar concentrators. The most important design methods of nonimaging optics are the Winston–Welford (flow line) and the Miñano–Benitez (or simultaneous multiple surface—SMS) [7–10]. The first nonimaging optics were initially mainly developed for low concentration solar thermal applications. An example of these optics is the CPC (compound parabolic concentrator). The latest developments in nonimaging optics, however, are increasingly occurring in systems designed for high concentration photovoltaic applications.

In the last decade the efficiency of multi-junction solar cells has experimented a rapid increase, from about 30% in 1999 to about 40% in 2009. This opened the possibility to develop concentration photovoltaic (CPV) systems that are competitive in terms of efficiency with crystalline silicon solar cell based modules. However, the price of these high efficiency multi-junction cells is still very high, making it difficult to develop a cost-effective concentration system. To reduce the impact of the cell cost, concentration may be increased, replacing a large area of very expensive solar cell with a potentially much cheaper large optic. However, adding a concentration optic also adds complexity and manufacturing and installation costs to the photovoltaic systems.

Minimizing the cost per kilowatt-hour of the energy produced is a fundamental condition to guarantee the success of concentration photovoltaics. Key to minimizing this cost is to have an efficient, tolerant to errors and low-cost optical concentrator. This implies using an optical design with few elements and able to provide the maximum possible acceptance angle (also called tolerance angle).

This acceptance angle is not only the maximum allowable deviation angle from the perfect aiming of the concentrator to the sun, but is also a measure of the tolerance available for the remaining parts of the system. This tolerance angle must be able to accommodate: (i) sun tracking accuracy, (ii) concentrator mounting tolerances, (iii) array assembling tolerances, (iv) tracker structure finite stiffness,

Illumination Engineering: Design with Nonimaging Optics, First Edition. R. John Koshel.
© 2013 the Institute of Electrical and Electronics Engineers. Published 2013 by John Wiley & Sons, Inc.

(v) movements of the system due to wind, and (vi) shape and roughness errors of the optical surfaces. All these items can be expressed in terms of a decrease in the tolerance angle. The larger the acceptance angle of a concentrator, the larger the "tolerance budget" available for "spending" on all these negative, but inevitable effects. After all these effects are factored in, the concentrator(s) must still have enough acceptance angle "left" to able to accommodate the sun's angular extension and collect its light. A system with higher tolerance has the potential to handle less precise manufacturing and mounting of the components and less precise tracking. This translates to the potential of using lower cost, mass production components.

5.2 ACCEPTANCE ANGLE

A central concept in nonimaging optics is conservation of étendue. Figure 5.1 shows a general nonimaging concentrator with its entrance aperture having an area A_1 immersed in a medium of refractive index n_1 and an exit aperture of area A_2 immersed in a medium of refractive index n_2. The incoming radiation is confined to a cone of angular aperture θ_1 at the entrance aperture and to a cone of angular aperture θ_2 at the exit aperture. Conservation of étendue states that [1–2]:

$$C = \frac{A_1}{A_2} = \frac{n_2^2 \sin^2 \theta_2}{n_1^2 \sin^2 \theta_1}, \tag{5.1}$$

where C is the concentration the optic provides.

The entrance aperture of a solar concentrator is in contact with air, and therefore $n_1 = 1$. Angular aperture θ_2 verifies $\theta_2 \leq \pi/2$ and the concentration C a solar concentrator can provide is then

$$C \leq C_{\max} = \frac{n_2^2}{\sin^2 \theta_1}, \tag{5.2}$$

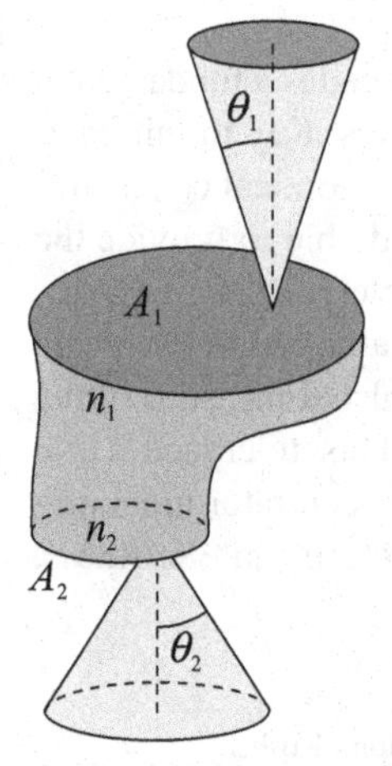

Figure 5.1 Conservation of étendue in a nonimaging concentrator.

where C_{max} is the maximum concentration possible obtained when $\theta_2 = \pi/2$ and assuming that all radiation entering through A_1 exits through A_2 (no radiation is lost or rejected by the optic). This expression can also be rewritten for the angular aperture θ_1 of the accepted light by the concentrator as

$$\theta_1 \leq \theta_{max} = \arcsin\left(\frac{n_2}{\sqrt{C}}\right), \tag{5.3}$$

where θ_{max} is the maximum angular aperture possible for the radiation entering through aperture A_1 and accepted by the optic, which has a concentration C. By accepted radiation, we mean that entering through A_1 and exiting through A_2 (it is not lost nor rejected by the optic). Yet another way to look at Equation (5.2) is to write it as $\sin^2 \theta_1 \, C \leq n_2^2$. So, if θ_1 increases, C must decrease, and vice versa.

Nonimaging optics may be designed to be ideal in two-dimensional (2D) geometry and to approach ideality in three-dimensional (3D) geometry. Therefore, for incoming radiation of a given angular aperture θ_1, they approach the maximum possible concentration C_{max}, according to Equation (5.2). Also, for a given concentration C, these optics approach the maximum angular aperture θ_{max} possible to accept, given by Equation (5.3).

The angular aperture of the solar radiation is very small (confined to a cone of angular aperture θ_S of about $\pm 0.26°$), and, therefore, the maximum concentration possible for solar energy is very high, as derived from Equation (5.2). In practical solar concentrator designs, we choose a value for the concentration (typically depending on the materials and other properties of the receiver) and try to maximize the angular aperture of the cone of radiation accepted by the optic, according to Equation (5.3).

In order to be able to evaluate the performance of different solar concentrators, we need to first characterize them. Figure 5.2 shows a concentrator and the incident light, here represented by a set of random rays tilted by an angle α (a large number of random rays would have to be generated to simulate a real light source). For different incidence angles α, different percentages of light that hit the entrance aperture

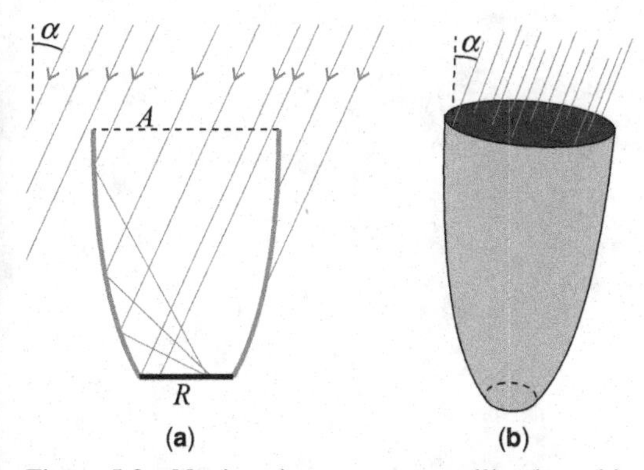

Figure 5.2 Nonimaging concentrator illuminated by light incident at an angle α. (a) 2D concentrator and (b) 3D concentrator with circular symmetry.

A will eventually reach the receiver *R*. This allows us to define an angular efficiency as a function of angle α.

If $\Phi_A(\alpha)$ is the flux falling on the entrance aperture *A* for light incident at an angle α and $\Phi_R(\alpha)$ is the corresponding flux transferred by the optic to the receiver *R*, the angular efficiency of the optic is given by

$$\eta(\alpha)=\frac{\Phi_R(\alpha)}{\Phi_A(\alpha)}=\frac{1}{\Phi_{A0}}\frac{\Phi_R(\alpha)}{\cos\alpha}, \tag{5.4}$$

where $\Phi_{A0} = \Phi_A(0)$ is the flux falling on aperture *A* for $\alpha = 0$. Function $\eta(\alpha)$ has a maximum η_{max}, and the normalized function $\tau(\alpha) = \eta(\alpha)/\eta_{max}$ is the relative angular transmission of the concentrator as a function of angle α. The angular efficiency and the relative angular transmission are then both proportional to the flux falling on the receiver divided by $\cos(\alpha)$. In the case in which the incident light is perfectly collimated, we obtain the relative angular transmission for collimated light. If this light has the sun's angular aperture, we obtain the relative angular transmission for sunlight. Examples will be given below.

A simpler way to characterize a concentrator is, instead of giving its concentration and the angular efficiency function, to give three values: concentration, efficiency, and acceptance angle. The acceptance angle is typically defined as the value of α for which the relative angular transmission drops to 0.9. The efficiency η_0 of the concentrator is given as the portion of light that reaches the receiver for $\alpha = 0$, that is, $\eta_0 = \Phi_{R0}/\Phi_{A0}$, where $\Phi_{R0} = \Phi_R(0)$ is the flux falling on the receiver for $\alpha = 0$.

Figure 5.3 shows an example of a low concentration nonimaging concentrator. It is a CPC with entrance aperture **CD**, receiver **AB**, and a concentration given by C = [**C**,**D**]/[**A**,**B**], where [**X**,**Y**] represents the distance between points **X** and **Y**. It

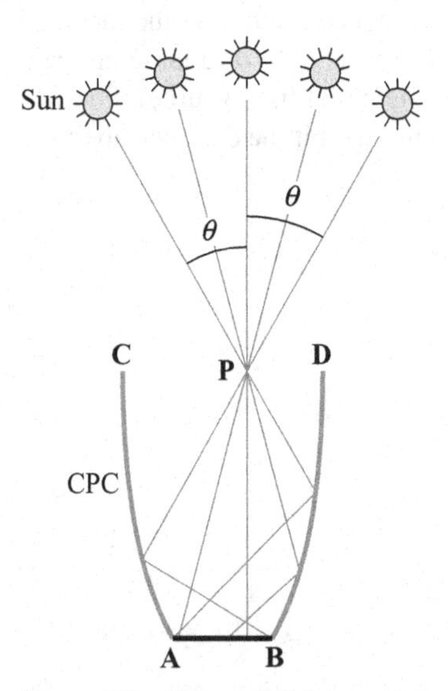

Figure 5.3 By maximizing acceptance $\pm\theta$, nonimaging optics (such as the CPC) are able to maximize sunlight collection while stationary. This reduces the need to track the sun.

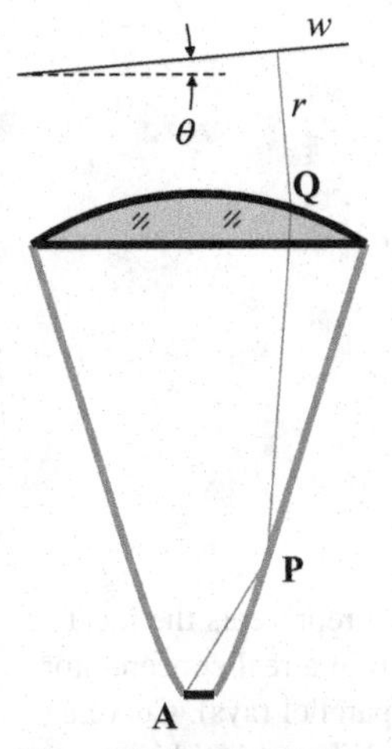

Figure 5.4 Lens-mirror nonimaging concentrator.

has an acceptance angle θ (the total acceptance of this optic is then $\pm\theta$), and it is therefore designed to collect radiation with total angular aperture 2θ. As long as the sun moves in the sky inside this angle, its light will be collected and concentrated onto the receiver. Figure 5.3 illustrates acceptance for a particular point **P** on the entrance aperture **CD**, but all points of **CD** have the same acceptance. This means that a set of parallel rays impinging on **CD** will be concentrated onto **AB** if these rays make an angle to the vertical smaller than the acceptance angle θ. In 3D geometry, the acceptance would be a cone with angular aperture θ, and sunlight would be collected as long as the sun moved inside that cone.

Figure 5.4 shows another example of a nonimaging concentrator. It has a smaller acceptance $\pm\theta$ and is a combination of a lens and mirrors. This optic is designed to collect radiation with total angular aperture 2θ. Its design starts by choosing the shape of the lens. Then, the shape of the mirror can be calculated by constant optical path length between flat wave front *w* and edge **A** of the receiver. The design process is similar to the one used to design DTIRC (dielectric total internal reflection concentrators) optics [2, 11].

By maximizing the acceptance angle, the CPC of Figure 5.3 or the optic of Figure 5.4 allow sunlight to be captured and concentrated for the longest period of time possible, diminishing the need to track the sun. Also, they maximize the capture of diffuse radiation that comes from the sky.

The relative angular transmission of an ideal concentrator for different incidence angles of perfectly collimated light (parallel rays) is flat (and equal to unity) up to an angle θ, and then it falls sharply to zero, as shown by curve *a* in Figure 5.5. This is, for example, the case of a 2D CPC as shown in Figure 5.2a. In practice, however, this is generally not the case, and the relative angular transmission starts to fall before and extends beyond angle θ, as shown by curve *b* in Figure 5.5. This is, for example, the case of a 3D CPC as shown in Figure 5.2b, which is not ideal [2]. In this case, the acceptance angle in defined as the incidence angle θ_R of the

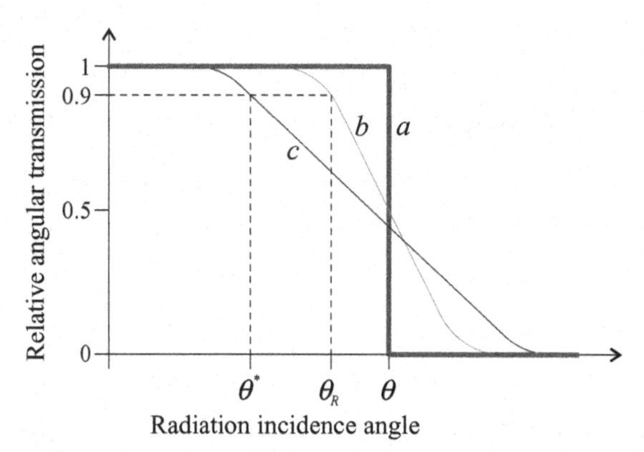

Figure 5.5 Relative angular transmission of a concentrator. Curve *a* represents the ideal case and curve *b* the typical case of the relative angular transmission in a real concentrator, both under different incidence angles of perfectly collimated light (parallel rays). Curve *c* represents the relative angular transmission of a concentrator under different incidence angles of sunlight.

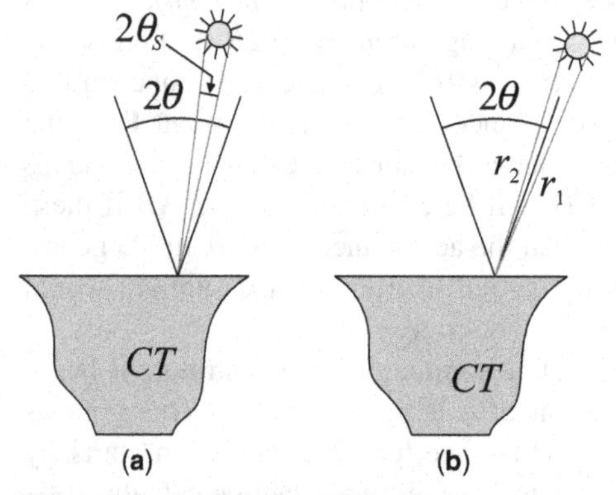

Figure 5.6 (a) When sunlight hits the entrance aperture of solar concentrator CT inside its acceptance angle, it is accepted and concentrated onto its receiver. (b) When sunlight hits a point at the entrance aperture at the limit of acceptance, some light is accepted (ray r_1), while other is lost (ray r_2).

radiation for which the relative angular transmission falls to 90% of its maximum value.

There is, however, yet another factor that affects the acceptance angle: the angular aperture of the sun. The relative angular transmission curve obtained for different incidence angles of sunlight (curve *c* in Fig. 5.5) does not fall as sharply as that obtained for perfectly collimated light (curve *b* in Fig. 5.5), and the acceptance angle has a lower value θ^* under sunlight. This is due to the total angular aperture $\pm\theta_S$ of sunlight, as shown in Figure 5.6. When sunlight hits a point on the entrance aperture of a solar concentrator CT confined inside its acceptance $\pm\theta$ all its light is captured, as in Figure 5.6a. However, as the direction of incidence of sunlight crosses the acceptance of the concentrator (Fig. 5.6b), there are rays (such as r_2) that are still accepted because they are still inside the acceptance angle, while others (such as r_1) are rejected (or lost) since they are already outside the acceptance angle. This smoothens the relative angular transmission curve relative to that obtained with collimated light.

Typically, the goal of solar concentrators is to provide a high concentration and a wide acceptance angle. Two concentrators may be thought of as "equivalent" if they have the same concentration and the same acceptance angle (same overall tolerances). Concentrators may then be thought of as devices that aim at maximizing the Concentration acceptance product (CAP), which may be defined as [12]

$$\mathrm{CAP} = \sqrt{C}\sin\theta, \tag{5.5}$$

where C is the geometrical concentration and θ the acceptance angle of a concentrator. With this definition, the CAP is limited to a maximum value of n, the refractive index in which the receiver is immersed. Therefore, a possibility to increase the CAP of a concentrator is to increase the refractive index of the material covering the receiver. The acceptance angle θ in Equation (5.5) is obtained for parallel rays. If the angular aperture of sunlight is considered, the acceptance angle will have a lower value θ^*, and the definition is now [12]

$$\mathrm{CAP}^* = \sqrt{C}\sin\theta^*. \tag{5.6}$$

The CAP* is now called the effective CAP, and θ^* is called the effective acceptance angle (since it considers the finite angular aperture of the sun). The CAP* can be measured experimentally.

A wide acceptance is important, for example, when we have an array of concentrators pointing at the sun. This is an interesting approach since, instead of a large concentrator, we can have many small ones that collect the same amount of sunlight. Many small concentrators use less space and materials than the large one they replace. When assembling an array, however, there are unavoidable errors in the positioning of the individual optics. Figure 5.7 shows a few concentrators in one such array. In this example, each one of these concentrators has an acceptance $\pm\theta$ and is similar to the one in Figure 5.4. Each one of these optics is therefore designed to collect radiation with total angular aperture 2θ.

When assembled, they have small pointing errors γ and that makes them point in slightly different directions. The overall acceptance of the array is then diminished from the original value of $\pm\theta$ of the individual optics to a smaller value of $\pm\beta$ for the array. The result of this is that when the array is pointed at the sun, some of the individual optics may be pointing correctly and collecting sunlight, while others are out of alignment and, therefore, are not collecting sunlight.

The result is that the units well aimed at the sun will produce a higher current than the ones where the pointing errors are larger. Since the module is composed of series and parallel connected units, as shown in Figure 5.8, the current of the series interconnection is limited by the unit that generates the lowest current. This current mismatch diminishes the power generated by the array. This effect will be more significant for higher concentration since it implies a smaller acceptance angle and, therefore, higher sensitivity to errors.

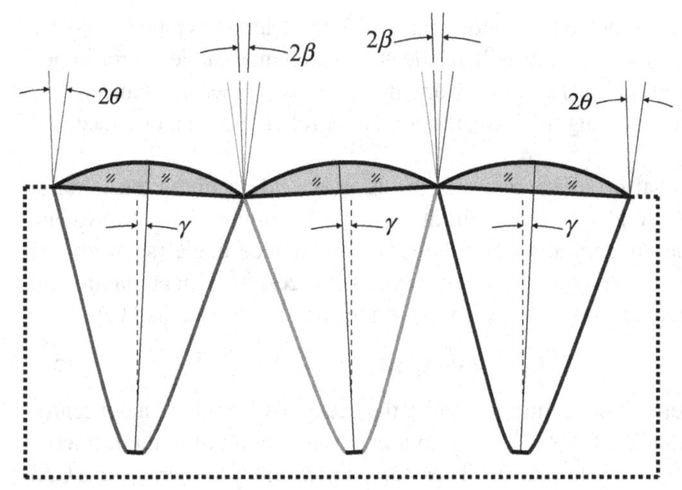

Figure 5.7 Several nonimaging concentrators assembled in an array. Misalignments of angle γ between the concentrators reduces the acceptance angle from $\pm\theta$ for each optic to $\pm\beta$ for the array.

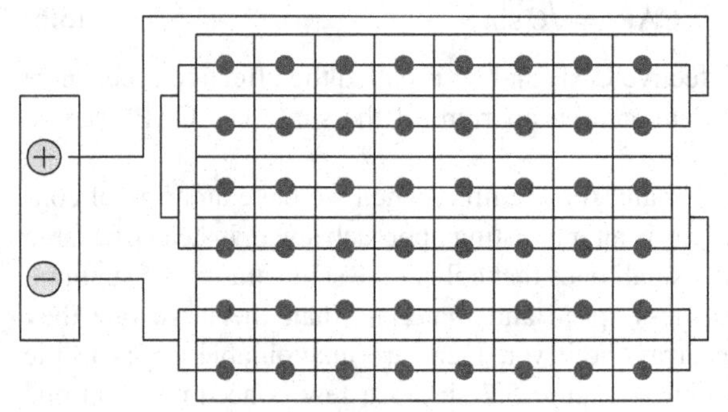

Figure 5.8 Schematic example of the interconnections of several concentration units in an array.

Assembly errors may be reduced by having a more elaborate and sturdier assembly of the concentrators. This, however, means more complex manufacturing. Also, a sturdier array will be heavier and requires a stronger tracking mechanism, adding to the cost. Also, small oscillations of the trackers due to wind will reduce the effective acceptance of the assembly, adding to the need of a wide acceptance angle.

In other applications, sunlight is concentrated using Fresnel mirrors in which many mirrors (called heliostats) concentrate radiation onto a single receiver. In these applications, maximizing acceptance relaxes the tracking precision needed for the heliostats.

Another advantage of a wider acceptance angle is tolerance to errors in the manufacturing of the optical surfaces. A good example of the manufacturing advantages introduced by having wide acceptance angle was demonstrated with the development of the DSMTS linear concentrator [13, 14]. Figure 5.9 shows its 2D section. It is a linear concentrator designed for a geometrical concentration of 30×, which implies the use of a one-axis tracking system.

The mirrors of this concentrator were made by mechanically flexing a metal sheet and imposing the derivatives at the mirror edges. This method results in a shape for the mirrors that is not the ideal one. However, the results obtained with prototypes show that given the high theoretical acceptance angle of this concentrator (1.88°), it can easily accommodate the acceptance angle reduction resulting from this simple manufacturing method, still maintaining a large acceptance angle of (1.63°).

Figure 5.10 shows a photo of a prototype DSMTS concentrator.

In this prototype used for testing the concept, the secondaries and receivers *R* do not run the full length of the mirrors *m*.

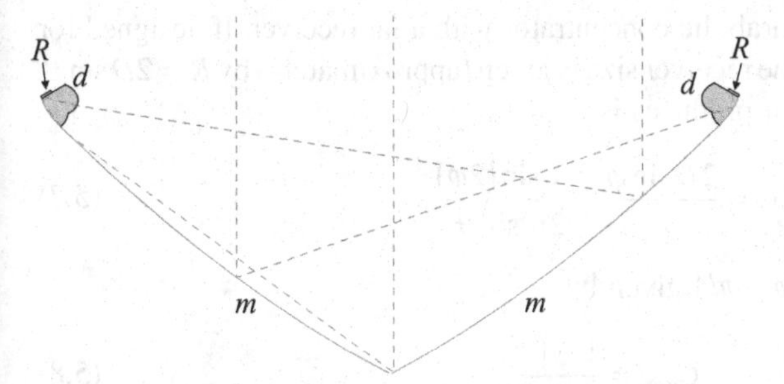

Figure 5.9 DSMTS concentrator with mirrors *m* and receivers *R*, in contact with dielectric parts *d*. Making the mirrors by a simple method, such as flexing a metal sheet, reduces the acceptance, but this is easily accommodated, giving the high theoretical acceptance angle of the concentrator.

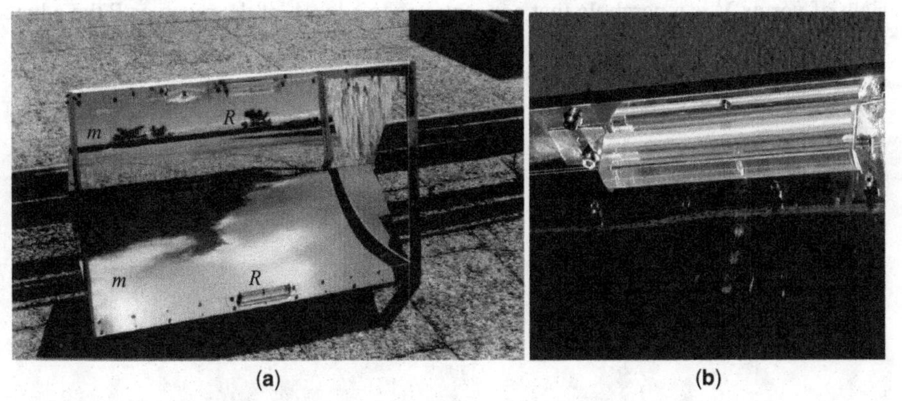

Figure 5.10 (a) Prototype DSMTS concentrator with mirrors *m* and short receivers *R*. In this prototype, the secondaries and receivers do not run the full length of the mirrors. (b) Detail of a secondary and receiver.

5.3 IMAGING AND NONIMAGING CONCENTRATORS

Nonimaging optics typically have two (or more) optical surfaces, and this allows the design of many types of configurations and also makes them flexible so they can be adjusted to the geometrical constraints of specific situations. Examples of nonimaging solar concentrator are CPCs, DTIRCs, combinations of parabolic primaries with TERC secondaries, XR, RXI, XX, or TIR lenses with secondaries and other configurations [1–3].

For the case of very wide acceptance angle, low concentration applications there is no parallel in imaging optics for nonimaging optics devices, such as the CPC illustrated in Figure 5.3, which is a very simple and highly efficient concentrator.

In the case of high concentration, however, traditional imaging optics may also be used as solar concentrators. A good example are the parabolic mirrors. The acceptance angle of these optics, however, falls short of the maximum possible. Also, these optics cannot attain the very high concentrations that nonimaging optics can deliver.

Figure 5.11 shows a parabolic concentrator with a fat receiver. If designed for a given acceptance angle θ, the receiver size is given (approximately) by $R = 2D\sin\theta/\cos\varphi$ and the concentration it produces is

$$C = \frac{2D\sin\varphi}{R} = \frac{1}{2}\frac{\sin(2\varphi)}{\sin\theta}, \tag{5.7}$$

which has a maximum for $\varphi = \pi/4$ given by

$$C_{\max} = \frac{1}{2}\frac{1}{\sin\theta}, \tag{5.8}$$

and is half the theoretical maximum in the 2D case (linear symmetry). In the case of circular symmetry, the maximum concentration is 1/4 of the theoretical maximum.

Figure 5.12 shows a combination of a compound parabolic primary and a TERC secondary [2, 15]. The portion of the compound parabolic primary p to the right of vertex **V** is a parabola that concentrates to edge **A** of the receiver R the edge

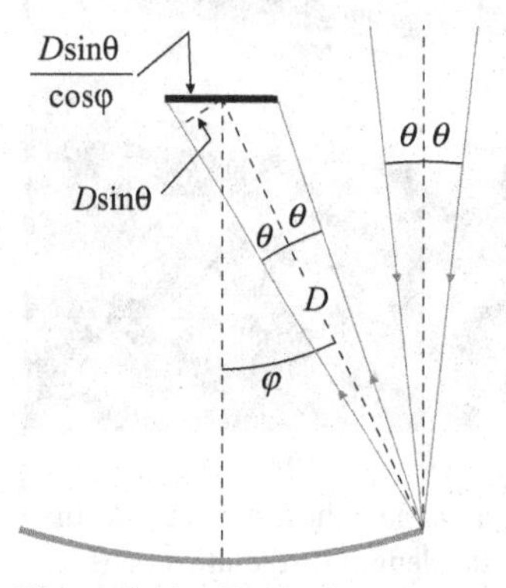

Figure 5.11 Parabolic concentrator.

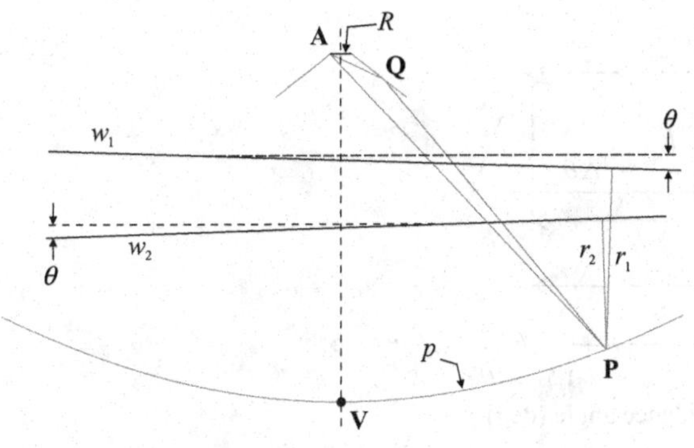

Figure 5.12 Combination of a compound parabolic primary and a TERC secondary.

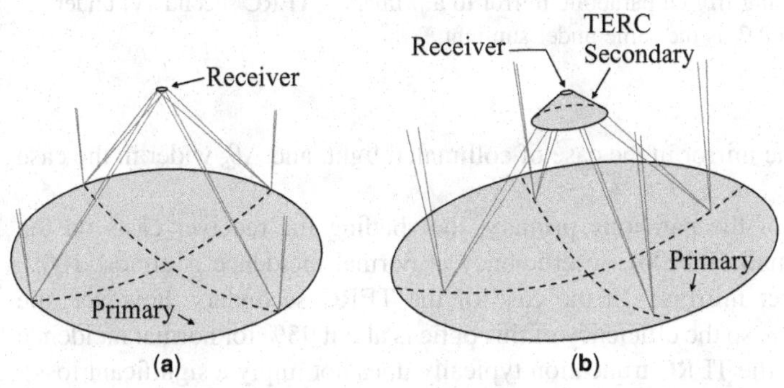

Figure 5.13 (a) Parabolic mirror with receiver. (b) Compound parabolic primary mirror with TERC secondary. Both optics have a concentration of 1000×.

rays perpendicular to flat wave front w_1 tilted down by an angle θ to the horizontal. Points **Q** on the TERC secondary are obtained by constant optical path length for rays r_2, perpendicular to wave front w_2 (tilted up by an angle θ to the horizontal), between w_2 and edge **A** of the receiver. A complete TERC secondary would completely shade the primary and, therefore, it is truncated to let light through to the primary.

We now compare the concentrators in Figure 5.11 and 5.12, when designed for a concentration of 1000× and given circular symmetry, as shown in Figure 5.13. The ideal acceptance angle for a concentration $C = 1000$ when the receiver is in air is given by Equation (5.3) as $\theta_I = \arcsin\left(1/\sqrt{C}\right) = 1.81°$.

Figure 5.14 shows the relative angular transmission curves for these optics. There are four curves in this figure: *a* is for the parabolic mirror under different incidence angles of collimated light (parallel rays), *b* is for the parabolic mirror under different incidence angles of sunlight (which has a small angular aperture), *c* is for the primary + TERC under different incidence angles of collimated light and *d* is for the primary + TERC under different incidence angles of sunlight. In the case of Figure 5.14, the acceptance angle of the primary + TERC optic is $\Delta\theta_P$ wider than

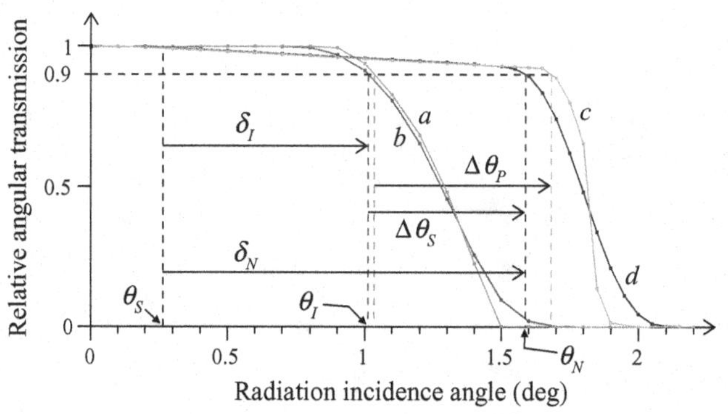

Figure 5.14 Relative angular transmissions for *a*: parabolic mirror under collimated light (parallel rays), *b*: parabolic mirror under sunlight, *c*: primary + TERC secondary under collimated light, *d*: primary + TERC secondary under sunlight. Angle $\Delta\theta_P$ is the increase in acceptance angle (going from a parabolic mirror to a primary + TERC secondary) under collimated light, and $\Delta\theta_S$ is the same under sunlight.

that of the parabolic mirror in the case of collimated light, and $\Delta\theta_S$ wider in the case of sunlight.

In the case of the parabolic primary, the shading the receiver casts on the primary is very small (1/1000), so efficiency at normal incidence is almost 100% (considering perfect mirrors). In the case of the TERC secondary, however, the shading is about 5%, so the efficiency of this optic is about 95% for normal incidence of light (for which the TERC truncation typically does not imply a significant loss).

When pointing at the sun, the nonimaging optic has an angular tolerance of $\pm\delta_N$ (given by the difference between its acceptance angle θ_N under sunlight and the angular aperture of the sun θ_S). The imaging parabolic mirror has a lower angular tolerance $\pm\delta_I$ (given by the difference between its acceptance angle θ_I under sunlight and the angular aperture of the sun θ_S). In practice, these tolerances are reduced due to errors, such as imperfect shapes of the optics, misalignment of the receiver,or imperfections in the optical surfaces. This reduction, however, affects the imaging optics more severely since they start with a much smaller tolerance.

It is possible to increase the acceptance angle of a given concentrator by increasing the refractive index in which the receiver is immersed, as seen from Equation (5.3). Figure 5.15 shows that possibility for a parabolic mirror (a) [16] and a compound primary + TERC secondary (b). The basic geometry of the optics in Figure 5.15 is the same as before, so when given circular symmetry, they will also attain a geometric concentration of 1000 (as was the case of the optics in Figure 5.13). Now, however, in both cases, light reaches the receiver within a dielectric material. The primary + TERC combination is more compact than the corresponding parabolic mirror. That was also the case in the concentrators of Figure 5.13.

In Figure 5.15a, the rays reflected at a point **N** closer to the center of the mirror produce a smaller sport than those rays reflected at the outer points **M**. This makes points **N** have a wider acceptance angle than points **M** and gives the acceptance curve of the parabola its gradual fall, as shown in Figure 5.14. The acceptance of

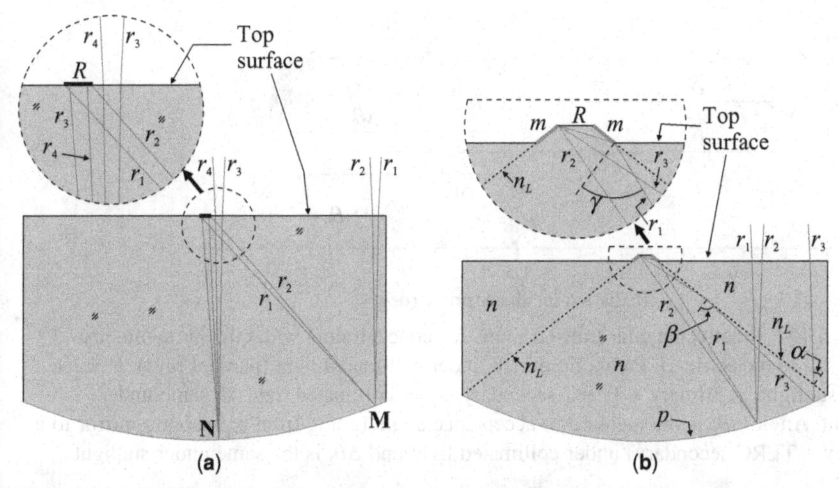

Figure 5.15 (a) Parabolic mirror immersed in dielectric material. Rays r_3 and r_4 closer to the center of the concentrator form a smaller spot than r_1 and r_2 incident closer to the edge. (b) Combination of a compound parabolic primary and TERC secondary, immersed in dielectric. The TERC extends all the way to the primary and is implemented as a mirror next to the receiver and as a low refractive index layer n_L further away from it.

the nonimaging optic falls off more steeply, a characteristic of the optics that reach closer to the maximum possible acceptance for a given concentration.

The TERC secondary in Figure 5.15b is designed using the same method shown in Figure 5.12. However, it is now implemented as two small mirrors m next to the receiver R and as a low refractive index layer of refractive index n_L inside a material of refractive index n. The secondary now extends all the way to the primary, contrary to what happened in Figure 5.12, in which it was truncated to prevent it from completely shading the compound parabolic primary p. An edge ray r_3, as it crosses the top surface of the concentrator, decreases its angle to the vertical according to the law of refraction. This ray then crosses the layer n_L, making a small angle α to its normal. It refracts from the material of index n into the layer of index n_L and then refracts back into the material of index n. If this layer is thin enough, the ray continues its path as if the layer did not exist. The ray then hits the primary mirror and is reflected back toward the TERC secondary, hitting it at a much wider angle β. It suffers total internal reflection (TIR), and is reflected toward the edge of the receiver R. Another ray r_1 has a similar path, but reaches the secondary at a point closer to the receiver R and making a smaller angle γ to the secondary normal. Now TIR no longer occurs, and this light ray is reflected toward the edge of the receiver by mirror m. A ray r_2 has a similar path initially, but is reflected by the primary directly toward the edge of the receiver.

The smaller the mirrors m are, the larger the difference in refractive indices between n and n_L for the concentrator to work as described. The reason for this is that the edge rays hit the secondary at smaller angles to its normal at points closer to the receiver. In this example, the size of mirrors m was chosen such that they

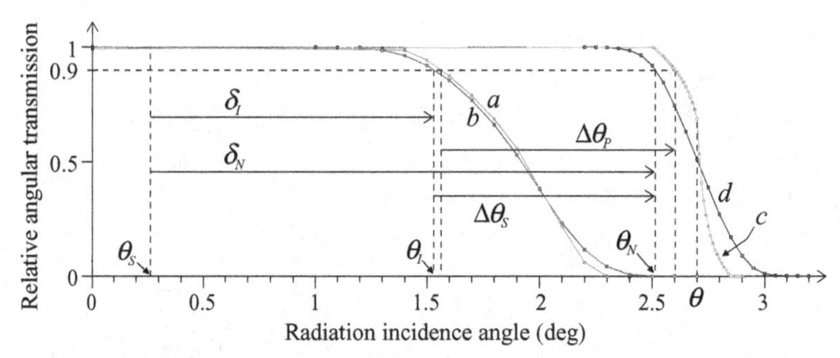

Figure 5.16 Relative angular transmissions for concentrators with circular symmetry immersed in dielectric. *a*: Parabolic mirror under collimated light (parallel rays), *b*: same under sunlight, *c*: primary + TERC secondary under collimated light, *d*: same under sunlight. Angle $\Delta\theta_P$ is the increase in acceptance angle (going from a parabolic mirror to a primary + TERC secondary) under collimated light and $\Delta\theta_S$ is the same under sunlight.

produce, together with the receiver *R*, a shading of 0.5% on the primary. The geometrical efficiency of this optic (considering a perfect system) is then 99.5% at normal incidence of light.

If the optics in Figure 5.15 are given circular symmetry (as was the case of the optics in Fig. 5.13), their relative angular transmissions are as shown in Figure 5.16.

When comparing these curves with the relative angular transmission curves in Figure 5.14, we can see that the acceptance angles are now much wider. As referred above, this is due to having the receiver immersed in dielectric (for this example, we chose $n = 1.49$, that of PMMA plastic, resulting in an ideal acceptance angle of $\theta = 2.7°$). Also, the relative angular transmission of the primary + TERC secondary combination (curves *c* and *d*) is now constant (equal to 1) all the way to a wide angle. This is due to the fact that the TERC secondary now extends all the way to the primary, reducing losses. The acceptance angle of the primary + TERC optic is $\Delta\theta_P$ wider than that of the parabolic mirror in the case of collimated light, and $\Delta\theta_S$ wider in the case of sunlight. When pointing at the sun, the nonimaging optic is able to tolerate an angular error $\pm\delta_N$ (where $\tau_N = \theta_N - \theta_S$), while the imaging parabolic mirror can only tolerate an angular error of $\pm\delta_I$ (where $\delta_I = \theta_I - \theta_S$).

If the optics in Figure 5.15 have linear symmetry, as shown in Figure 5.17, their relative angular transmission curves are as shown in Figure 5.18.

Now the relative angular transmission curve of the nonimaging optic under collimated light is a step function that falls off sharply at acceptance angle θ (matching the ideal maximum). For angles larger than θ, the relative angular transmission should be zero. In this particular configuration, this is not the case because of mirrors *m* on the sides of the receiver. These mirrors shade the primary, decreasing the efficiency of the optic, but they also reflect back toward the primary some light outside the acceptance angle. The primary then reflects this light back up and eventually it reaches the receiver. This is not typical of nonimaging optics, but happens in this particular configuration.

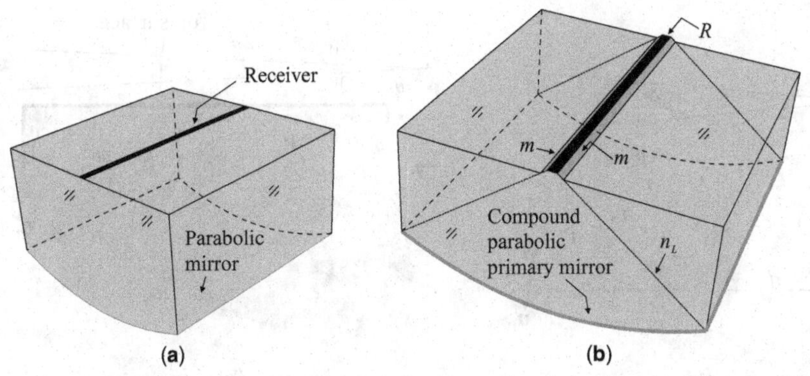

Figure 5.17 (a) Linear parabolic primary with receiver immersed in dielectric. (b) combination of a compound parabolic primary and TERC secondary. The TERC is implemented as a low refractive index layer n_L and two small mirrors m next to the receiver R.

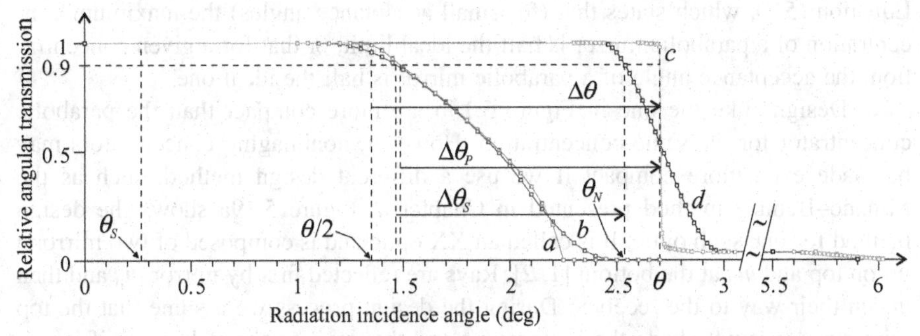

Figure 5.18 Relative angular transmission curves for concentrators with linear symmetry immersed in dielectric. *a*: parabolic mirror under collimated light (parallel rays), *b*: same under sunlight, *c*: primary + TERC secondary under collimated light, *d*: same under sunlight. Angle $\Delta\theta$ is the increase in ideal acceptance angle (defined here as the angle for which acceptance falls below 100%) for collimated light. In this case, the nonimaging optic has double the ideal acceptance angle of the imaging one.

In the linear configuration, the shading that the receiver casts on the primary for the geometry of Figure 5.17a is now 3%, and the shading that mirrors m and receiver cast on the primary in the case of Figure 5.17b is now 7%. The geometrical efficiency (considering perfect optics) of the parabolic mirror for normal incidence of light is then 97%, while that of the nonimaging optic is 93%.

The increase in acceptance angle, as shown in Figure 5.18 for the case of linear symmetry, when going from an imaging to a nonimaging optic is $\Delta\theta_P$ under collimated light and $\Delta\theta_S$ under sunlight.

If we defined the ideal acceptance angle as that for which the acceptance falls below 100% for collimated light, then the acceptance of the nonimaging optic would be θ, while that for the imaging optic would be $\theta/2$. This is in accordance with

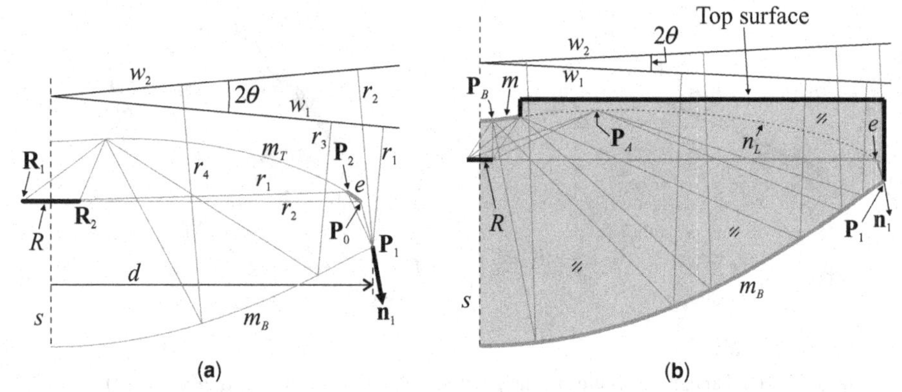

Figure 5.19 (a) Design of an XX SMS concentrator. (b) Practical implementation of an XX SMS concentrator.

Equation (5.8), which states that (for small acceptance angles) the maximum concentration of a parabolic mirror is half the ideal limit, or that for a given concentration, the acceptance angle of a parabolic mirror is half the ideal one.

Designs like the one in Figure 5.15b are more compact than the parabolic concentrator for the same concentration. However, nonimaging concentrators may be made even more compact if we use a different design method, such as the Miñano–Benitez method presented in Chapter 4. Figure 5.19a shows the design method for one such optic. It is called an XX optic and is composed of two mirrors: m_T on top and m_B at the bottom [1, 2]. Rays are reflected first by mirror m_B and then m_T on their way to the receiver. During the design process we assume that the top mirror m_T does not shade the bottom one and that rays go through m_T as if it was not there, as is shown by the paths of rays r_3 or r_4.

The design of the optic starts by specifying the size and position of the receiver R. Then one chooses a point $\mathbf{P}_1$ and its normal $\mathbf{n}_1$ at the edge of the bottom mirror m_B. Point $\mathbf{P}_1$ must be at a distance d from the optical axis that verifies the conservation of étendue: $2d\sin\theta = R$.

Next, reflect at $\mathbf{P}_1$ the edge rays r_1 and r_2 perpendicular to wave fronts w_1 and w_2, respectively. These two rays are directed toward edge $\mathbf{R}_1$ of the receiver using the elliptical arc e as the guiding curve. Edge $\mathbf{P}_0$ of this elliptical arc is on the horizontal line through $\mathbf{R}_1$ and $\mathbf{R}_2$ of the receiver. The next step is to use reverse ray tracing by reflecting from e the reversed rays coming from the edge $\mathbf{R}_2$ of the receiver and calculate a new portion of mirror m_B. These reversed rays are reflected from e in a direction perpendicular to wave front w_2. Finally, a new set of rays is reflected from this new portion of the bottom mirror, m_B, perpendicular to wave front w_1. From this, one calculates a new portion of the top mirror m_T to the left of $\mathbf{P}_2$ in such a way that these rays are reflected toward the edge $\mathbf{R}_1$ of the receiver. Ultimately, this process is repeated as the bottom and top mirror surfaces are calculated, until the optical surfaces reach the optical axis. The complete optic is obtained by symmetry about the optical axis s.

A practical implementation of this optic is shown in Figure 5.19b. The top mirror is implemented as a low refractive index layer n_L inside a higher refractive index material. At a point $\mathbf{P}_A$ on n_L light is reflected by total internal reflection (TIR), but for points $\mathbf{P}_B$ closer to the optical axis TIR fails and we must mirror this portion m of the top optical surface. The paths of light rays inside the optic then start by a refraction at the top surface, then they cross the low refraction index layer n_L, are reflected by the bottom mirror m_B, and then again by the top optical surface (n_L or m) toward the receiver. Again, the complete optic is obtained by mirror symmetry about the optical axis s.

The portion of the XX optic above the low refractive index layer n_L may be removed and the shape of the optical surfaces readjusted to obtain an RXI optic, as shown in Figure 5.20 [1, 2, 17]. This figure also shows the paths of two edge rays perpendicular to wave fronts w_1 and w_2 in their way to the edges $\mathbf{R}_1$ and $\mathbf{R}_2$ of the receiver R. Light now refracts as it enters the top surface of the optic, then it is reflected at the bottom surface and then TIRs at the top surface, being redirected toward the receiver R. The central portion m of the top surface also needs to be mirrored since TIR fails at those points.

Figure 5.21 shows a prototype RXI solar concentrator.

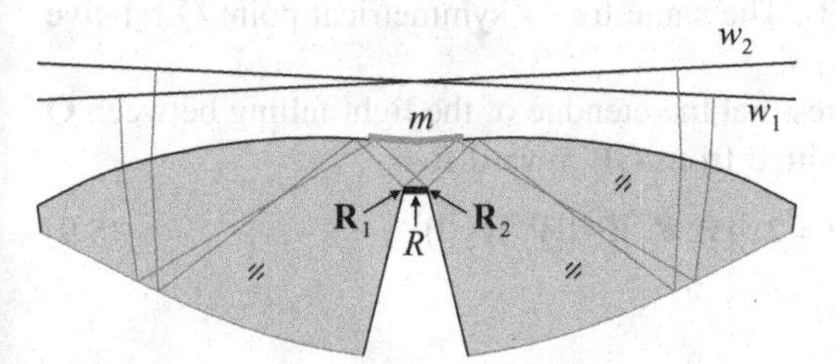

Figure 5.20 RXI concentrator.

Figure 5.21 Prototype RXI solar concentrator.

Different nonimaging designs are applicable to different situations. The configuration of Figure 5.13b may be used in large solar concentrators since it uses only mirrors, while the RXI in Figure 5.20 is limited to small optics since it is made of a single piece of dielectric material [18].

5.4 LIMIT CASE OF INFINITESIMAL ÉTENDUE: APLANATIC OPTICS

Nonimaging optics redirect the edge rays of the incoming radiation toward the edges of the receiver. This enables them to handle wide acceptance angles, which are very useful in solar concentration. Here, we go in the opposite direction and analyze the limit case in which the acceptance angle of a nonimaging optic goes to zero. This approach, although not optimal, may be applied in very high concentration in which the acceptance angles are naturally small.

Figure 5.22a shows a concentrating optic acceptance $\pm\theta$ (total acceptance angle 2θ), and a receiver of size $R = [\mathbf{R}_1, \mathbf{R}_2]$, whose edges $\mathbf{R}_1$ and $\mathbf{R}_2$ are immersed in a medium of refraction index n (where $[\mathbf{X}, \mathbf{Y}]$ represents the distance between points **X** and **Y**). Light enters the optic through A. Light that hits point **P** with angular aperture $\pm\theta$ is redirected onto $\mathbf{R}_1\mathbf{R}_2$. The same for its symmetrical point **Q** relative to the symmetry axis (optical axis) s.

Conservation of étendue states that the étendue of the light falling between **Q** and **P** matches that of the light emitted from **QP** toward $\mathbf{R}_1\mathbf{R}_2$

$$\xi = 4x\sin\theta = 2([[\mathbf{P}, \mathbf{R}_1]] - [[\mathbf{P}, \mathbf{R}_2]]), \tag{5.9}$$

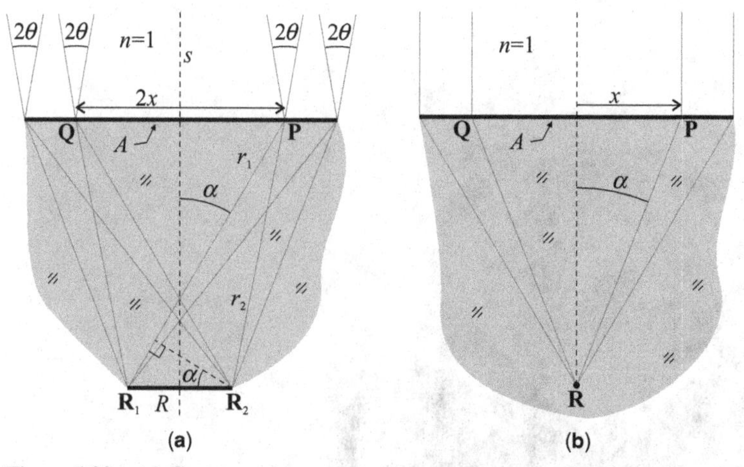

Figure 5.22 (a) Concentrating optic with entrance acceptance $\pm\theta$ and receiver $\mathbf{R}_1\mathbf{R}_2$. Light enters the optic through A. (b) Limit case of panel a in which the receiver size R and acceptance $\pm\theta$ goes to zero, that is, the limit case in which the étendue accepted be optic and redirected toward $\mathbf{R}_1\mathbf{R}_2$ goes to zero.

where [[**X**,**Y**]] represents the optical path length between **X** and **Y**. In general, the paths of the rays between **P** or **Q** and R will have the shape of broken lines, as the light is reflected and/or refracted in the optic. However, Equation (5.9) still holds in that case [2].

We now take the limit case in which receiver $\mathbf{R}_1\mathbf{R}_2$ becomes infinitely small, that is, the limiting case in which $R \to 0$. The size R of the receiver now becomes an infinitesimal quantity dR. The étendue of the light falling on dR is now also infinitesimal, and angle θ tends to an infinitesimal angle $d\theta$. Since, as $\theta \to 0$, then $\sin\theta \to \theta$, and we may write $\sin d\theta = d\theta$ (angle θ in radians). Also, ray r_2 tends to be parallel to r_1 and $[[\mathbf{P},\mathbf{R}_1]] - [[\mathbf{P},\mathbf{R}_2]] \to n dR \sin\alpha$, where α is the angle these rays make to the optical axis s. The above Equation (5.9) for étendue conservation now becomes $2x d\theta = n dR \sin\alpha$. Making

$$g = \frac{n\,dR}{2\sin\theta} \approx \frac{n\,dR}{2d\theta}, \tag{5.10}$$

we get

$$x = g\sin\alpha. \tag{5.11}$$

This equation is known in imaging optics as the condition of aplanatism and may be used to design solar concentrators in the limit case of small acceptance angles [19, 20]. The resulting optic is an imaging optic.

Figure 5.22b shows this limiting case in which $R \to 0$ and $\theta \to 0$, that is, the limiting case in which the étendue of the light entering the optic through A and redirected toward receiver $\mathbf{R}_1\mathbf{R}_2$ also goes to zero. Now we have a set of parallel rays incident upon A that verify two conditions: they are all redirected toward a point **R** (limiting case of an infinitesimal receiver), and they verify condition (Eq. 5.11). If at point **R** we place a receiver of size dR and want the optic to have an acceptance $\pm d\theta$, we may calculate g by Equation (5.10). These two conditions allow us to design two optical surfaces simultaneously. These optics may, therefore, be seen as limit cases of SMS optics when the coupled étendue goes to zero. All SMS optics, such as RR, RX, XR, XX, or RXI when taken to this limit, tend to aplanatic designs.

The condition of aplanatism must also be met when a concentrator has circular symmetry and the meridian and saggital acceptances are the same (a condition for the acceptance cone to be circular). Figure 5.23 shows a 2D vertical cut of a circularly symmetric optic O_C, where x is now the radial coordinate. Some meridian edge rays defining a meridian acceptance angle θ_M are also shown diagrammatically. Conservation of étendue states that

$$2\,dx\sin\theta_M = 2n\,dR\cos\alpha\sin\left(\frac{d\alpha}{2}\right), \tag{5.12}$$

where n is the refractive index of the material in which the receiver of diameter dR is immersed.

Now, making $\sin(d\alpha/2) = d\alpha/2$, we get $dx/d\alpha = ndR/(2\sin\theta_M)\cos\alpha$, or $x = ndR/(2\sin\theta_M)\sin\alpha + C$, where C is a constant of integration. Making, for example, $x = 0$ for $\alpha = 0$, we get $C = 0$, or

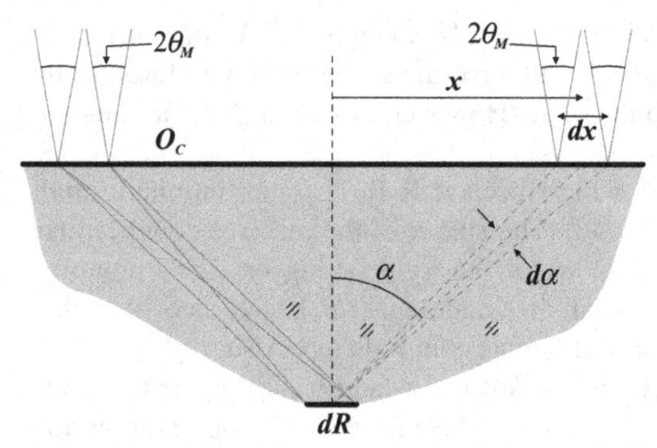

Figure 5.23 Meridian rays in a concentrator with circular symmetry.

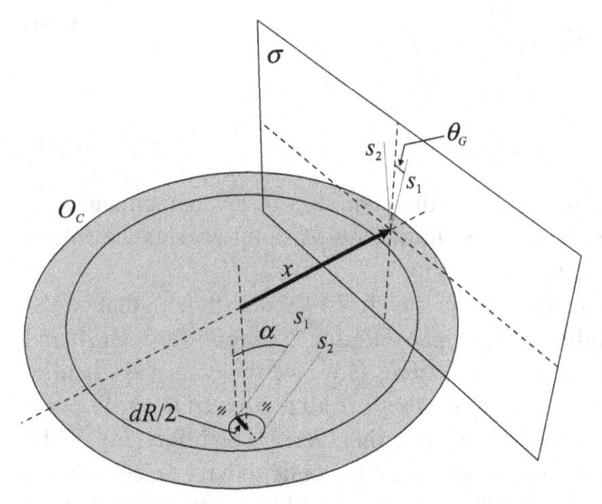

Figure 5.24 Saggital rays in a concentrator with circular symmetry.

$$x = \frac{n\,dR}{2\sin\theta_{\mathrm{M}}}\sin\alpha. \tag{5.13}$$

This is the case when the vertical ray through the receiver leaves the optic at $x = 0$. This is the same as Equation (5.11).

Figure 5.24 shows a 3D view of the same concentrator O_C with circular symmetry and also the paths of two sagittal rays s_1 and s_2 (shown diagrammatically) contained on sagittal plane σ. If these are edge rays of the incoming radiation, they define a sagittal acceptance angle θ_G. For a given point on the entrance aperture of the concentrator, these rays have the highest value of skewness. They must arrive at the receiver also with the highest value of skewness and, therefore, at its edge. In particular, ray s_1 arrives at the receiver, also as a sagittal ray, making an angle α to the vertical (parallel to the optical axis). Conservation of skewness is then

$$x\sin\theta_{\mathrm{G}} = n\frac{dR}{2}\sin\alpha \Leftrightarrow x = \frac{n\,dR}{2\sin\theta_{\mathrm{G}}}\sin\alpha, \tag{5.14}$$

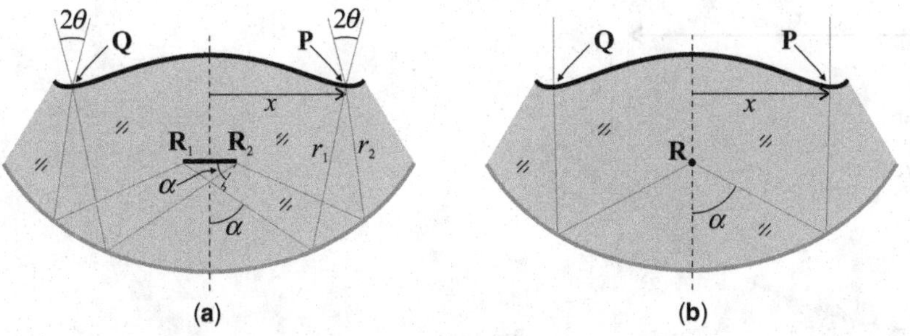

Figure 5.25 (a) RX SMS optic for a wide acceptance $\pm\theta$. As the acceptance $\pm\theta$ goes to zero, this optic tends to an aplanatic RX, as in panel b.

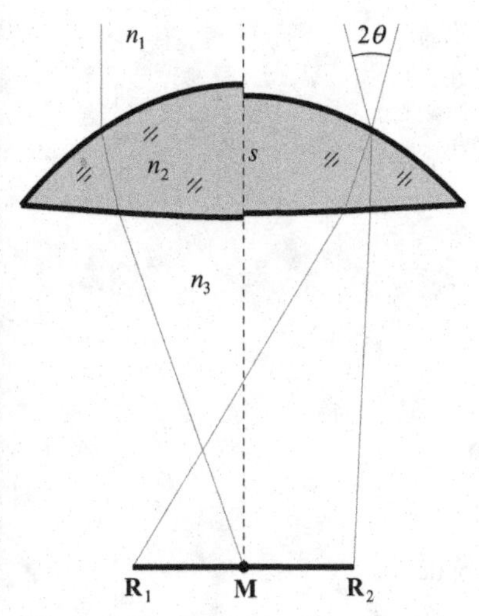

Figure 5.26 SMS lens for acceptance angle and receiver $\mathbf{R}_1\mathbf{R}_2$ to the right of s and aplanatic lens with receiver **M** to the left of s.

where x is again the radial coordinate. Equation (5.14) is the same as Equation (5.13) when the meridian acceptance θ_M and the sagittal acceptance θ_G have the same value. Therefore, an optic which verifies the condition of aplanatism has the same meridian and sagittal acceptance.

Figure 5.25a shows an RX SMS optic [1, 2, 21] designed for an acceptance $\pm\theta$. This is a particular case of the general situation shown in Figure 5.22a and for which Equation (5.9) also applies. Figure 5.25b shows the same device in the limit case in which $\theta \rightarrow 0$, so this new optic is aplanatic, for which Equation (5.11) applies. In the case of the RX, these two optics, although different, look quite similar.

Figure 5.26 shows another comparison, now between an RR SMS lens to the right of s and an aplanatic lens to the left of s. The SMS lens has acceptance $\pm\theta$ and receiver $\mathbf{R}_1\mathbf{R}_2$, while the aplanatic lens has focus **M**.

Aplanatic optics may be designed using a simple method in which we proceed in very small steps, building both optical surfaces at a time. Figure 5.27 shows an

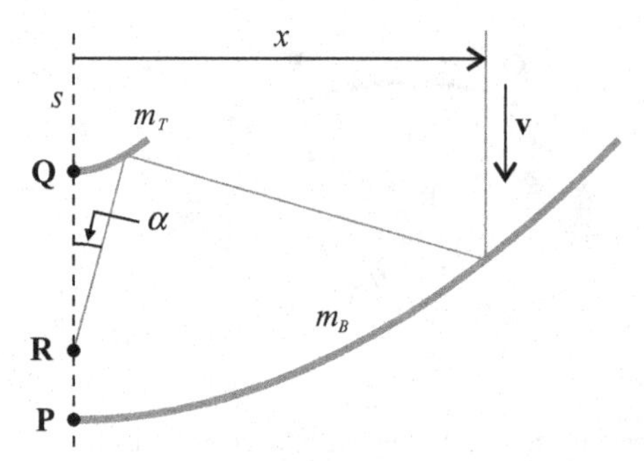

Figure 5.27 Aplanatic XX optic with two mirrors m_B and m_T and point receiver **R**.

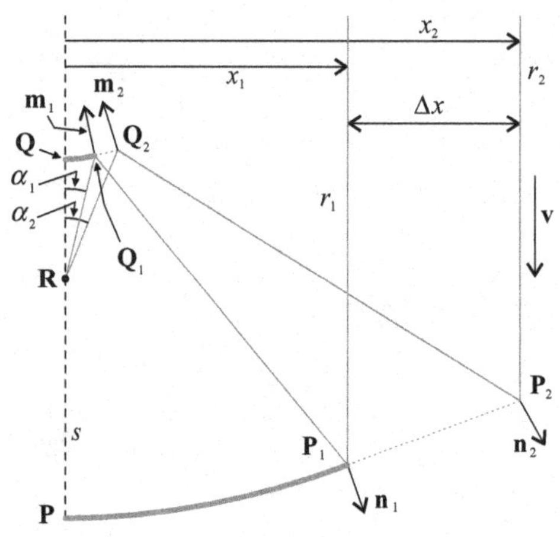

Figure 5.28 Construction method for aplanatic XX optic.

example of one of these optics for the particular case in which both optical surfaces are mirrors. An incoming ray with direction **v** (from an object at an infinite distance) is reflected by the first optical surface m_B, and then by the second m_T, toward focus **R** (where an image of the object is formed).

Referring now to Figure 5.28, and according to the above Equation (5.11), we can define a method for calculating the points of the mirrors of an aplanatic optic.

Suppose we already know the positions of points $\mathbf{P}_1$ and its normal $\mathbf{n}_1$ and $\mathbf{Q}_1$ and its normal $\mathbf{m}_1$, that is, the path of ray r_1 is already known. We now launch a ray r_2, shifted by a distance Δx away from the optical axis from ray r_1 and in the direction of vector **v**. We intersect ray r_2 with the tangent to m_B at point $\mathbf{P}_1$ (perpendicular to its normal $\mathbf{n}_1$) and determine the position of point $\mathbf{P}_2$. From Equation (5.11), we can determine $\alpha_2 = \arcsin(x_2/g)$.

Launching from **R** a ray tilted by an angle α_2 to the optical axis and intersecting it with the tangent to m_T at $\mathbf{Q}_1$ (perpendicular to its normal $\mathbf{m}_1$), we can determine the position of point $\mathbf{Q}_2$. We now have the complete path of ray r_2. From the direction of the incident and reflected rays at $\mathbf{P}_2$, we can determine its normal $\mathbf{n}_2$. Also, from the directions of the incident and reflected rays at $\mathbf{Q}_2$, we can determine its normal $\mathbf{m}_2$. Repeating the same process for another ray to the right of r_2, we can determine two new points on m_B and m_T to the right of $\mathbf{P}_2$ and $\mathbf{Q}_2$. This process continues as we determine more and more points and completely define both optical surfaces. The design starts by defining the position of points **P** and **Q** on the optical axis where the surfaces m_B and m_T start. The normals to **P** and **Q** are parallel to the optical axis *s*.

In general, this method allows us to design two optical surfaces in which n_1 is the refractive index before the first surface, n_2 is the refractive index between surfaces, and n_3 is the refractive index after the second surface (see Fig. 5.26). If $n_1 = n_2$, then the first surface is reflective, otherwise refractive. Also, if $n_2 = n_3$, the second surface is reflective, otherwise refractive. As seen above, as we step through the construction method, we define the path of a new ray using Equation (5.11) and determining two new points, one on each optical surface. The corresponding surface normals at these points can be calculated by [2]

$$\mathrm{dflnrm}(\mathbf{i}, \mathbf{r}, n_A, n_B) = \frac{n_A\mathbf{i} - n_B\mathbf{r}}{\|n_A\mathbf{i} - n_B\mathbf{r}\|}, \tag{5.15}$$

where $\|\mathbf{i}\| = \|\mathbf{r}\| = 1$ and **i** is the direction of the incident ray and **r** is the direction of the deflected ray, and n_A and n_B are the refractive indices before and after deflection. If $n_A \neq n_B$, the ray is refracted, and if $n_A = n_B$, the ray is reflected. This method can then be used to design different types of aplanatic optics, such as RR, XR, RX, or XX.

Figure 5.29 shows the same optic as in Figure 5.27, but now used as a solar concentrator, when combined with an inverted glass pyramid *p* attached to the solar cell *c*. This last glass optical element *p* homogenizes the illuminance on the solar cell while further increasing concentration. CPC-type optics *p* may also be used [22]. Practical implementations of concentrators using an aplanatic XX design have been optimized for a geometrical concentration of 500× and a 1-cm^2 square solar cell [23, 24].

Figure 5.30a shows another example where $n_1 = n_2 = 1$ and $n_3 = 1.5$. This is, therefore, an XR concentrator. Now the cell is immersed in a medium with refractive

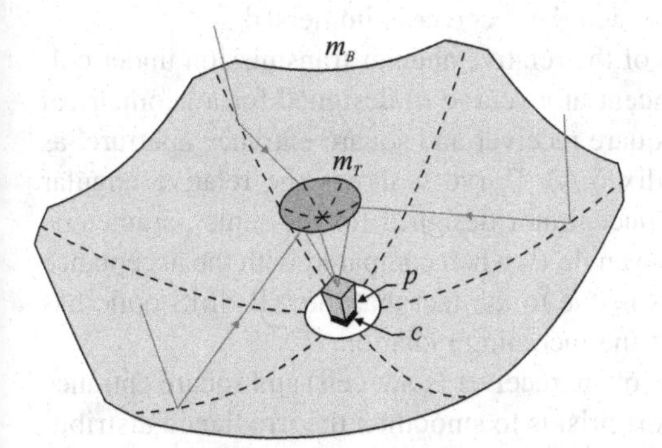

Figure 5.29 3D view of an aplanatic XX optic used as a concentrator when combined with a inverted glass pyramid *p* and receiver (solar cell) *c*.

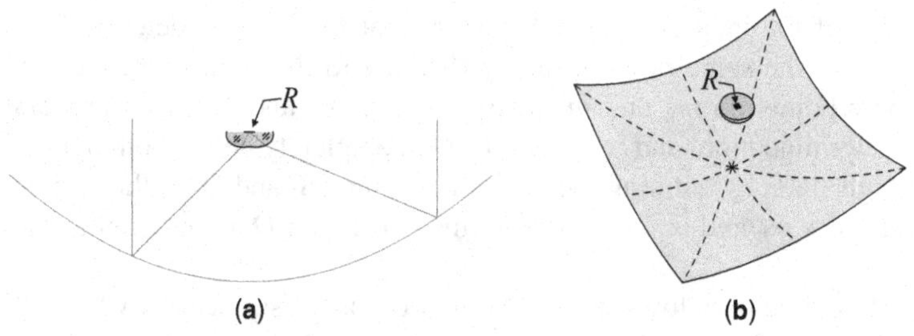

Figure 5.30 XR concentrator. (a) 2D section and (b) a 3D model resulting from giving the 2D section circular symmetry. The primary mirror is cut with a square shape, and the receiver is also given a square shape. Although the optical surfaces are different, the XR SMS and XR aplanatic look similar.

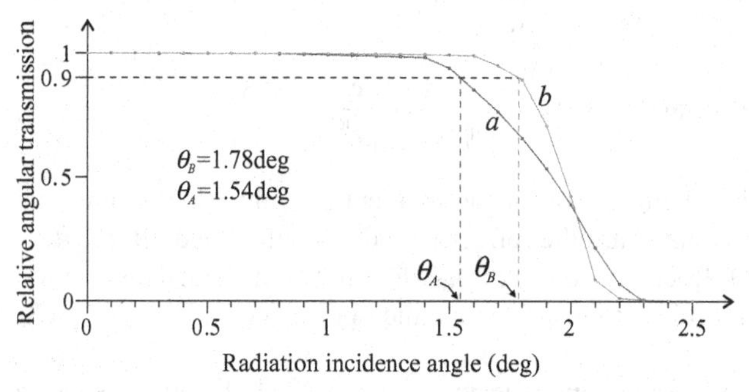

Figure 5.31 Relative angular transmissions for a XR aplanatic concentrator (curve *a*) and a XR SMS concentrator (curve *b*) under different incidence angles of collimated light. In this example, the optics were designed for a concentration of 1000× and the same parameters.

index $n_3 > 1$ that increases the acceptance angle by a factor n_3 (in the 2D case and n_3^2 in the 3D case). Note that for small acceptance angles, arcsin $\theta \approx \theta$ and, therefore, from Equation (5.3), the maximum acceptance angle increases (approximately) linearly with the refractive index in which the receiver is immersed.

Figure 5.31 shows an example of the relative angular transmission under collimated light of an XR aplanatic concentrator (curve *a*) designed for a geometrical concentration of 1000× and with a square receiver and square entrance aperture, as shown in Figure 5.30b (see Appendix 5.A). Curve *b* shows the relative angular transmission curve of an XR SMS concentrator designed for the same parameters. The SMS optic has a wider acceptance angle θ_B when compared with the acceptance angle θ_A for the aplanatic optic. This is due to the fact that the XR SMS optic has a better control over the edge rays of the incoming radiation.

These XR concentrators with a square receiver (solar cell) and square entrance apperture may be combined with short prisms to smoothen the irradiance distribution on the cell. For example, one of these SMS optics, when combined with an

(a) (b)

Figure 5.32 (a) Prototype module of XR solar concentrators with optics designed by LPI. (b) Shows a rendered view of a single concentrator with its heat sink: a heat pipe with radial fins. The heat sink shadow falls on the secondary and solar cell, not on the primary mirror. Therefore, the heat sink does not increase the shadow produced by the secondary.

ultra-short prism, optimized for a concentration of 890×, and manufactured, shows an acceptance of ±1.73°, obtained by measurements [25].

Figure 5.32a shows an array of prototype XR concentrators. Figure 5.32b shows a rendered view of a single unit with its heat sink: a heat pipe with radial fins.

The heat sink is dimensioned so that its shadow falls on the secondary refractive optic and, therefore, does not add to the shadow on the primary mirror.

5.5 3D MIÑANO–BENITEZ DESIGN METHOD APPLIED TO HIGH SOLAR CONCENTRATION

The most advanced method of nonimaging optics with practical applications is currently the Miñano-Benítez (or SMS) method, which allows the design of two (or more) simultaneous free-form surfaces [1, 2, 7–10].

As described in Chapter 4, the SMS design method couples two freeform input wave fronts into two output freeform wave fronts. These wave fronts are typically described mathematically as nonuniform rational b-spline (NURBS) surfaces, and the calculations are, in general, complex. However, in the case of solar concentration, the input wave fronts are typically flat and the output wave fronts are points, which simplifies the method considerably and allows for the analytical calculation of the points in the SMS chains. Below is a description of this simplified method, typically used in the design of solar concentrators.

Figure 5.33 shows an example of the basic element of an SMS calculation for flat input wave fronts and point output wave fronts. The light rays in this figure are

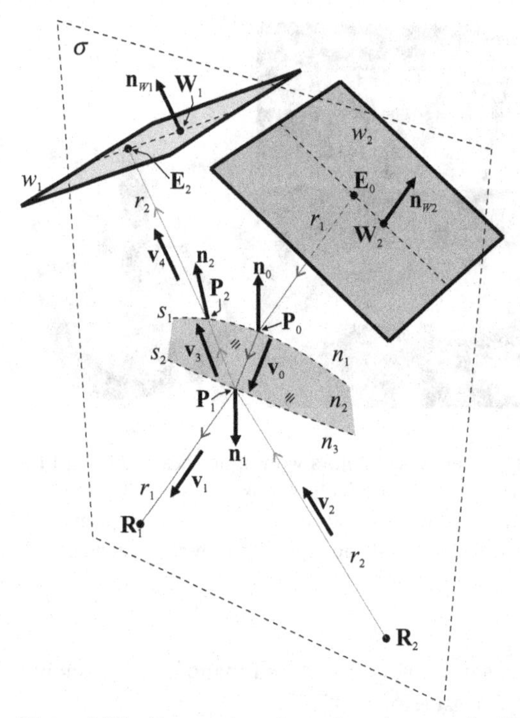

Figure 5.33 Perspective view of an element of a 2D-SMS chain.

2D and contained on plane σ. The rays perpendicular to flat wave fronts w_1 and w_2 are to be focused onto points $\mathbf{R}_2$ and $\mathbf{R}_1$. Flat wave front w_1 is defined by point $\mathbf{W}_1$ and normal $\mathbf{n}_{W1}$ and wave front w_2 is defined by point $\mathbf{W}_2$ and normal $\mathbf{n}_{W2}$. Both wave fronts are perpendicular to plane σ. The 2D-SMS method allows us to calculate two surfaces s_1 and s_2 in which s_1 separates two media of refractive indices n_1 and n_2 while surface s_2 separates two media of refractive indices n_2 and n_3. If $n_1 = n_2$, then s_1 is a mirror, otherwise it is a refractive surface. Also, if $n_2 = n_3$, then s_2 is a mirror, otherwise it is a refractive surface.

The calculation starts by choosing the position of point $\mathbf{P}_0$ (which will be on s_1) and the direction of its normal $\mathbf{n}_0$. Point $\mathbf{E}_0$ on wave front w_2 may be calculated by: $\mathbf{E}_0 = \text{islp}(\mathbf{P}_0,\mathbf{n}_{w2},\mathbf{W}_2,\mathbf{n}_{w2})$, where function islp($\mathbf{P},\mathbf{v},\mathbf{Q},\mathbf{n}$) returns the intersection between a straight line defined by a point $\mathbf{P}$ and a vector $\mathbf{v}$ and a plane defined by a point $\mathbf{Q}$ and normal $\mathbf{n}$ as:

$$\text{islp}(\mathbf{P}, \mathbf{v}, \mathbf{Q}, \mathbf{n}) = \mathbf{P} + \frac{(\mathbf{Q}-\mathbf{P})\cdot\mathbf{n}}{\mathbf{v}\cdot\mathbf{n}}\mathbf{v}s. \tag{5.16}$$

We calculate the direction of the deflected ray at $\mathbf{P}_0$ as $\mathbf{v}_0 = \text{dfl}(-\mathbf{n}_{w2},\mathbf{n}_0,n_1,n_2)$. Function dfl($\mathbf{i},\mathbf{n},n_A,n_B$) returns the direction of the deflected ray, given the incidence direction $\mathbf{i}$, the normal $\mathbf{n}$ to the surface, and the refractive indices n_A before and n_B after deflection. The ray is refracted if $n_A \neq n_B$ or reflected if $n_A = n_B$

$$\text{dfl}(\mathbf{i}, \mathbf{n}, n_A, n_B) = \begin{cases} \text{rfx}(\mathbf{i}, \mathbf{n}) & \text{if} \quad n_A = n_B \\ \text{rfr}(\mathbf{i}, \mathbf{n}, n_A, n_B) & \text{if} \quad n_A \neq n_B \end{cases}, \tag{5.17}$$

in which rfx(**i**,**n**) returns the direction of a reflected ray, given the direction **i** of the incident ray, and the normal **n** to the surface as [2]:

$$\text{rfx}(\mathbf{i}, \mathbf{n}) = \mathbf{i} - 2(\mathbf{i} \cdot \mathbf{n})\mathbf{n} \tag{5.18}$$

Function rfx(**i**,**n**,n_A,n_B) returns the direction of the refracted ray, given the direction **i** of the incident ray, the normal **n** to the surface, and the refractive indices n_A before and n_B after refraction as [2]:

$$\text{rfr}(\mathbf{i}, \mathbf{n}, n_A, n_B) = \begin{cases} \dfrac{n_A}{n_B}\mathbf{i} + \left(-(\mathbf{i} \cdot \mathbf{n}_S)\dfrac{n_A}{n_B} + \sqrt{\Delta} \right)\mathbf{n}_S & \text{if} \quad \Delta > 0 \\ \text{rfx}(\mathbf{i}, \mathbf{n}) & \text{if} \quad \Delta \leq 0 \end{cases}, \tag{5.19}$$

in which

$$\mathbf{n}_S = \begin{cases} \mathbf{n} & \text{if} \quad \mathbf{i} \cdot \mathbf{n} \geq 0 \\ -\mathbf{n} & \text{if} \quad \mathbf{i} \cdot \mathbf{n} < 0 \end{cases} \quad \text{and} \quad \Delta = 1 - \left(\frac{n_A}{n_B} \right)^2 \left[1 - (\mathbf{i} \cdot \mathbf{n}_S)^2 \right] \tag{5.20}$$

In the above expressions, $\|\mathbf{i}\| = \|\mathbf{n}\| = 1$.

Now we choose a value for the optical path length S_{W2R1} between wave front w_2 and point $\mathbf{R}_1$. The optical path length S_{01} between point $\mathbf{P}_0$ and $\mathbf{R}_1$ is $S_{01} = S_{w2R1} - n_1[\mathbf{E}_0,\mathbf{P}_0]$ in which $[\mathbf{X},\mathbf{Y}]$ is the distance between points **X** and **Y** given by $[\mathbf{X},\mathbf{Y}] = \|\mathbf{Y} - \mathbf{X}\|$ and where $\|\mathbf{v}\| = \sqrt{\mathbf{v} \cdot \mathbf{v}}$ is the magnitude of a vector.

Since we now have the position of point $\mathbf{P}_0$, direction $\mathbf{v}_0$ of the ray deflected at this point and the optical path length between $\mathbf{P}_0$ and $\mathbf{R}_1$, we can calculate a new point $\mathbf{P}_1$ on the bottom surface as $\mathbf{P}_1 = \text{coptpt}(\mathbf{P}_0,n_2,\mathbf{v}_0,\mathbf{R}_1,n_3,S_{01})$.

Function coptpt(**F**,n_A,**v**,**G**,n_B,S) returns the point (on a Cartesian oval) at which deflection of a light ray occurs, given the emission point **F**, the refractive index n_A at **F**, the direction **v** of the emitted ray, the point **G** onto which the light ray is deflected, the refractive index n_B at **G**, and the optical path length S between **F** and **G** as [2]:

$$\text{coptpt}(\mathbf{F}, n_A, \mathbf{v}, \mathbf{G}, n_B, S) = \begin{cases} \mathbf{F} + \dfrac{C_1 + \delta\sqrt{C_2\left(n_B^2 - n_A^2\right) + C_1^2}}{n_A^2 - n_B^2}\mathbf{v} & \text{if} \quad n_A \neq n_B \\ \mathbf{F} + \dfrac{(S/n_A)^2 - (\mathbf{F}-\mathbf{G})\cdot(\mathbf{F}-\mathbf{G})}{2(S/n_A + (\mathbf{F}-\mathbf{G})\cdot\mathbf{v})}\mathbf{v} & \text{if} \quad n_A = n_B \end{cases}, \tag{5.21}$$

with

$$\begin{aligned} C_1 &= n_A S + n_B^2(\mathbf{F}-\mathbf{G})\cdot\mathbf{v} \\ C_2 &= S^2 - n_B^2(\mathbf{F}-\mathbf{G})\cdot(\mathbf{F}-\mathbf{G}), \end{aligned} \tag{5.22}$$

and

$$\begin{aligned} \delta &= -1 \quad \text{for} \quad n_A > n_B \\ \delta &= 1 \quad \text{for} \quad n_A < n_B \end{aligned}. \tag{5.23}$$

Since we now have the position of $\mathbf{P}_1$, we can calculate the unit vector $\mathbf{v}_1$ from $\mathbf{P}_1$ to $\mathbf{R}_1$ as $\mathbf{v}_1 = \text{nrm}(\mathbf{R}_1 - \mathbf{P}_1)$, where

$$\text{nrm}(\mathbf{v}) = \frac{\mathbf{v}}{\|\mathbf{v}\|} = \frac{\mathbf{v}}{\sqrt{\mathbf{v} \cdot \mathbf{v}}}. \tag{5.24}$$

The normal at point $\mathbf{P}_1$ can now be calculated by $\mathbf{n}_1 = \text{dflnrm}(\mathbf{v}_0,\mathbf{v}_1,n_2,n_3)$, where function $\text{dflnrm}(\mathbf{i},\mathbf{r},n_A,n_B)$ is defined by Equation (5.15). At this point, we have calculated the path of ray $\mathbf{E}_0$–$\mathbf{P}_0$–$\mathbf{P}_1$–$\mathbf{R}_1$.

We can now calculate vector $\mathbf{v}_2 = \text{nrm}(\mathbf{P}_1 - \mathbf{R}_2)$ and deflect this ray at $\mathbf{P}_1$, obtaining the direction of $\mathbf{v}_3$ as $\mathbf{v}_3 = \text{dfl}(\mathbf{v}_2,\mathbf{n}_1,n_3,n_2)$. We now choose a value S_{W1R2} for the optical path length between flat wave front w_1 and point $\mathbf{R}_2$. The optical path length S_{11} between point $\mathbf{P}_1$ and wave front w_1 is $S_{11} = S_{W1R2} - n_3[\mathbf{P}_1,\mathbf{R}_2]$. We can now calculate a new point $\mathbf{P}_2$ on the top surface as $\mathbf{P}_2 = \text{coptsl}(\mathbf{P}_1,n_2,\mathbf{v}_3,\mathbf{W}_1,n_1,\mathbf{n}_{W1},S_{11})$.

Function $\text{coptsl}(\mathbf{F},n_A,\mathbf{v},\mathbf{Q},n_B,\mathbf{n},S)$ returns the point (on a Cartesian oval) at which deflection of a light ray occurs, given the emission point $\mathbf{F}$, the refractive index n_A at $\mathbf{F}$, the direction $\mathbf{v}$ of the emitted ray, a point $\mathbf{Q}$ on a flat wave front onto which the light ray is headed, the refractive index n_B at the flat wave front, its normal $\mathbf{n}$, and the optical path length S between $\mathbf{F}$ and the flat wave front as [2]:

$$\text{coptsl}(\mathbf{F}, n_A, \mathbf{v}, \mathbf{Q}, n_B, \mathbf{n}, S) = \mathbf{F} + \frac{S - n_B(\mathbf{Q} - \mathbf{F}) \cdot \mathbf{n}}{n_A - n_B \mathbf{v} \cdot \mathbf{n}} \mathbf{v}. \tag{5.25}$$

Vector $\mathbf{v}_4 = \mathbf{n}_{W1}$ points in the direction perpendicular to wave front w_1. Point $\mathbf{E}_2$ is given by $\mathbf{E}_2 = \text{islp}(\mathbf{P}_2,\mathbf{v}_4,\mathbf{W}_1,\mathbf{n}_{W1})$. The normal to the surface at point $\mathbf{P}_2$ is given by $\mathbf{n}_2 = \text{dflnrm}(\mathbf{v}_3,\mathbf{v}_4,n_2,n_1)$.

This process can now be repeated by deflecting at point $\mathbf{P}_2$ a ray r_3 coming from wave front w_2, as shown in Figure 5.34. This allows us to calculate a new point $\mathbf{P}_3$ and its normal on the bottom surface. Deflecting at this new point $\mathbf{P}_3$ a ray r_4 coming from $\mathbf{R}_2$ allows us to calculate a new point $\mathbf{P}_4$ on the top surface, and so on. This procedure defines an SMS chain.

Going back to initial point $\mathbf{P}_0$ and deflecting there a ray r_A coming from w_1, we can calculate a new point $\mathbf{P}_A$ and its normal on the bottom surface. Deflecting there a ray r_B coming from $\mathbf{R}_1$, we can calculate a new point $\mathbf{P}_B$ and its normal on the top surface. This process now continues as we calculate the points of an SMS chain to the right of $\mathbf{P}_0$. This process defines the points on two curves, one on the top optical surface and one on the bottom optical surface.

The calculation above was 2D, since all the points and light rays were contained in plane σ. Choosing an initial point $\mathbf{Q}_0$ and its normal $\mathbf{m}_0$ away from plane σ results in an SMS chain that is 3D. The calculations for the SMS chain are the same as before. Figure 5.35 shows one such example.

Just like we did for the case of the two-dimensional geometry, we deflect at point $\mathbf{Q}_0$ a ray t_1 perpendicular to wave front w_2. This allows us to calculate point

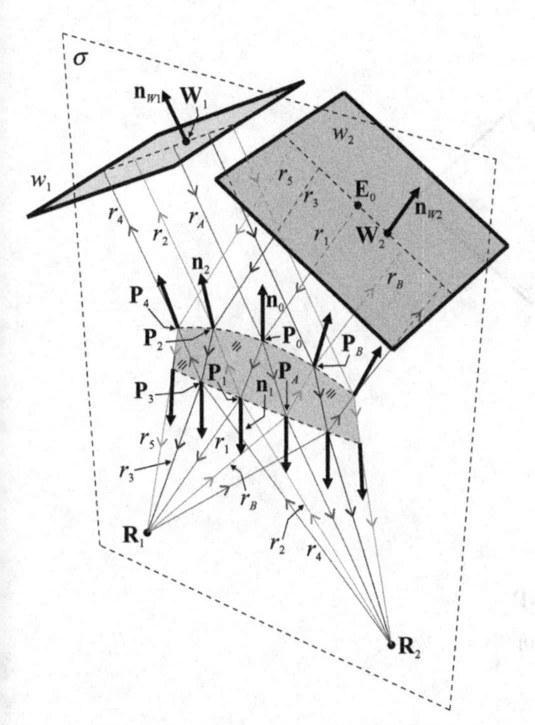

Figure 5.34 Perspective view of a 2D-SMS chain.

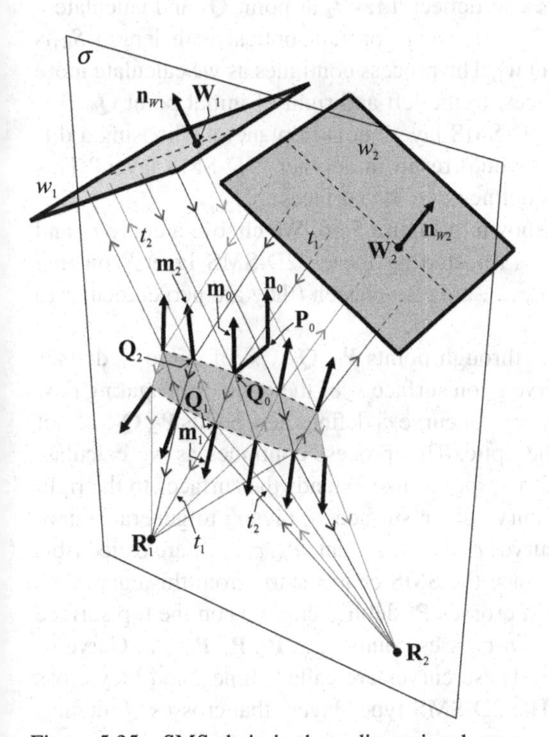

Figure 5.35 SMS chain in three-dimensional space.

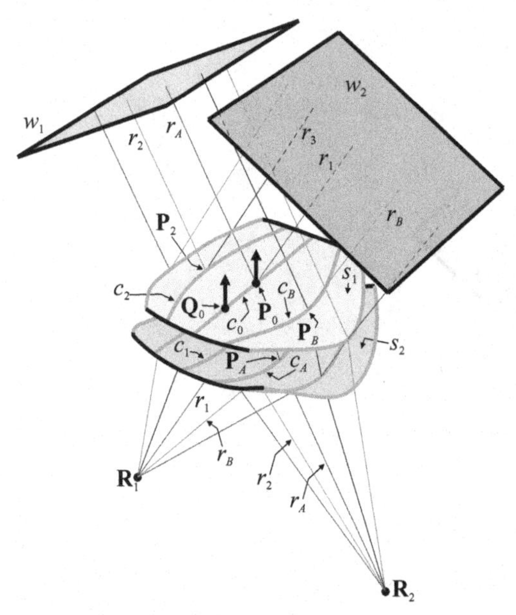

Figure 5.36 Three-dimensional SMS optic.

$\mathbf{Q}_1$ and its normal $\mathbf{m}_1$ on the second surface, by imposing constant optical path length S_{W2R1} between w_2 and $\mathbf{R}_1$. Now, we can deflect a ray t_2 at point $\mathbf{Q}_1$ and calculate a new point $\mathbf{Q}_2$ on the top surface by imposing constant optical path length S_{W1R2} between point $\mathbf{R}_2$ and flat wave front w_1. The process continues as we calculate more points on the first and second surfaces, to the left and right of initial point $\mathbf{Q}_0$.

This process creates a new "2D-SMS layer" next to plane σ. Choosing a different point $\mathbf{Q}_0$ away from plane σ would result in another "2D-SMS layer." This "layering" of 2D-SMS-type chains defines two 3D surfaces.

The result of this process is shown in Figure 5.36. We choose a curve c_0 and choose the initial points $\mathbf{P}_0$, $\mathbf{Q}_0$, . . . for starting these "2D-SMS layers" on that curve. The normals to points $\mathbf{P}_0$, $\mathbf{Q}_0$, . . . are also chosen (they are perpendicular to curve c_0).

Propagation of rays r_1, t_1, . . . through points $\mathbf{P}_0$, $\mathbf{Q}_0$, . . . of curve c_0 defines new points $\mathbf{P}_1$, $\mathbf{Q}_1$, . . . of a new curve c_1 on surface s_2 of the optic. Propagating now rays r_2, t_2, . . . through points $\mathbf{P}_1$, $\mathbf{Q}_1$, . . . of curve c_1 defines new points $\mathbf{P}_2$, $\mathbf{Q}_2$, . . . of a new curve c_2 on surface s_1 of the optic. This process continues as we calculate more curves on surfaces s_2 and s_1. The process also extends the surfaces to the right of c_0 by using rays r_A, to generate curve c_A on surface s_2, rays r_B to generate curve c_B on surface s_1, and so on. These curves c_0, c_1, c_2, . . . and c_A, c_B, . . . are called ribs. The initial curve c_0 is the seed rib since the SMS chains starts from this curve.

The 2D-SMS-type "layer" that crosses $\mathbf{P}_0$ defines curve a_P on the top surface and b_P on the bottom surface. Curve a_P crosses points . . . , $\mathbf{P}_2$, $\mathbf{P}_0$, $\mathbf{P}_B$, . . . Curve b_P crosses points . . . , $\mathbf{P}_3$, $\mathbf{P}_1$, $\mathbf{P}_A$, ... These curves are called spines, and they cross the ribs, as shown in Figure 5.37. The 2D-SMS-type "layer" that crosses $\mathbf{Q}_0$ defines

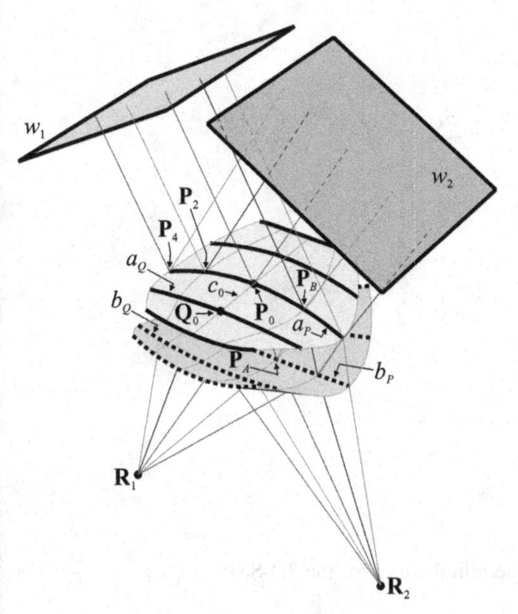

Figure 5.37 3D-SMS ribs and spines.

spine a_Q on the top and b_Q on the bottom surface. Other points on the seed rib c_0 define other spines on the top and bottom surfaces.

There is no specific rule to defining the initial curve (seed rib) c_0. In the simple example in Figure 5.36, it is simply a straight line perpendicular to plane σ. A possibility, however, is to use the SMS method and two other wave fronts to define it. Figure 5.38 shows two input wave fronts w_{C1} and w_{C2} and two output wave fronts (in this case two points) $\mathbf{R}_{C2}$ and $\mathbf{R}_{C1}$. With these wave fronts, we calculate points on two optical curves c_0 and b_0 using the SMS method (as in Figs. 5.33 and 5.34) and take points P_I (ant their normals) c_0 for the 3D-SMS calculation.

Using this seed rib, we can now calculate the optical surfaces using the method in Figure 5.35 and 5.36. The result is shown in Figure 5.39, in which we couple input wave fronts w_1 and w_2 onto output wave fronts (in this particular case two points) $\mathbf{R}_2$ and $\mathbf{R}_1$.

This method allows us to have some control over the light emitted perpendicularly to wave fronts w_{C1} and w_{C2} that we want concentrated onto $\mathbf{R}_{C2}$ and $\mathbf{R}_{C1}$. In this example, the normals to input wave fronts w_{C1} and w_{C2} for the initial curve and also input wave fronts w_1 and w_2 for calculating the 3D-SMS curves all make an angle θ to the vertical (the direction of the incoming light). The optic is, therefore, designed for an acceptance $\pm\theta$.

The same concentrator can be used for different acceptance angles (within a certain range) simply by changing the receiver dimensions [1]. By designing for a small receiver and acceptance angle, the calculated SMS points are closer together, giving a better definition of the optical surfaces. In particular, we can design a concentrator for a small receiver defined by points $\mathbf{R}_{C1}$, $\mathbf{R}_{C2}$, $\mathbf{R}_1$, and $\mathbf{R}_2$ (and

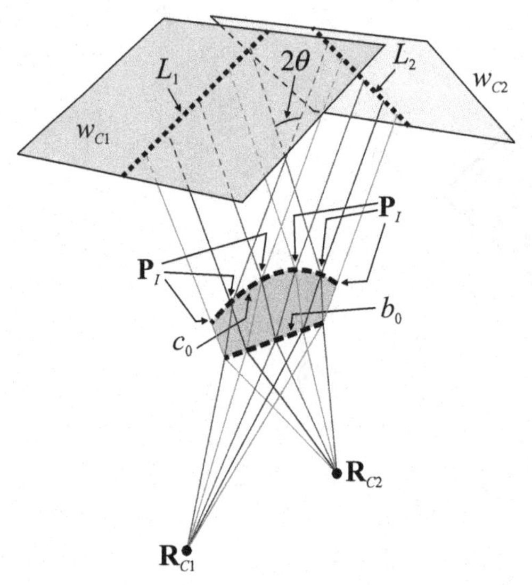

Figure 5.38 Wave fronts for defining the initial curve for the 3D-SMS.

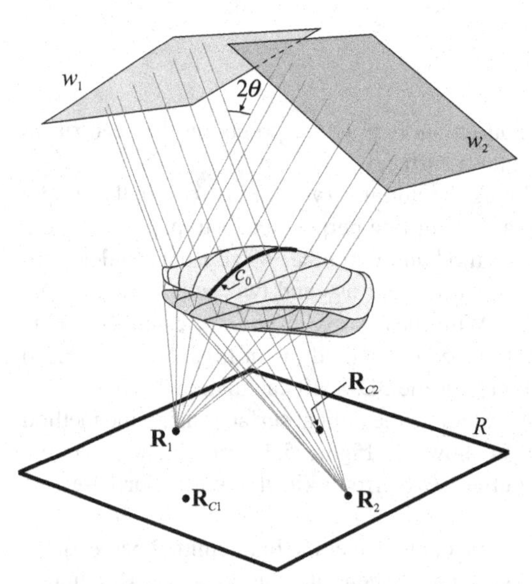

Figure 5.39 3D-SMS design based on an initial curve also designed using the SMS design method.

corresponding small acceptance $\pm\theta$), and then use it with a larger receiver R, resulting in a wider acceptance angle, since incoming light at wider angles will still intersect a larger receiver (Fig. 5.39).

The calculation method described above can also be used to design optics with other geometries. Figure 5.40 shows an XR configuration [26], described in detail in Chapter 4. The 2D-SMS design starts by defining the positions of the foci $\mathbf{R}_{C1}$

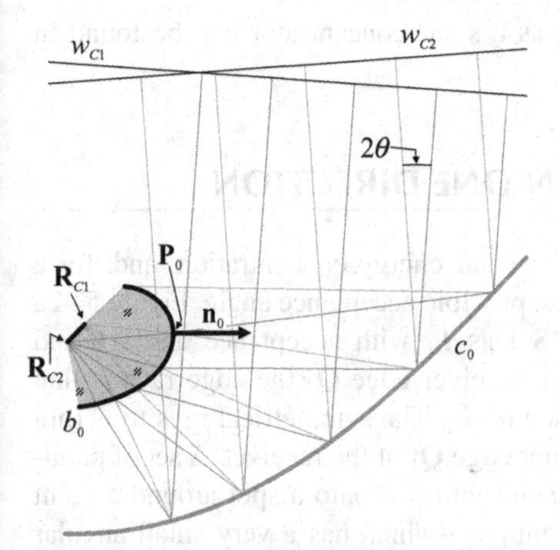

Figure 5.40 Construction of an XR 2D-SMS concentrator. The points of curve c_0 (or b_0) can later be used for the design of a 3D free-form XR SMS optic.

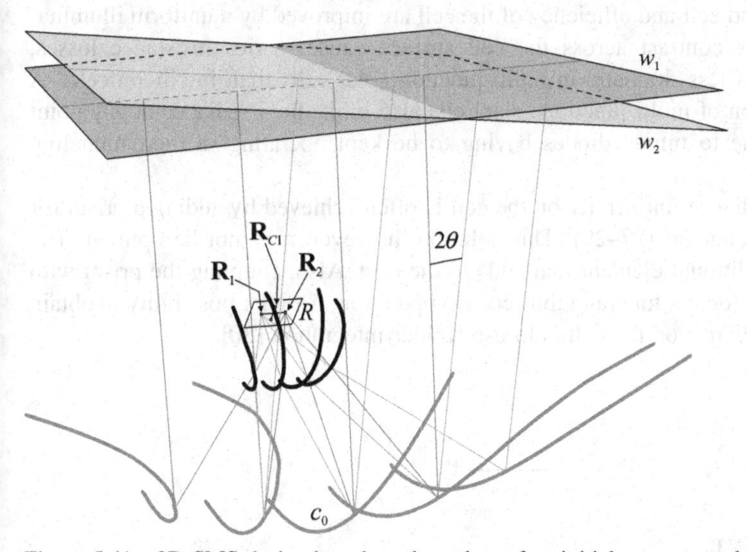

Figure 5.41 3D-SMS design based on the points of an initial curve c_0 and input wave fronts w_1 and w_2 and two output point wave fronts $\mathbf{R}_1$ and $\mathbf{R}_2$.

and $\mathbf{R}_{C2}$ and the input wave fronts w_{C1} and w_{C2}. We also choose the position of point $\mathbf{P}_0$ and its normal $\mathbf{n}_0$ and the refractive index of the secondary optic. With these elements, we can calculate the mirror c_0 and refractive curve b_0 using the SMS method.

We now take the points of curve c_0 as initial curve for the 3D-SMS design, as shown in Figure 5.41 (we could also take the points of curve b_0 on the refractive curve as the initial curve for the 3D-SMS design). This design will focus wave fronts w_1 and w_2 onto points $\mathbf{R}_2$ and $\mathbf{R}_1$ contained in the larger receiver R. The design has an acceptance $\pm\theta$ defined by the input wave fronts.

A detailed description of this optic as a solar concentrator can be found in Chapter 4.

5.6 KÖHLER INTEGRATION IN ONE DIRECTION

Nonimaging optics are very well suited for solar energy concentration, and, for a given concentration, they provide the widest possible acceptance angle. Figure 5.42a shows an example of a nonimaging SMS lens L_1 with acceptance angle θ and receiver $\mathbf{Q}_1\mathbf{Q}_2$ [2]. It concentrates onto the receiver edge $\mathbf{Q}_2$ the edge rays r_1 that make an angle θ to the vertical before reaching L_1. The symmetrical rays to r_1 (not shown) would be concentrated onto the other edge $\mathbf{Q}_1$ of the receiver. A set of parallel rays r_2 inside the acceptance angle are concentrated onto a spot around a point $\mathbf{Q}$ on the receiver. This means that also sunlight (which has a very small angular aperture) will be concentrated onto a spot on the receiver. This may be a problem for some applications in which a uniform illumination of the receiver is important.

When concentrating light onto solar cells, the long-term durability of the concentrator and cell and efficiency of the cell are improved by a uniform illumination. High flux contrast across the cell surface causes series resistance losses, although this is less dramatic in multi-junction solar cells than in silicon cells. A proper operation of multi-junction solar cells also limits the irradiance at any point of the cell (due to tunnel diodes having to be kept operating in their tunneling regime) [12].

The irradiance uniformity on the cell is often achieved by adding a prism (or kaleidoscope) element [27–29]. This solution, however, may not be optimal. The prism is an additional element that adds to the cost. Also, coupling the prism with the cell may become a fine (and thus costly) operation. Another possibility to obtain a uniform irradiance on the cell is to use Köhler integration [30].

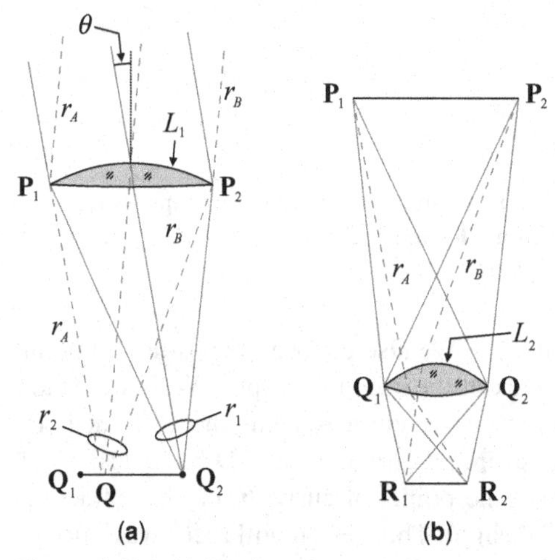

Figure 5.42 (a) Nonimaging lens for an acceptance angle θ and receiver $\mathbf{Q}_1\mathbf{Q}_2$. (b) Nonimaging lens for a source $\mathbf{P}_1\mathbf{P}_2$ and receiver $\mathbf{R}_1\mathbf{R}_2$.

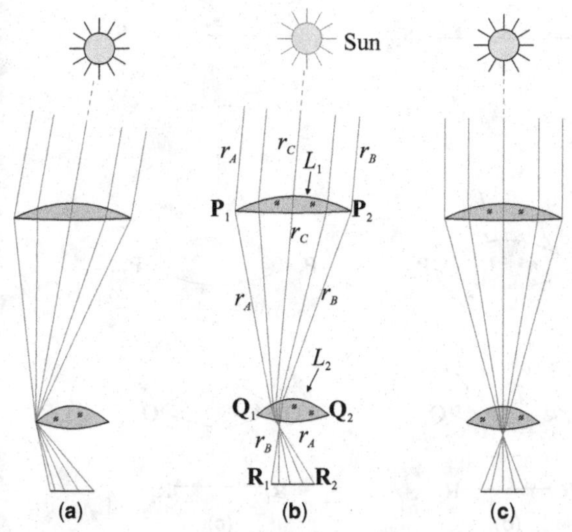

Figure 5.43 Integrator composed of two nonimaging optics placed in series. Panels a, b, and c show the behavior of the concentrator for different incidence angles of the sunlight.

A way around this problem is to consider another nonimaging SMS lens L_2 as shown in Figure 5.42b. It is designed for a finite source $\mathbf{P}_1\mathbf{P}_2$ and receiver $\mathbf{R}_1\mathbf{R}_2$. The edge rays coming from the edges $\mathbf{P}_1$ and $\mathbf{P}_2$ of the source are concentrated onto the edges $\mathbf{R}_2$ and $\mathbf{R}_1$ of the receiver, respectively.

These two lenses may now be combined as shown in Figure 5.43b, in which the source $\mathbf{P}_1\mathbf{P}_2$ for lens L_2 is now lens L_1, and the receiver $\mathbf{Q}_1\mathbf{Q}_2$ of lens L_1 is now lens L_2. The resulting optical train has the acceptance angle θ of lens L_1 and the receiver $\mathbf{R}_1\mathbf{R}_2$ of lens L_2.

We now consider some rays r_A, r_B, r_C, . . . which are parallel before reaching lens L_1. Incoming ray r_A is refracted at the edge $\mathbf{P}_1$ of the top lens L_1. This is also an edge ray for lens L_2 (since it comes from the edge $\mathbf{P}_1$ of its source) and, therefore, is redirected toward the edge $\mathbf{R}_2$ of its receiver. Also, incoming ray r_B is refracted at the edge $\mathbf{P}_2$ of the top lens L_1. Again, this is an edge ray for lens L_2 (since it comes from the edge $\mathbf{P}_2$ of its source) and, therefore, is redirected toward the edge $\mathbf{R}_1$ of its receiver. Since rays r_A and r_B are redirected toward edges $\mathbf{R}_2$ and $\mathbf{R}_1$ of the receiver, those rays r_C in between them must be redirected toward points in between $\mathbf{R}_2$ and $\mathbf{R}_1$.

The result of combining lenses L_1 and L_2 is then an optical system that produces a quite uniform illuminance on the receiver, even for a set of parallel rays inside the acceptance angle of the optical train. This means that as the sun moves inside the acceptance angle of the optic, the illuminance on the receiver will remain very uniform, as shown in Figure 5.43.

A similar configuration can also be used in the case of a finite source $\mathbf{S}_1\mathbf{S}_2$ at a finite distance (instead of an acceptance angle θ), as shown in Figure 5.44. Now top lens L_1 has source $\mathbf{S}_1\mathbf{S}_2$ and target $\mathbf{Q}_1\mathbf{Q}_2$, while bottom lens L_2 has source $\mathbf{P}_1\mathbf{P}_2$ and target $\mathbf{R}_1\mathbf{R}_2$. The light emitted from different points $\mathbf{S}_1$, $\mathbf{S}_2$, $\mathbf{S}_A$, $\mathbf{S}_B$, . . . on the

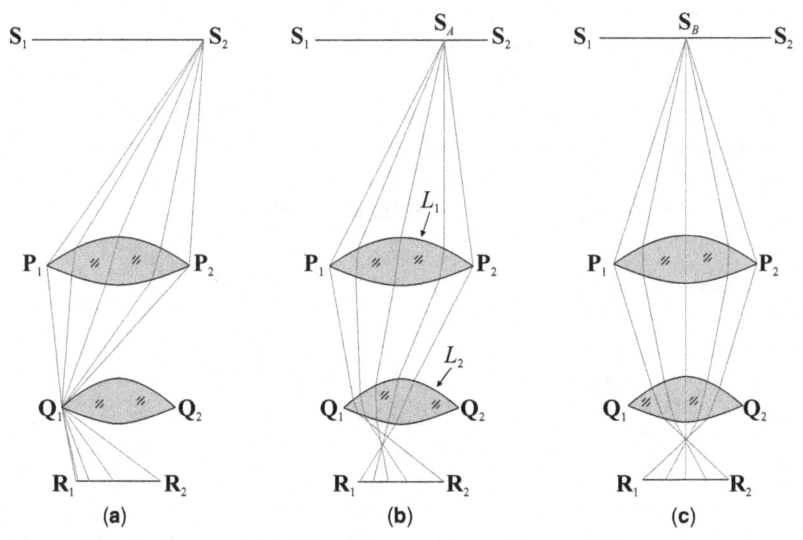

Figure 5.44 Integrator for a finite source.

source is spread over the entire receiver $\mathbf{R}_1\mathbf{R}_2$, greatly improving illuminance uniformity over what a single nonimaging lens could do. Conservation of étendue in this system implies that the étendue of the light from $\mathbf{S}_1\mathbf{S}_2$ to $\mathbf{P}_1\mathbf{P}_2$ equals that from $\mathbf{P}_1\mathbf{P}_2$ to $\mathbf{Q}_1\mathbf{Q}_2$, and also equals that from $\mathbf{Q}_1\mathbf{Q}_2$ to $\mathbf{R}_1\mathbf{R}_2$.

The nonimaging lenses used above only guarantee sharp focusing of the edges of the source onto the edges of the receiver. In general, they will not form a good image of the source on the receiver. Figure 5.45 shows a diagrammatic representation of what an ideal imaging system would do. If optics L_1 and L_2 were perfect imaging devices, L_1 would produce on $\mathbf{Q}_1\mathbf{Q}_2$ an image of object $\mathbf{S}_1\mathbf{S}_2$, and optic L_2 would then produce on $\mathbf{R}_1\mathbf{R}_2$ an image of $\mathbf{P}_1\mathbf{P}_2$. Since L_1 is ideal, it focuses every point of object $\mathbf{S}_1\mathbf{S}_2$ onto the corresponding point of the image $\mathbf{Q}_1\mathbf{Q}_2$. In particular, it focuses $\mathbf{S}_1$ onto $\mathbf{Q}_2$ and $\mathbf{S}_2$ onto $\mathbf{Q}_1$ (and also mid-point $\mathbf{M}_S$ onto mid-point $\mathbf{M}_Q$). This ensures that all the light emitted by $\mathbf{S}_1\mathbf{S}_2$ and captured by $\mathbf{P}_1\mathbf{P}_2$ will hit L_2 and will not be lost (a nonimaging lens can also accomplish this). Optic L_2 will then distribute this light on $\mathbf{R}_1\mathbf{R}_2$. On the other hand, since L_2 is ideal, it focuses every point of its object $\mathbf{P}_1\mathbf{P}_2$ onto the corresponding point of the image $\mathbf{R}_1\mathbf{R}_2$. In particular, it focuses $\mathbf{P}_1$ onto $\mathbf{R}_2$ and $\mathbf{P}_2$ onto $\mathbf{R}_1$ and guarantees that all the light that crosses L_1 toward L_2 ends up on $\mathbf{R}_1\mathbf{R}_2$, without losses (a nonimaging lens can also accomplish this). However, an ideal imaging optic L_2 also concentrates onto $\mathbf{R}$ all the light that reaches point $\mathbf{P}$ of L_1. This means that the illuminance at $\mathbf{R}$ will be proportional to the illuminance that $\mathbf{S}_1\mathbf{S}_2$ produces at $\mathbf{P}$. Therefore, if aperture $\mathbf{P}_1\mathbf{P}_2$ is uniformly illuminated by the source $\mathbf{S}_1\mathbf{S}_2$, then the image of $\mathbf{P}_1\mathbf{P}_2$ on $\mathbf{R}_1\mathbf{R}_2$ will also be uniformly illuminated.

A possible implementation of these principles is the concentrator in Figure 5.46 [31], called SILO (SIngLe Optical surface). A Fresnel lens primary images the sun as it "moves" in the sky onto a secondary refractive optic. This optic creates an

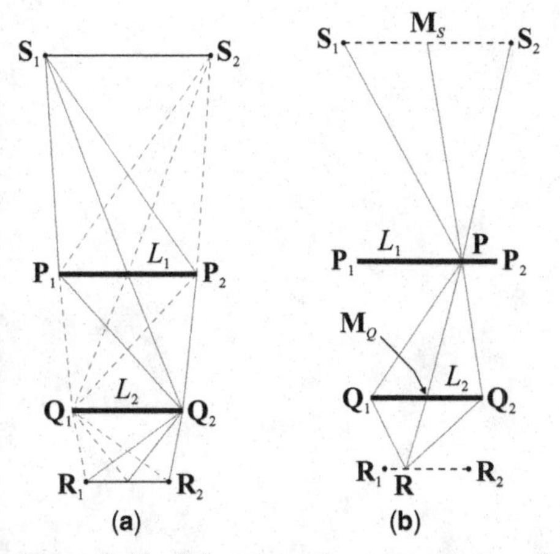

Figure 5.45 Diagrammatic representation of an ideal imaging integrator.

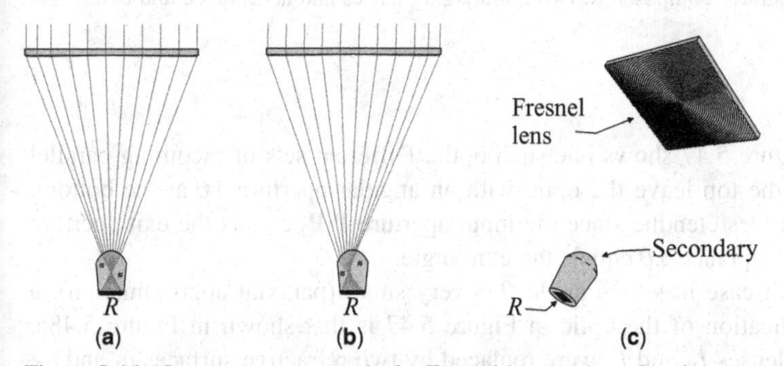

Figure 5.46 Integrator composed of a Fresnel lens and a refractive secondary. (a) Normal incidence of light on the lens. (b) Light incidence at an angle. (c) Three-dimensional view of the same.

image of the Fresnel lens primary onto the receiver *R*. Figure 5.46a shows the behavior of the system for normal incidence of light on the Fresnel lens. Figure 5.46b shows the same when light hits the Fresnel lens at an angle (similarly to what is shown in Fig. 5.43b).

Figure 5.46c shows a 3D view of the same optic, with a square Fresnel lens and square receiver *R*. Since the secondary optic images the Fresnel lens onto the receiver *R* and the lens is cut with a square shape, its image on the receiver will also be square, matching the shape of the receiver. This is an important characteristic when using square solar cells as receivers, as it improves efficiency.

The optics shown above have only one lens as primary and another lens as secondary. It is, however, possible to design other optics with several elements in the primary and a corresponding number of elements in the secondary.

A limit case of the optic shown in Figure 5.43 is when receiver $\mathbf{R}_1\mathbf{R}_2$ moves away from the optic increasing in size in such a way that in the limit case it subtends

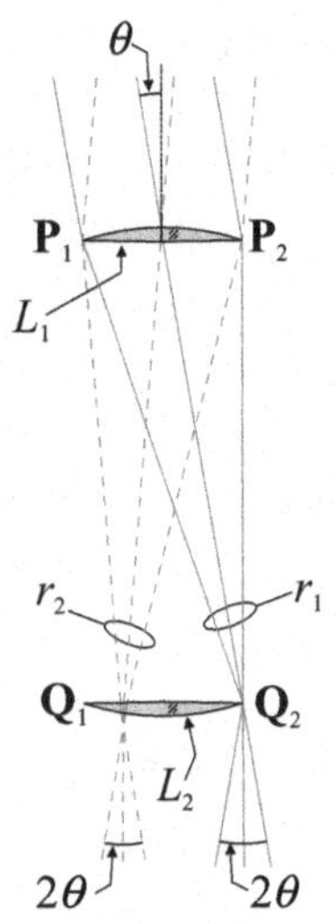

Figure 5.47 Integrator composed of two nonimaging lenses and acceptance and exit angles $\pm\theta$.

an angle $\pm\theta$. Figure 5.47 shows one such optic. Different sets of incoming parallel rays r_1 or r_2 at the top leave the optic with an angular aperture $\pm\theta$ at the bottom. This optic conserves étendue since the input aperture $\mathbf{P}_1\mathbf{P}_2$ equals the exit aperture $\mathbf{Q}_1\mathbf{Q}_2$ and the acceptance $\pm\theta$ equals the exit angle.

In the limit case in which angle θ is very small (paraxial approximation), a possible simplification of the optic in Figure 5.47 is that shown in Figure 5.48a. Here, the SMS lenses L_1 and L_2 were replaced by two refractive surfaces a_1 and a_2. Surface a_1 images an infinite object subtending an angle $\pm\theta$ onto a_2, and a_2 images a_1 onto an infinite image, also subtending an angle $\pm\theta$.

Figure 5.48b shows an array of integrators like the one in Figure 5.48a. These produce an output with angular aperture $\pm\theta$ for any set of parallel rays coming from any direction inside the acceptance $\pm\theta$.

These arrays of integrators may also be designed for finite size sources and targets. Figure 5.49 shows one such optic with a finite source $\mathbf{S}_1\mathbf{S}_2$ and a finite receiver $\mathbf{R}_1\mathbf{R}_2$. Just like the optic in Figure 5.48b, this optic is composed of several elements, but now having variable geometry. One of these elements is formed by surfaces a_1 and a_2. Refractive surface a_1 focuses to $\mathbf{Q}$ (mid-point of $\mathbf{Q}_1\mathbf{Q}_2$) the rays coming from $\mathbf{M}_S$ (mid-point of $\mathbf{S}_1\mathbf{S}_2$). Refractive surface a_2 focuses to $\mathbf{M}_R$ (mid-point of $\mathbf{R}_1\mathbf{R}_2$) the rays coming from $\mathbf{P}$ (mid point of $\mathbf{P}_1\mathbf{P}_2$) (Fig. 5.49a). We must then have for surface a_1:

$$[\mathbf{M}_S, \mathbf{P}_1] + n[\mathbf{P}_1, \mathbf{Q}] = [\mathbf{M}_S, \mathbf{P}_2] + n[\mathbf{P}_2, \mathbf{Q}], \tag{5.26}$$

where n is the refractive index of the optic. And for surface a_2:

$$n[\mathbf{P}, \mathbf{Q}_1] + [\mathbf{Q}_1, \mathbf{M}_R] = n[\mathbf{P}, \mathbf{Q}_2] + [\mathbf{Q}_2, \mathbf{M}_R]. \tag{5.27}$$

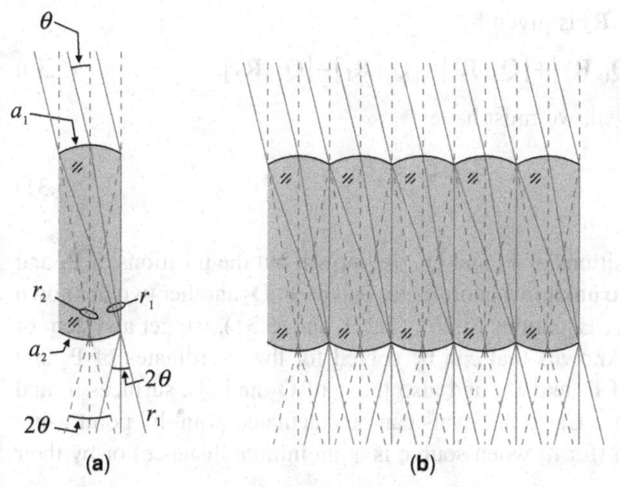

Figure 5.48 (a) Imaging integrator. (b) Array of imaging integrators.

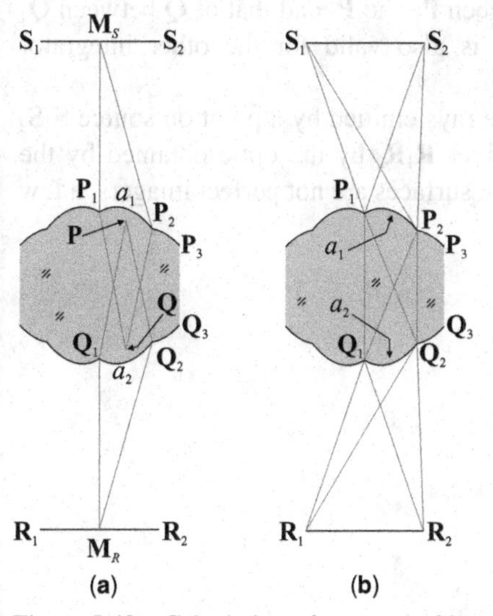

Figure 5.49 Calculation of an array of integrator elements for finite source $\mathbf{S}_1\mathbf{S}_2$ and receiver $\mathbf{R}_1\mathbf{R}_2$.

This element formed by a_1 and a_2 must also conserve étendue (Fig. 5.49b). The étendue from $\mathbf{S}_1\mathbf{S}_2$ to a_1 bound by points $\mathbf{P}_1$ and $\mathbf{P}_2$ is given by:

$$\xi = 4x\sin\theta = 2([[\mathbf{P}, \mathbf{R}_1]] - [[\mathbf{P}, \mathbf{R}_2]]). \tag{5.28}$$

The étendue from a_1 to a_2 bound by points $\mathbf{Q}_1$ and $\mathbf{Q}_2$ is given by:

$$\xi_2 = n([\mathbf{P}_1, \mathbf{Q}_2] + [\mathbf{P}_2, \mathbf{Q}_1] - [\mathbf{P}_1, \mathbf{Q}_1] - [\mathbf{P}_2, \mathbf{Q}_2]). \tag{5.29}$$

The étendue from a_2 to $\mathbf{R}_1\mathbf{R}_2$ is given by:

$$\xi_3 = [\mathbf{Q}_1, \mathbf{R}_2] + [\mathbf{Q}_2, \mathbf{R}_1] - [\mathbf{Q}_1, \mathbf{R}_1] - [\mathbf{Q}_2, \mathbf{R}_2]. \tag{5.30}$$

For étendue to be conserved, we must have:

$$\begin{aligned} \xi_1 &= \xi_2 \\ \xi_2 &= \xi_3 \end{aligned}. \tag{5.31}$$

Let us assume that the positions of $\mathbf{P}_1$ and $\mathbf{Q}_1$ are known, but the positions of $\mathbf{P}_2$ and $\mathbf{Q}_2$ are not. Point $\mathbf{P}_2$ has two unknown coordinates and point $\mathbf{Q}_2$ another two unknown coordinates. Using the four Equations (5.26), (5.27), and (5.31), we get a system of four equations in four unknowns that can be solved for the coordinates of $\mathbf{P}_2$ and $\mathbf{Q}_2$. Given the positions of $\mathbf{P}_1$ and $\mathbf{P}_2$, and also those of $\mathbf{Q}_1$ and $\mathbf{Q}_2$, surfaces a_1 and a_2 can be determined. These Cartesian ovals can be calculated point by point using Equations (5.21) or (5.25) (for a_1 when source is at an infinite distance) or by their parametric equations [2].

Repeating the same process we can now calculate the positions of points $\mathbf{P}_3$ and $\mathbf{Q}_3$ and two other surfaces extending from $\mathbf{P}_2$ to $\mathbf{P}_3$ and from $\mathbf{Q}_2$ to $\mathbf{Q}_3$.

In real designs, the position of $\mathbf{P}$ between $\mathbf{P}_1$ and $\mathbf{P}_2$ and that of $\mathbf{Q}$ between $\mathbf{Q}_1$ and $\mathbf{Q}_2$ may have to be optimized. This is also valid for the other integrator elements.

Figure 5.50 shows a ray tracing of the rays emitted by a point on source $\mathbf{S}_1\mathbf{S}_2$ and how that light is spread over the receiver $\mathbf{R}_1\mathbf{R}_2$ by the optic obtained by the method described above. Since the refractive surfaces are not perfect imagers, a few rays miss the receiver $\mathbf{R}_1\mathbf{R}_2$.

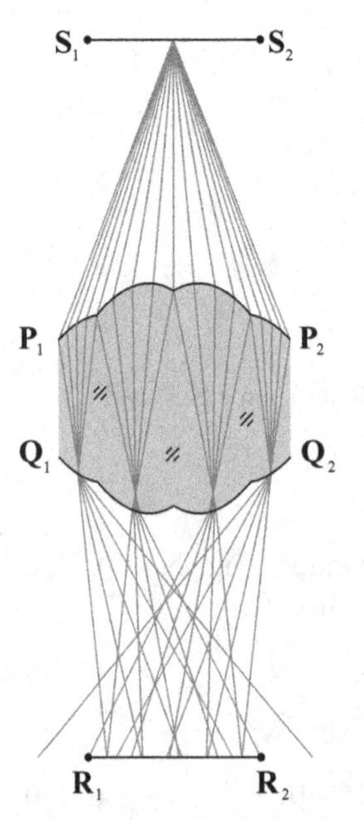

Figure 5.50 Array of integrator elements for finite source and target.

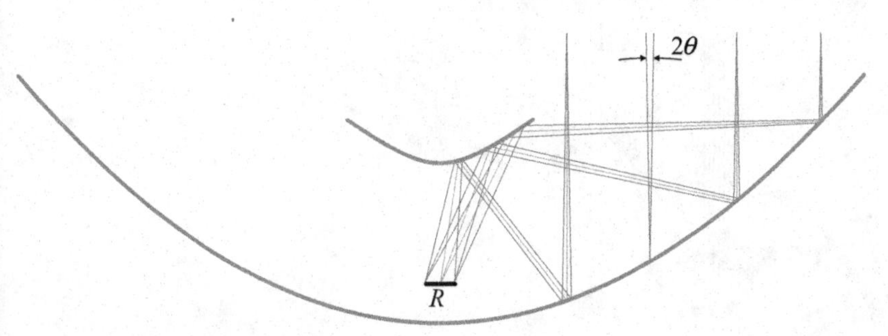

Figure 5.51 XX aplanatic concentrator.

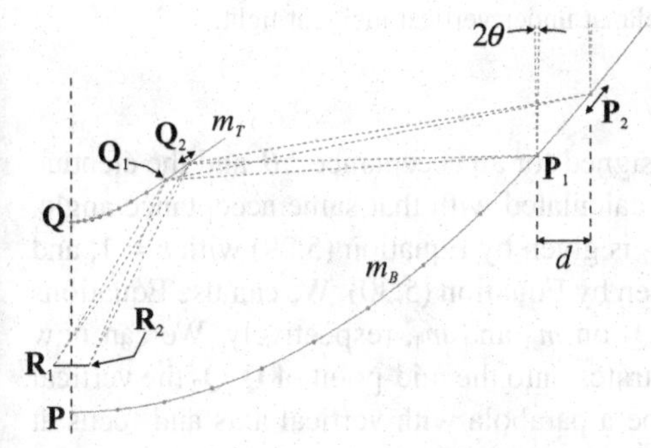

Figure 5.52 Calculation of the points along the XX aplanatic mirrors defining the edges of the integrator elements. The calculation of each new pair of points is based on the two previous ones. The process starts at points **P** and **Q** on the symmetry axis.

Another possible way to design one of these optics is to start with an existing concentrator and modify it to make it an integrator. Figure 5.51 shows an aplanatic XX (two mirrors) concentrator. As seen from the figure, the vertical rays are concentrated at the center of the receiver, and the edge rays are (approximately) concentrated onto its edges. The acceptance of the concentrator is $\pm\theta$.

Let us then assume we already calculated the edges of the integrator elements on the mirrors of the XX concentrator up to points $\mathbf{P}_1$ on mirror m_B and point $\mathbf{Q}_1$ on mirror m_T. We now want to calculate new points $\mathbf{P}_2$ on m_B and $\mathbf{Q}_2$ on m_T that will bind the next integrator element, as shown in Figure 5.52.

Now we only have two degrees of freedom: the parameter (or position) of point $\mathbf{P}_2$ along m_B and the parameter (or position) of $\mathbf{Q}_2$ along m_T. For that reason, we cannot impose the four Equations (5.31), (5.26), and (5.27), but must choose two of them. We calculate the new points $\mathbf{P}_2$ and $\mathbf{Q}_2$ by imposing the conservation of étendue. In this case, the source is at an infinite distance and subtends an angle $\pm\theta$, and all the optical surfaces are mirrors in air ($n = 1$). The étendue ξ_1 of the incoming light falling on $\mathbf{P}_1\mathbf{P}_2$ is:

$$\xi_1 = 2d\sin\theta. \tag{5.32}$$

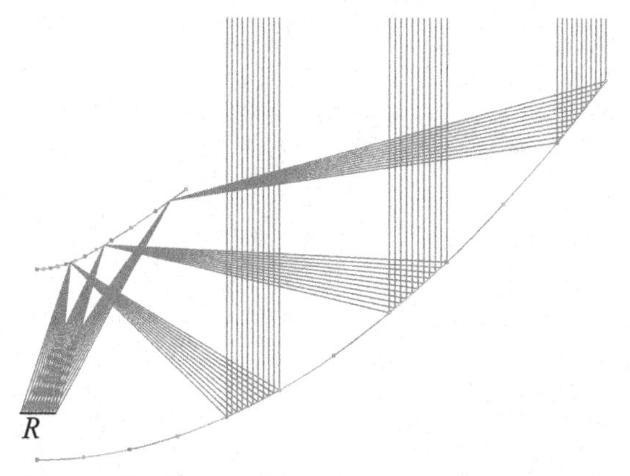

Figure 5.53 Integrator based on an XX aplanat under vertical incident light.

The concentrator in Figure 5.51 was designed for an acceptance $\pm\theta$, and the étendue balance for the integrator is also being calculated with that same acceptance angle.

The étendue ξ_2 from $\mathbf{P}_1\mathbf{P}_2$ to $\mathbf{Q}_1\mathbf{Q}_2$ is given by Equation (5.29) with $n = 1$, and the étendue ξ_3 from $\mathbf{Q}_1\mathbf{Q}_2$ to $\mathbf{R}_1\mathbf{R}_2$ is given by Equation (5.30). We can use Equations (5.31) to find the positions of $\mathbf{P}_2$ and $\mathbf{Q}_2$ on m_B and m_T, respectively. We can now calculate the Cartesian oval that concentrates onto the mid-point of $\mathbf{Q}_1\mathbf{Q}_2$ the vertical rays that hit $\mathbf{P}_1\mathbf{P}_2$ (in this case, it will be a parabola with vertical axis and focus at the midpoint of $\mathbf{Q}_1\mathbf{Q}_2$). Also, we can calculate the Cartesian oval that concentrates onto the center of the receiver the mid-point of $\mathbf{P}_1\mathbf{P}_2$ (in this case, it will be an ellipse with foci at the center of the receiver and mid-point of $\mathbf{P}_1\mathbf{P}_2$). Since we could not force condition (5.27) (with $n = 1$ in this case), if we force the ellipse to go through $\mathbf{Q}_1$, it (in general) will not go through point $\mathbf{Q}_2$, and there will be a discontinuity on the surface at $\mathbf{Q}_2$, since the next ellipse (to the right of $\mathbf{Q}_2$) will start at point $\mathbf{Q}_2$. The same happens for the mirror between $\mathbf{P}_1$ and $\mathbf{P}_2$, and there will also be discontinuities on that surface. This process starts at points $\mathbf{P}$ and $\mathbf{Q}$ on the symmetry axis of the concentrator.

Figure 5.53 shows the resulting integrator based on the XX aplanatic concentrator of Figure 5.51 and its behavior under vertical incident light [32]. The discontinuities at the optical surfaces are very small compared with the size of the concentrator.

The method described above matches the étendue ξ_1 of the incoming light falling on $\mathbf{P}_1\mathbf{P}_2$ with ξ_2 from $\mathbf{P}_1\mathbf{P}_2$ to $\mathbf{Q}_1\mathbf{Q}_2$ and with ξ_3 from $\mathbf{Q}_1\mathbf{Q}_2$ to $\mathbf{R}_1\mathbf{R}_2$, that is, $\xi_1 = \xi_2 = \xi_3$, according to Equation (5.31). Ideally, the mirror between $\mathbf{P}_1$ and $\mathbf{P}_2$ would focus one set of edge rays onto $\mathbf{Q}_1$ and the other set of edge rays onto $\mathbf{Q}_2$, so that all intermediate rays would end up between $\mathbf{Q}_1$ and $\mathbf{Q}_2$. This, however, will not be the case in practice, resulting in a decrease in efficiency. One way to reduce this decrease in efficiency is to recalculate the integrator elements in such a way that $\xi_1 < \xi_2 < \xi_3$. This way, even it the mirror at $\mathbf{P}_1\mathbf{P}_2$ is not perfect, it will still be able to couple the incoming light onto $\mathbf{Q}_1\mathbf{Q}_2$, since the étendue ξ_1 of the light it collects is smaller than the étendue ξ_2 available between $\mathbf{P}_1\mathbf{P}_2$ and $\mathbf{Q}_1\mathbf{Q}_2$. The same for the

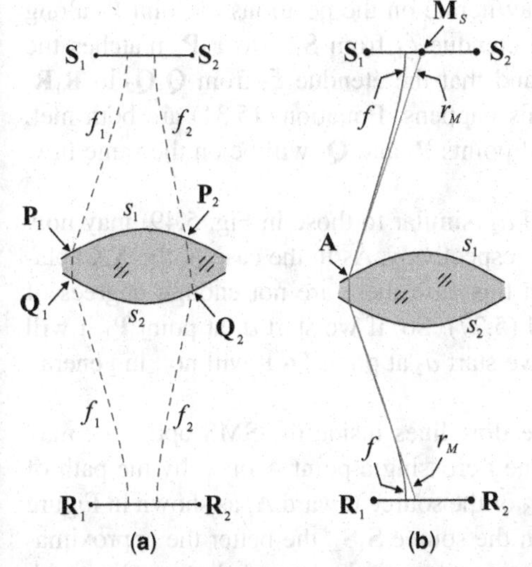

Figure 5.54 (a) SMS lens and flow lines from the source $\mathbf{S}_1\mathbf{S}_2$ toward the receiver $\mathbf{R}_1\mathbf{R}_2$. (b) Comparison of the paths of a flow line f trough a point **A** of the SMS optic with that of a light ray r_M emitted from the mid-point $\mathbf{M}_S$ of the source toward **A**.

optic at $\mathbf{Q}_1\mathbf{Q}_2$ which even if not perfect, it will still be able to couple the incoming light onto $\mathbf{R}_1\mathbf{R}_2$, since the étendue ξ_2 of the light it collects is smaller than the étendue ξ_3 available between $\mathbf{Q}_1\mathbf{Q}_2$ and $\mathbf{R}_1\mathbf{R}_2$.

Another possibility for designing an integrator is to start with an SMS optic and calculate the integrator elements on the SMS surfaces the way we did earlier for the XX concentrator in Figure 5.52. This is a more general situation since the aplanatic designs are the limit case of the SMS when the collected étendue goes to zero. Figure 5.54 shows an SMS lens for source $\mathbf{S}_1\mathbf{S}_2$ and receiver $\mathbf{R}_1\mathbf{R}_2$. Also shown are two flow lines f_1 and f_2 that cross the top surface s_1 of the lens at points $\mathbf{P}_1$ and $\mathbf{P}_2$ and the bottom surface s_2 at points $\mathbf{Q}_1$ and $\mathbf{Q}_2$.

The portion of curve s_1 with end points $\mathbf{P}_1$ and $\mathbf{P}_2$ is illuminated by light whose edge rays come from (point wave fronts) $\mathbf{S}_1$ and $\mathbf{S}_2$ (each point of s_1 between $\mathbf{P}_1$ and $\mathbf{P}_2$ is illuminated by the whole source $\mathbf{S}_1\mathbf{S}_2$). The étendue from $\mathbf{S}_1\mathbf{S}_2$ to $\mathbf{P}_1\mathbf{P}_2$ is then given by Equation (5.28) [2]. Also, the étendue from $\mathbf{Q}_1\mathbf{Q}_2$ to $\mathbf{R}_1\mathbf{R}_2$ is given by Equation (5.30). Étendue is conserved between flow lines [2], and the étendue from $\mathbf{S}_1\mathbf{S}_2$ to $\mathbf{P}_1\mathbf{P}_2$ equals that from $\mathbf{Q}_1\mathbf{Q}_2$ to $\mathbf{R}_1\mathbf{R}_2$. That is, by choosing points $\mathbf{P}_1$ and $\mathbf{Q}_1$ on the same flow line f_1 and also $\mathbf{P}_2$ and $\mathbf{Q}_2$ on the same flow line f_2 we ensure that $\xi_1 = \xi_3$.

Given the positions of $\mathbf{P}_1$, $\mathbf{P}_2$ and $\mathbf{Q}_1$, $\mathbf{Q}_2$ we can calculate a value for ξ_2 from Equation (5.29), but in general, this value ξ_2 does not match the étendue transferred by the SMS lens between $\mathbf{P}_1\mathbf{P}_2$ and $\mathbf{Q}_1\mathbf{Q}_2$. Equation (5.29) assumes that each point of s_2 between $\mathbf{Q}_1$ and $\mathbf{Q}_2$ is illuminated by light whose edge rays come from (point wave fronts) $\mathbf{P}_1$ and $\mathbf{P}_2$ (it assumes that $\mathbf{P}_1\mathbf{P}_2$ emits light to all the points of s_2 between $\mathbf{Q}_1$ and $\mathbf{Q}_2$), and that is not the case inside the lens. Still, from the positions of points

$\mathbf{P}_1$ and $\mathbf{Q}_1$ on the same flow line, we may iterate on the positions of point $\mathbf{P}_2$ along s_1 and of point $\mathbf{Q}_2$ along s_2 so that the étendue ξ_1 from $\mathbf{S}_1\mathbf{S}_2$ to $\mathbf{P}_1\mathbf{P}_2$ matches the value of ξ_2 given by Equation (5.29), and that the étendue ξ_3 from $\mathbf{Q}_1\mathbf{Q}_2$ to $\mathbf{R}_1\mathbf{R}_2$ also matches this value of ξ_2. When this happens, Equations (5.31) are both met. Since this also makes $\xi_1 = \xi_3$, calculated points $\mathbf{P}_2$ and $\mathbf{Q}_2$ will be on the same flow line.

Integrator element surfaces a_1 and a_2 (similar to those in Fig. 5.49) may now be constructed between $\mathbf{P}_1\mathbf{P}_2$ and $\mathbf{Q}_1\mathbf{Q}_2$, respectively. As in the case of the XX aplanatic integrator analyzed above, also in this case there are not enough degrees of freedom to verify conditions (5.26) and (5.27). So, if we start a_1 at point $\mathbf{P}_1$ it will not, in general, cross point $\mathbf{P}_2$. Also, if we start a_2 at point $\mathbf{Q}_2$ it will not, in general, cross point $\mathbf{Q}_2$.

When calculating the paths of the flow lines inside the SMS optic, we may approximate the path of a given flow line f crossing a point $\mathbf{A}$ on s_1 by the path of the ray r_M emitted from the mid point $\mathbf{M}_S$ of the source toward $\mathbf{A}$, as shown in Figure 5.54b. The further away point $\mathbf{A}$ is from the source $\mathbf{S}_1\mathbf{S}_2$, the better the approximation, since r_M more closely matches the asymptote to the hyperbolic flow line with foci $\mathbf{S}_1$ and $\mathbf{S}_2$.

Another possible way to calculate the boundaries of the integrator elements starting from an SMS optic is to use the construction shown in Figure 5.55. We start with an SMS optic (in this case it is a lens) and choose a point $\mathbf{P}_1$ on the top surface and determine the corresponding point $\mathbf{Q}_1$ on the bottom surface on the same flow line f_1. Now, determine the normal $\mathbf{n}_1$ at point $\mathbf{P}_1$ so that ray r_2 emitted from $\mathbf{S}_2$ is refracted at $\mathbf{P}_1$ toward $\mathbf{Q}_1$. This would be the normal that an ideal integrator element would have at point $\mathbf{P}_1$. Ray trace ray r_1 emitted from $\mathbf{S}_1$ through point $\mathbf{P}_1$ with normal $\mathbf{n}_1$ and intersect this ray with the bottom surface, determining point $\mathbf{Q}_2$. Point $\mathbf{P}_2$ on the top surface is determined as being on the same flow line f_2 as $\mathbf{Q}_2$. Points $\mathbf{P}_1$ and $\mathbf{P}_2$ on the top surface and $\mathbf{Q}_1$ and $\mathbf{Q}_2$ on the bottom surface can now be taken as the

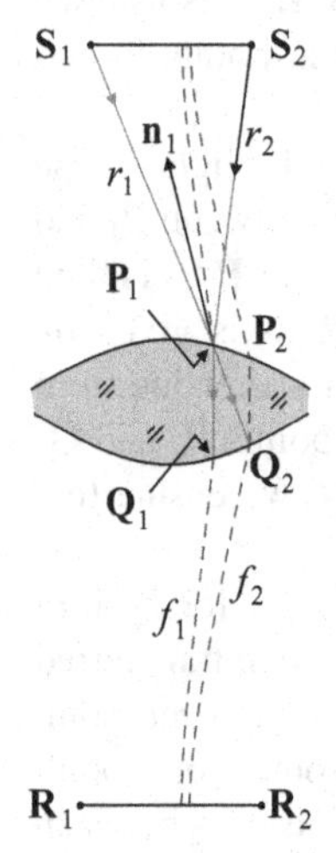

Figure 5.55 Calculation of the boundaries of integrator elements based on an SMS optic.

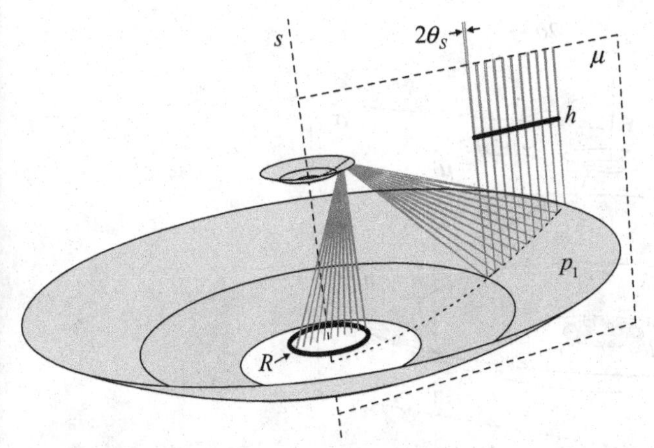

Figure 5.56 Radial Köhler concentrator (with circular symmetry). Receiver R is illuminated by light passing through a narrow slit hole h (on meridian plane μ) illuminated by light with small angular aperture $\pm\theta_S$ and contained in plane μ.

boundaries of an integrator element. Repeating this same procedure for the new points $\mathbf{P}_2$ and $\mathbf{Q}_2$, new points $\mathbf{P}_3$ and $\mathbf{Q}_3$ to the right (not shown) can be determined and the boundaries of a new integrator element defined.

In this method, the étendue from $\mathbf{S}_1\mathbf{S}_2$ toward $\mathbf{P}_1\mathbf{P}_2$ is the same as that from $\mathbf{Q}_1\mathbf{Q}_2$ to $\mathbf{R}_2\mathbf{R}_2$ since points $\mathbf{P}_1$ and $\mathbf{Q}_1$ are on the same flow line f_1 and points $\mathbf{Q}_2$ and $\mathbf{P}_2$ are also on the same flow line f_2. However, fulfillment of Equations (5.26), (5.27), and (5.31) is not guaranteed. This method can also be extended to 3D geometry [33, 34].

The optics presented above can be made into 3D concentrators by either linear symmetry (resulting in trough optics) or circular symmetry, resulting in the geometry shown in Figure 5.56. We now estimate the irradiance pattern on the receiver that one of these optics produces if it has circular symmetry. To that end, we consider the case shown in Figure 5.56, in which the hole entrance aperture of the concentrator is covered except for a narrow radial slit hole h covering element p_1 of the integrator. This slit hole h is contained in meridian plane μ. Through h, light contained in plane μ and confined to a narrow angular aperture $\pm\theta_S$ (say the solar angular aperture) illuminates p_1, and from there it is redirected to the receiver R, where it illuminates a line covering the diameter of circular receiver R.

Figure 5.57 shows another situation in which a point $\mathbf{H}$ of h is illuminated by incident light contained in the sagittal plane σ (perpendicular to meridian plane μ). If this light has angular aperture $\pm\theta$ (the acceptance of the concentrator), and is contained between edge rays r_1 and r_2, these rays will spread out over the entire diameter of the receive R. However, light with a narrower angular aperture $\pm\theta_S$ will illuminate a small line i at the receiver of receiver R. The length of i is related to the diameter d of R by $i/d = 2\theta_S/2\theta$.

If we now combine the effects shown in Figure 5.56 and 5.57 and illuminate slit hole h with light confined angularly within a cone of total angular aperture $\pm\theta_S$

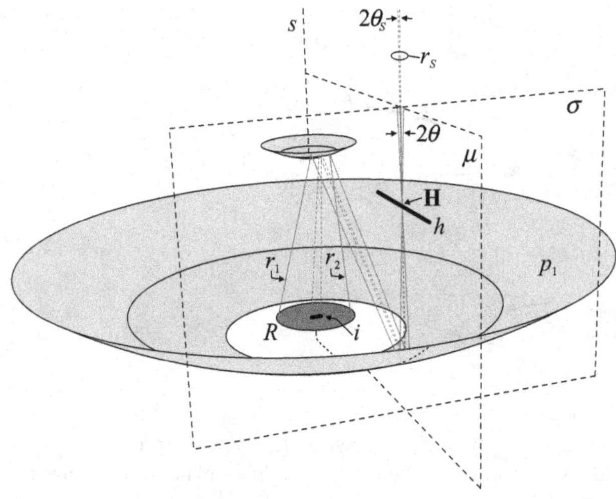

Figure 5.57 A point **H** of a slit hole h covering element p_1 of the integrator is illuminated by light contained in the sagittal plane σ (perpendicular to meridian plane μ). Shown are two rays r_1 and r_2 impinging on **H** with an angular aperture corresponding to the acceptance $\pm\theta$ of the concentrator and also two other rays r_S with a smaller angular aperture $\pm\theta_S$.

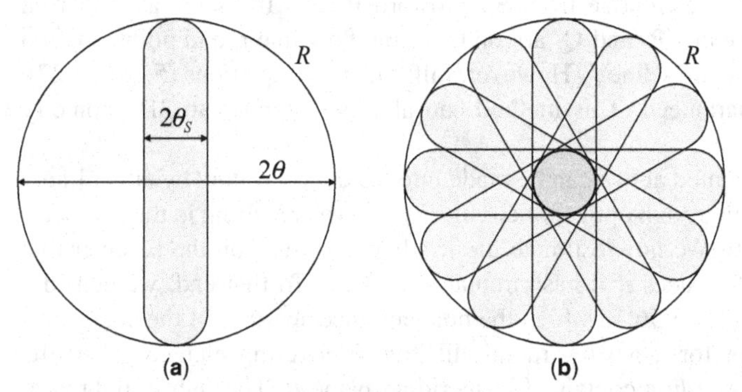

Figure 5.58 (a) Receiver illuminance through a slit hole by light confined to a cone of angular aperture $\pm\theta_S$. (b) Effect on the receiver illuminance as the slit hole rotates around the symmetry axis of the optic.

for every point **H** of h, the resulting illumination of the receiver is as shown in Figure 5.58a. As the slit hole rotates around the symmetry axis s of the concentrator, the illuminated area also rotates on the receiver, resulting in the pattern shown in Figure 5.58b. All these patterns superimpose at the center, creating a hotspot (shown as a dark circle).

This situation, however, is still much better than not integrating. The pattern created by a concentrating optic (such as the one in Fig. 5.51) is an image of the

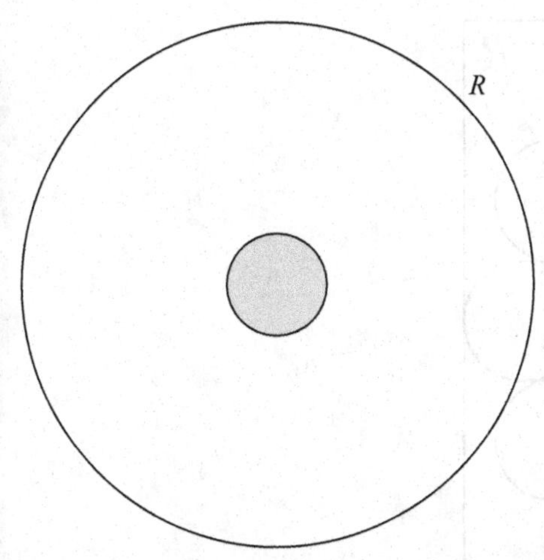

Figure 5.59 Image of the sun on the receiver produced by a nonintegrating concentrator.

sun on the receiver, resulting in a much higher hotspot, as shown in Figure 5.59. As the sun moves in the sky inside the cone defined by the acceptance angle θ, this image moves around inside receiver R.

The difference in the hotspot intensity between the cases of Figure 5.58b and 5.59 may now be estimated. We start with the situation in Figure 5.58a. If dE is the illuminance produced in the illuminated area by the slit hole h in Figure 5.58a, the total flux $d\Phi_T$ in the pattern it creates is proportional to $4dE\theta\theta_S$, that is, $d\Phi_T \propto 4dE\theta\theta_S$. The area A_R of the receiver is proportional to $\pi\theta^2$, that is, $A_R \propto \pi\theta^2$, and the average illuminance on the receiver is $dE_R = d\Phi_T/A_R \propto dE\theta_S/\theta$. The illuminance on the center circle of radius θ_S is $dE_C = dE$, and therefore, the ratio of illuminances between the center and the whole receiver is proportional to $dE_C/dE_R \propto \theta/\theta_S$. As the slit hole rotates around the symmetry axis s, it generates more illuminance patterns as the one in Figure 5.58a that superimpose, generating a hotspot in the center. However, each one of them has the same ratio between illuminance at the center and overall illuminance of the receiver, so this ratio $dE_C/dE_R \propto \theta/\theta_S$ is still valid in the case of Figure 5.58b. The hotspot illuminance E_C is therefore proportional to θ/θ_S times the average illuminance E_R on the receiver.

On the other hand, in the case of Figure 5.59, if $d\Phi$ is the flux falling on the hotspot, the hotspot illuminance is $E_C = d\Phi/\pi\theta_S^2$ while the average illuminance of the receiver is $E_R = d\Phi/\pi\theta^2$. The hotspot illuminance E_C is therefore proportional to θ^2/θ_S^2 times the average illuminance E_R on the receiver. The hotspot created by the integrator optic with circular symmetry is therefore much lower that that generated by a focusing optic concentrator.

Similar conclusions can be obtained for square receivers R (such as square solar cells). In that case, the hotspot illuminance E_C is proportional to $4/\pi\ \theta/\theta_S$ times the average illuminance E_R on the receiver for the circular integrator and $4/\pi\ \theta^2/\theta_S^2$ for the focusing optic [35]. For the optic in Figure 5.56 and 5.57 to be used with a square receiver and maintain its acceptance angle θ, the square receiver must circumscribe the slit hole images on it, as shown in Figure 5.60 (see Appendix 5.A).

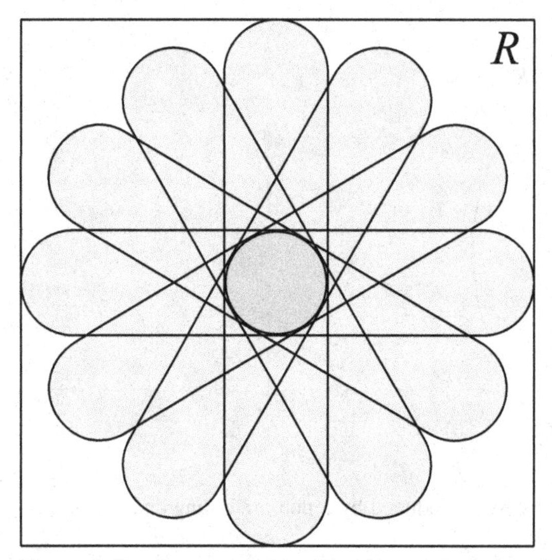

Figure 5.60 Square receiver illuminance through a slit hole as it rotates around the symmetry axis of the optic.

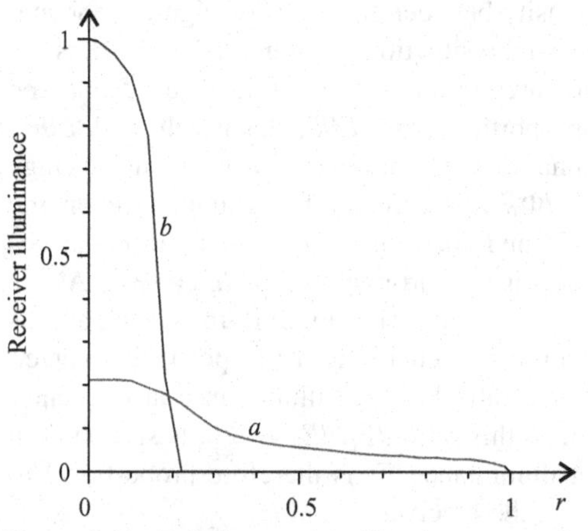

Figure 5.61 Relative receiver illuminance as a function of distance r to the center of the receiver for an integrator concentrator (curve a) and an aplanatic concentrator (curve b), both with circular symmetry.

In this case, dE_C is still the same, but dE_R is now lower since the receiver area increased from a value proportional to $\pi\theta^2$ to a value proportional to $4\theta^2$, and dE_R decreased by a factor $\pi\theta^2/4\theta^2 = \pi/4$, increasing E_C/E_R by a factor $4/\pi$.

Figure 5.61 shows the comparison of the irradiance distribution on the cell of the two examples (XX aplanatic and a radial Köhler XX) in Figure 5.51 and 5.53 as a function of the radial distance to the center of the receiver. The Köhler optic (curve a) shows a much lower irradiance peak at the center of the receiver than the aplanatic optic (curve b). Implementation of radial Köhler XX designs have been used as solar concentrators [36].

One option to reduce the hotspot in the center for integrator optics with circular symmetry is to modify the optic in order to change the pattern generated by each slit hole on the receiver. The shape of the optical surfaces are modified so that they generate a higher illuminance toward the outside of the receiver, compensating for the hotspot when the slit hole rotates around the symmetry axis of the optic [37]. Although improving over a "conventional" circular integrating optic, the illuminance on the receiver is typically still not completely uniform.

5.7 KÖHLER INTEGRATION IN TWO DIRECTIONS

Earlier we have seen how integration in one direction (radial direction) improves illumination uniformity on the receiver. To further increase receiver illuminance uniformity, it is possible to design integrators that "integrate" in two directions (radial and sagittal). To that end, the same principle of microlenses on the optical surfaces can be applied in 3D geometry. Figure 5.62 shows one such possibility [35]. The primary is a mirror composed of four parabolic sections p_1, p_2, p_3, and p_4. The secondary is composed of four microlenses. Figure 5.62a shows the complete primary and one of the microlenses of the secondary.

Figure 5.63 shows a diagonal, vertical cut through points **E**, **V**, and **G** of the optic in Figure 5.62.

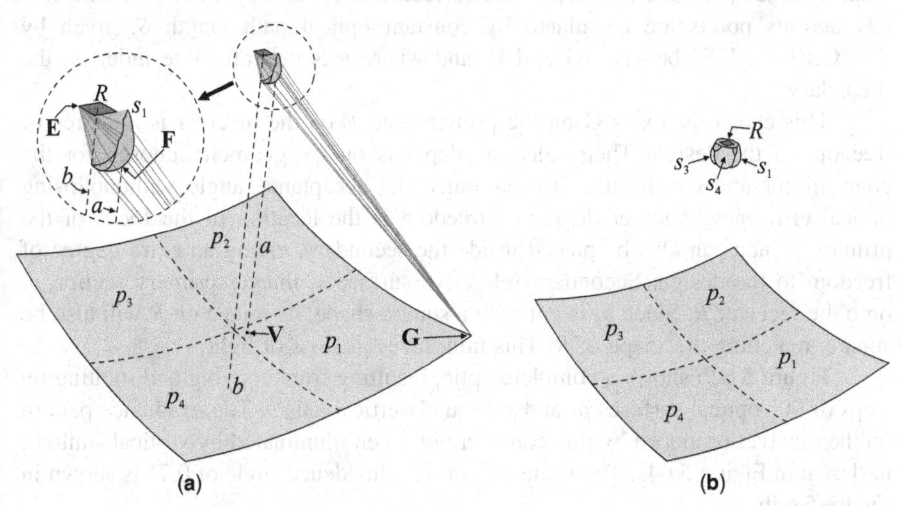

Figure 5.62 (a) Parabolic primary section p_1 has vertical axis a, vertex **V** and focus **F**. Secondary refractive section s_1 has foci **G** and **E** on the opposing edges of p_1 and receiver R and goes through point F. (b) A complete optic resulting from copying and rotating in steps of 90° p_1 and s_1 around vertical axis b. Axis b crosses the center of the primary and receiver.

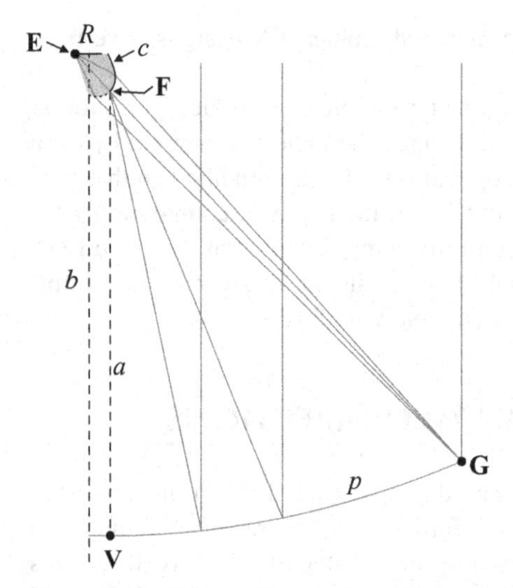

Figure 5.63 Diagonal, vertical cut of the optic in Figure 5.62 through points **E**, **V**, and **G**.

Parabolic section p_1 in Figure 5.62 has circular symmetry around vertical axis a; it has vertex **V** and focus **F** and results from giving parabola p in Figure 5.63 circular symmetry around its axis a. Secondary microlens s_1 is a Cartesian oval that focuses edge **G** of p_1 onto the opposing edge **E** of the receiver and goes through the focus **F** of the primary and results from giving Cartesian oval c in Figure 5.63 circular symmetry around line **EG**. This surface has circular symmetry around line **EG** and its points are calculated by constant optical path length S, given by $S = [\mathbf{G},\mathbf{F}] + n[\mathbf{F},\mathbf{E}]$ between **G** and **E**, and where n is the refractive index of the secondary.

This choice of focus **G** on the primary and **E** on the receiver is a degree of freedom of the design. Their selection depends on the geometry chosen for the concentrator and can be used to maximize the acceptance angle without losing optical efficiency. Another degree of freedom is the location of the focus of the primary, which can also be placed inside the secondary, giving an extra degree of freedom to the design. Secondary refractive surface s_1 images primary section p_1 onto the receiver R. Since p_1 is cut with a square shape, its image on R will also be square, matching the shape of R. This minimizes the loss of light.

Figure 5.62b shows a complete optic, resulting from copying and rotating by steps of 90° optical surfaces p_1 and s_1 around vertical axis b. The irradiance pattern on the receiver produced by this concentrator when illuminated by vertical sunlight is shown in Figure 5.64a. The same for sunlight incidence angle of 0.7° is shown in Figure 5.64b.

The same principle may be applied to the case in which the primary is a Fresnel lens combined with a refractive surface secondary. Figure 5.65 shows one of those concentrators, the Fresnel–R Köhler (FK) [38–40].

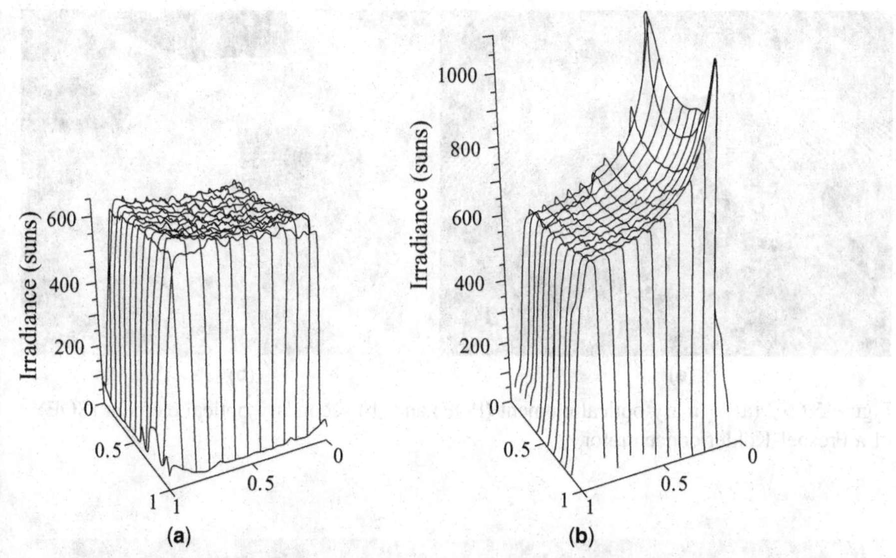

Figure 5.64 Irradiance on the receiver generated by the concentrator in Figure 5.62. (a) For normal incidence of sunlight; (b) for sunlight incidence angle of 0.7deg.

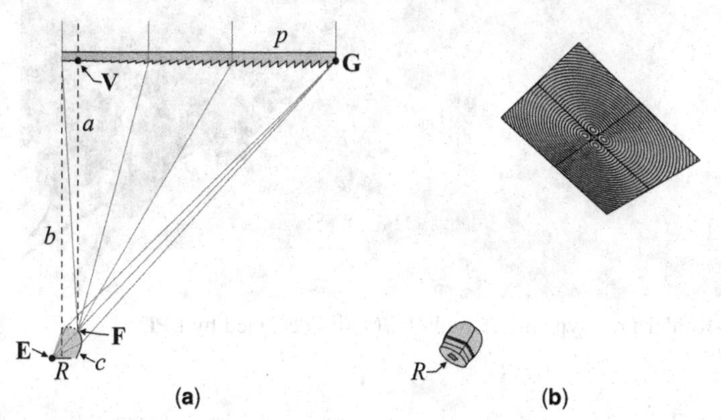

Figure 5.65 (a) Diagonal, vertical cut of a Fresnel-R Köhler integrator. (b) Fresnel-R Köhler integrator with square receiver R.

Figure 5.65a is the equivalent of Figure 5.63, only now with the parabolic primary reflector replaced by a Fresnel lens. Again, the primary element (Fresnel lens) focuses the incoming light onto focus **F** on the surface of the secondary element. Cartesian oval c focuses the edge **G** of the primary onto the opposite edge **E** of the receiver R. Figure 5.65b is the equivalent of Figure 5.62, showing the final 3D optic with square entrance aperture and square receiver R. Each one of the quadrants in Figure 5.65b results from giving circular symmetry to the Fresnel lens in Figure 5.65a around axis a and giving Cartesian oval c circular symmetry around line **EG**.

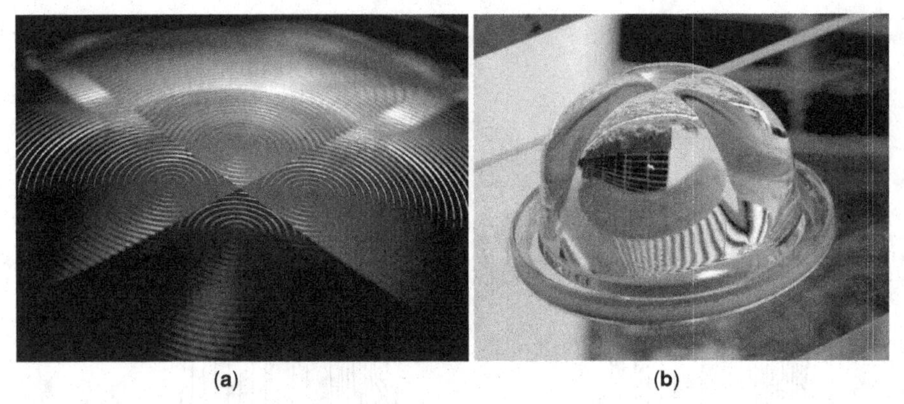

Figure 5.66 (a) Primary optical element (POE) and (b) secondary optical element (SOE) of a Fresnel-Köhler concentrator.

Figure 5.67 Fresnel-Köhler prototype under sunlight. Optics designed by LPI.

Like before, additional degrees of freedom may be gained by varying the geometry of the system. One such possibility is to place the focus **F** of the Fresnel lens inside the secondary.

Figure 5.66a shows a detail of the center of the Fresnel lens primary and Figure 5.66b the corresponding secondary of a Fresnel-Köhler concentrator.

Figure 5.67 shows a Fresnel-Köhler prototype being tested under sunlight.

The FK concentrator may be compared with other concentrators based on Fresnel lenses [12]. Figure 5.68 shows several of these concentrators (from left to right): single Fresnel lens (no secondary optical element—SOE), Fresnel lens combined with a spherical dome, the SILO (SIngLe Optical surface, presented above in Fig. 5.46), the XTP (reflective truncated pyramid), the RTP (dielectric truncated pyramid), and the FK. Both the Fresnel lenses and the solar cells are square. Most of these configurations either are or have been used in commercial CPV products.

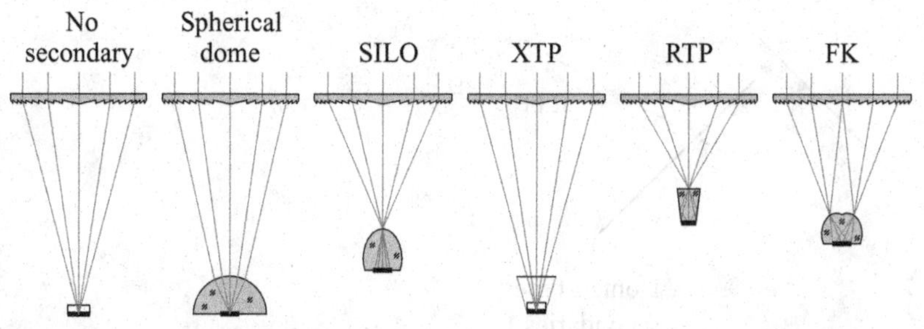

Figure 5.68 Concentration of the Fresnel-R Köhler (FK) with other Fresnel lens-based concentrators.

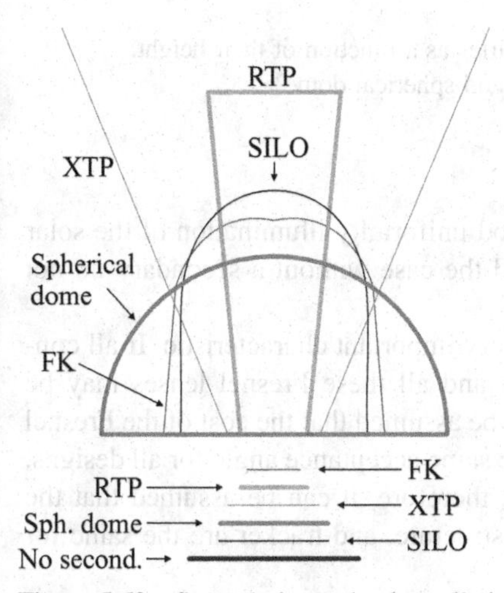

Figure 5.69 Secondaries and solar cell sizes for the concentrator configurations in Figure 5.68.

In this comparison, all concentrators in Figure 5.68 have the same Fresnel lens area ($625\,cm^2$) and the same acceptance angle ($\theta = \pm 1°$).

The secondaries and solar cells are not to scale, but the ratio between system height and Fresnel lens diagonal (*f*-number) are. The corresponding secondaries and solar cell sizes are shown in Figure 5.69.

The optical efficiency of these different configurations is similar, except when the focal distance of the Fresnel lens is very short. This is the case of the RTP, which has an efficiency about 4% lower than the others [12].

The FK is the geometry that has the highest CAP*, even as the focal distance of the Fresnel lens varies. The RTP is the second concentrator in the CAP* ranking [12]. However, the average ray path inside the RTP is about twice as long as that in the FK, reducing the material candidates to be used in the RTP to a few highly transparent glasses (such as BK7) when compared with the FK which can use a wider range of materials.

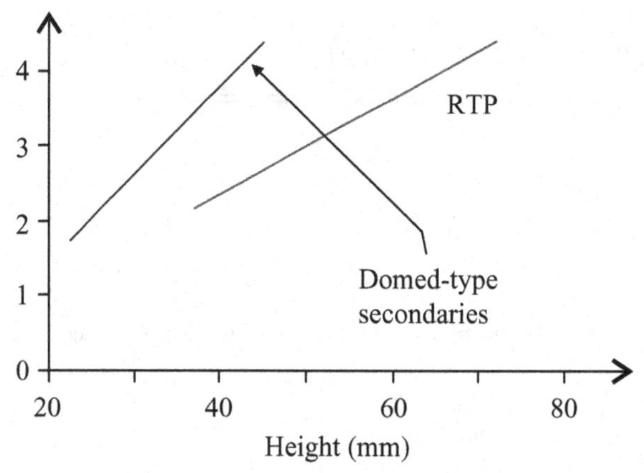

Figure 5.70 Estimated cost of dielectric secondaries as a function of their height. Domed-type secondaries refers to the FK, SILO, and spherical dome.

The FK, RTP, and SILO produce a good uniformity illumination of the solar cell, while the XTP, the spherical dome, and the case without a secondary do not [12].

The cost of these several options is a very important characteristic. If all configurations have the same Fresnel lens area and all these Fresnel lenses may be manufactured using the same methods, it can be assumed that the cost of the Fresnel lens is the same for all of them. By having the same acceptance angle for all designs, they all have similar overall tolerances and, therefore, it can be assumed that the cost of module assembly, array installation, structure, and tracker are the same for all cases.

The differences in cost will then be due to the different costs of each secondary and the corresponding solar cell. For a given acceptance angle the concentration attainable by each configuration varies and, therefore, the solar cell sizes will be different.

The price of the solar cell was considered to be \$7/cm^2, the cost of the reflective secondary was considered proportional to its mirrored area, at a cost of \$0.004/cm^2, and the cost of the dielectric secondaries (no antireflective coatings) as shown in Figure 5.70 (domed-type secondaries refers to the FK, SILO, and spherical dome). In the case of the spherical dome, its outer portion (not optically active) was removed to reduce its volume.

The FK is the configuration that has the lowest cost for secondary plus cell among those in Figure 5.68. This is still true for varying acceptance angles, as shown in Figure 5.71, showing the secondary plus cell cost as a function of the effective acceptance angle θ^* of the different concentrators [12].

The comparison between the different concentrators should be made at the same effective acceptance angle θ^* (as indicated, e.g., by the vertical dashed line for $\theta^* = 1$) since this corresponds to the same overall tolerances for the different concentrators. The price advantage of the FK relative to the second least expensive

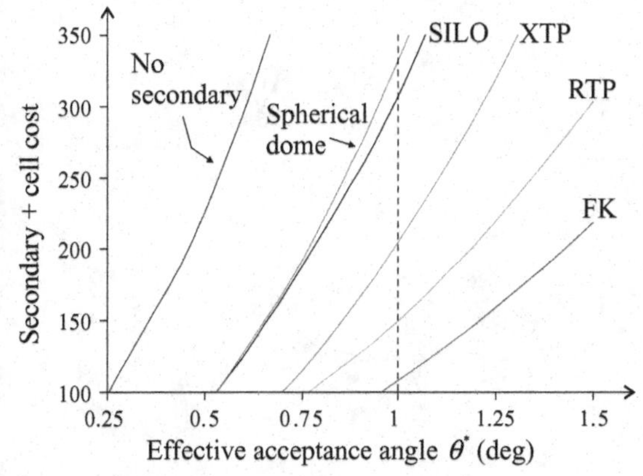

Figure 5.71 Secondary optical element (SOE) plus cell cost per unit entrance (primary lens) aperture area in $/m^2 as a function of the effective acceptance angle θ^*.

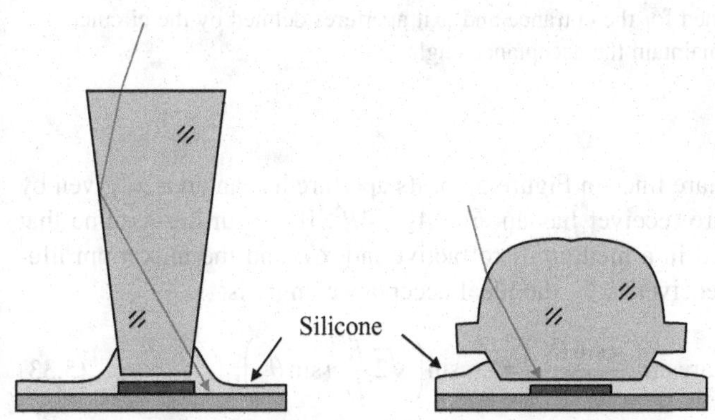

Figure 5.72 Practical aspects of assembling an RTP onto the solar cell when compared with an FK.

(the RTP) is mainly due to the smaller solar cell of the FK, which attains a higher concentration.

There is another practical aspect of the FK when compared with the RTP, as shown in Figure 5.72. Coupling the secondary optic to the solar cell may be done with a silicone. This silicone, however, may spill over the surfaces of the secondary, changing their shape and producing leakage (RTP secondary, Fig. 5.72 left). This is not an issue in the case of the FK since these surfaces (to the sides of the solar cell) are not optically active (Fig. 5.72 right) [12].

APPENDIX 5.A ACCEPTANCE ANGLE OF SQUARE CONCENTRATORS

The ideal acceptance angle for a square concentrator can be obtained from Equation (5.1) and Figure 5.73. If the entrance and exit apertures of the concentrator are

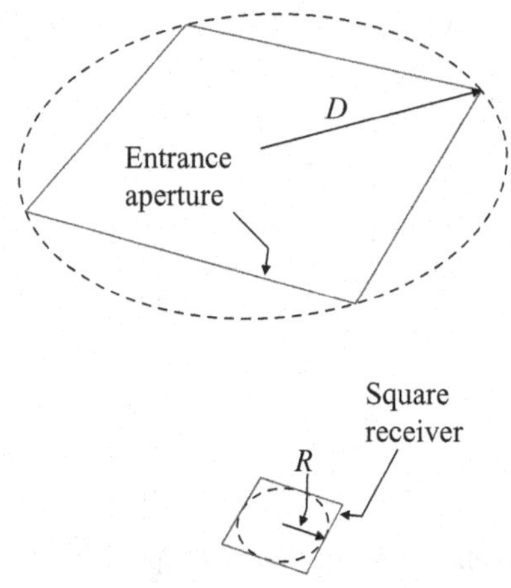

Figure 5.73 Square entrance aperture and square exit apertures of a concentrator (square solid lines). If this concentrator is obtained from another optic with circular symmetry, this other optic must be designed for the entrance and exit apertures defined by the circular dashed lines, in order to maintain the acceptance angle.

defined by the solid square lines in Figure 5.73, its aperture has an area A_1 given by $A_1 = 2D^2$, and the square receiver has an area $A_2 = 4R^2$. If we further assume that the receiver is immersed in a medium if refractive index n and the maximum illumination angle at the receiver is θ_2, the ideal acceptance angle is:

$$\theta_1 = \arcsin\left(\frac{n\sin\theta_2}{\sqrt{C_S}}\right) = \arcsin\left(\sqrt{2}\frac{R}{D}n\sin\theta_2\right), \tag{5.33}$$

where $C_S = A_1/A_2 = D^2/2R^2$ is the concentration of the ideal square optic.

Concentrators with square entrance apertures and square receivers are often obtained by designing a concentrator with circular symmetry and then cutting the entrance aperture with a square shape and also giving the receiver a square shape, as is the case of the concentrator in Figure 5.30. The resulting concentrators have an acceptance angle that is lower than ideal.

In that case, the optic is designed with circular symmetry for entrance and exit apertures indicated by the circular dashed lines in Figure 5.73. The radius of the entrance aperture of the optic with circular symmetry is D, the diagonal of the desired square entrance aperture. The radius of the exit aperture of the optic with circular symmetry is R, half the side of the square receiver.

Once the optic with circular symmetry is designed (for the dashed lines), we replace the entrance aperture by the inscribed square and the exit aperture by the circumscribed square. Cutting the entrance aperture square reduces the amount of captured light, but does not (necessarily) affect the acceptance. This is shown, for example, in Figure 5.3, which illustrates the acceptance angle for a given point **P** of the entrance aperture **AB**. Ideally, the acceptance angle is the same for all points of

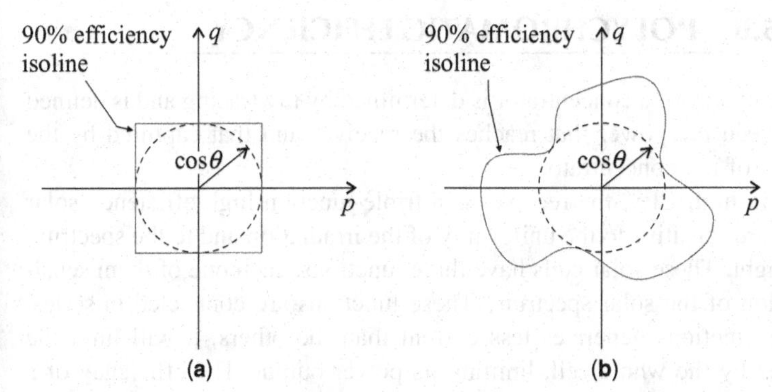

Figure 5.74 (a) Acceptance angle θ in the case of a square acceptance (on the plane of the direction cosines p,q). (b) Acceptance angle θ in the case of an arbitrary shape acceptance (also on the plane of the direction cosines p,q).

AB, so blocking part of **AB** results in less captured light, but it does not change the acceptance angle. On the other hand, increasing the exit aperture to the circumscribed square (receiver) increases the efficiency (some light that before fell out of the receiver is now captured by a larger receiver), but does not affect the acceptance. Note that the acceptance in the diagonal plane of the receiver does increase, but this does not affect the overall acceptance, which is defined as the minimum angle to the vertical (axis of the concentrator) for which the efficiency is greater than 90% (see Fig. 5.74a). Figure 5.74b shows the general definition of acceptance angle θ. If the aiming of the concentrator toward the sun is rotated by an angle smaller than θ in any direction, its efficiency is still greater than 90%.

The theoretical maximum acceptance angle of a concentrator designed with circular symmetry and then cut with a square aperture and combined with a square receiver is then

$$\theta = \arcsin\left(\frac{n \sin\theta_2}{\sqrt{C_C}}\right) = \arcsin\left(\frac{R}{D} n \sin\theta_2\right), \tag{5.34}$$

where $C_C = \pi D^2/\pi R^2 = D^2/R^2$ is the concentration of the ideal circular optic. When compared with the ideal square concentrator, the circular optic was designed for a larger area entrance aperture and a smaller area exit aperture and, therefore, has a smaller acceptance angle. Then, this value for θ is smaller than the ideal acceptance angle θ_1 given by Equation (5.33). For small values of θ (as is the case of high concentration), $\arcsin\theta \approx \theta$ and $\theta \approx \theta_1/\sqrt{2}$.

The CAP was defined above as $\mathrm{CAP} = \sqrt{C}\sin\theta$. Since designing the optical surfaces with circular symmetry reduces the maximum acceptance angle by a factor $1/\sqrt{2}$, the maximum CAP is reduced by the same factor.

Ideally, if the receiver is immersed in a medium of refractive index n and has an acceptance angle θ, from expression (8.2) we get $CAP = \sqrt{C}\sin\theta < n$. However, in the case of a square receiver, we get $CAP = \sqrt{C}\sin\theta < n/\sqrt{2}$.

APPENDIX 5.B POLYCHROMATIC EFFICIENCY

Typically, the efficiency of a concentrator is determined by ray tracing and is defined as the ratio between the power that reaches the receiver and that captured by the entrance aperture of the concentrator.

However, in high CPV, the receiver is a triple-junction high efficiency solar cell. These cells are sensitive to the uniformity of the irradiation and to the spectrum of the incident light. These solar cells have three junctions, each one of them sensitive to one portion of the solar spectrum. These junctions are connected in series, so if one of the junctions generates less current than the others, it will limit the current generated by the whole cell, limiting its power output. The efficiency of a concentrator coupled to one of these solar cells is not only a function of how much power the concentrator transfers toward the solar cell, but also a function of how it changes the spectrum of the light. Different irradiation distributions for different wavelengths may significantly affect the efficiency of the cell due to current mismatch between the top and middle junctions.

The efficiency of a solar cell is described by its external quantum efficiency Q_E, which is the number of electrons generated and collected for each photon incident on the cell, that is, $Q_E = n_E/n_p$, where n_E is the number of electrons generated and collected and n_P the number of incident photons, both per unit time (the electrons generated but then recombined do not count; this would be the internal quantum efficiency, which is higher). The number of electrons produced by the incident light at a wavelength λ is then given by $n_E = Q_E(\lambda)n_p$. On the other hand, the electrical current is given by $I = n_E q$, where n_E is the number of electrons per second (generated and collected as current) and q the charge of the electron.

The spectral irradiance of the incoming light is given by $I_D = dP/(Ad\lambda)$, defined as the light power (dP, energy per unit time) per unit surface area (A) and unit wavelength ($d\lambda$) and it therefore has units of $\mathrm{Wm^{-2}\ nm^{-1}}$. The energy of each photon at a wavelength λ is given by $E = h\nu = hc/\lambda$, where h is the Planck constant, ν is the photon frequency, and c the speed of light. The power associated with these photons is $dP = n_p E$, where n_P is the number of photons per unit time per unit wavelength at wavelength λ. Introducing this into the equation for I_D, we get $I_D = n_p E/(Ad\lambda)$. The number of photons per unit time in the range λ to $\lambda + d\lambda$ can then be obtained from $n_p = I_D A d\lambda/E$. The current is then given by:

$$I_{\lambda,\lambda+d\lambda} = n_E q = Q_E(\lambda) n_P q = A\frac{q}{h\nu} I_D(\lambda) Q_E(\lambda) d\lambda \tag{5.35}$$

where $q = 1.602176462 \times 10^{-19}$ C is the electron charge, $h = 6.62606876 \times 10^{-34}$ J·s is the Planck constant, A is the solar cell area, I_D the spectral irradiance, Q_E the external quantum efficiency and ν the photon frequency. This last equation can now be integrated for all wavelengths, resulting in the total current produced by the solar cell [41]:

$$I = A\frac{q}{hc_n}\int \lambda I_D(\lambda) Q_E(\lambda) d\lambda = \frac{A}{h_C}\int \lambda I_D(\lambda) Q_E(\lambda) d\lambda, \tag{5.36}$$

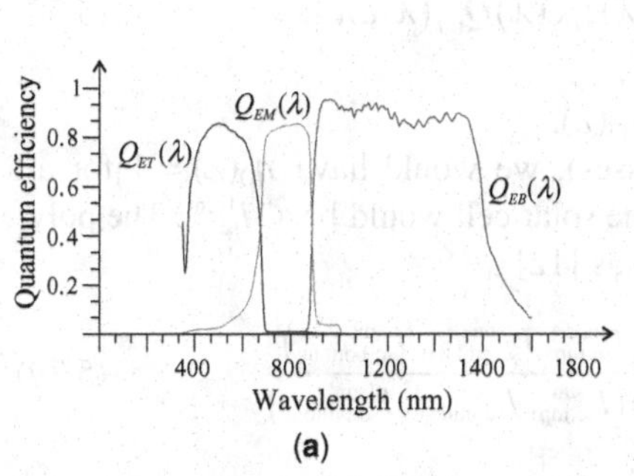

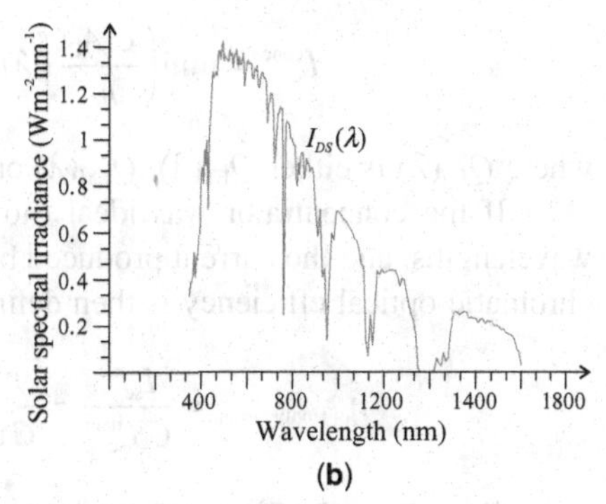

Figure 5.75 (a) Triple junction top, middle and bottom cell's external quantum efficiency. (b) Solar spectral irradiance.

where $c_n = 299{,}792{,}458 \times 10^9$ nm/s is the speed of light in nanometers per second (the solar irradiance is typically given as a function of wavelength in nanometers, so the speed of light is also be given in nanometers per second). Constant $h_C = hc_n/q = 1240$ and A is the area of the solar cell.

Figure 5.75a shows three curves exemplifying the external quantum efficiency of the top cell $Q_{ET}(\lambda)$, of the middle cell $Q_{EM}(\lambda)$, and of the bottom cell $Q_{EB}(\lambda)$. Figure 5.75b shows the solar spectral irradiance $I_{DS}(\lambda)$, according to the standard Air Mass 1.5 direct (AM1.5d).

Replacing in Equation (5.36) $Q_E(\lambda)$ by $Q_{ET}(\lambda)$ and $I_D(\lambda)$ by $I_{DS}(\lambda)$, we get the (short circuit) current $I_{\mathrm{sc,top}}^{\mathrm{1sun}}$ generated by the top cell when exposed to direct sunlight. The same procedure gives us the current $I_{\mathrm{sc,middle}}^{\mathrm{1sun}}$ generated by the middle cell (now calculated using $Q_{EM}(\lambda)$) and $I_{\mathrm{sc,bottom}}^{\mathrm{1sun}}$ generated by the bottom cell (calculated using $Q_{EB}(\lambda)$). The short circuit current generated by the solar cell may be estimated as the minimum of these three: $I_{\mathrm{sc}}^{\mathrm{1sun}} = \min\left(I_{\mathrm{sc,top}}^{\mathrm{1sun}}, I_{\mathrm{sc,middle}}^{\mathrm{1sun}}, I_{\mathrm{sc,bottom}}^{\mathrm{1sun}}\right)$, or:

$$I_{\mathrm{sc}}^{\mathrm{1sun}} = \min\left(\frac{A_C}{h_C}\int \lambda I_{\mathrm{DS}}(\lambda) Q_{\mathrm{E}k}(\lambda)\, d\lambda\right), \tag{5.37}$$

where $Q_{\mathrm{E}k}(\lambda)$ is either $Q_{ET}(\lambda)$, $Q_{EM}(\lambda)$, or $Q_{EB}(\lambda)$, and A_C is the solar cell area.

Optics used to concentrate sunlight onto solar cells increase the irradiance on the cell. However, these optics may, for example, refract different wavelengths differently or absorb light of different wavelengths differently, changing the spectral irradiance of the sunlight reaching the cell. The behavior of the optics for different wavelengths may be obtained by ray tracing, resulting in a spectral optical efficiency $\eta_O(\lambda)$ that, for each wavelength, measures what fraction of the light crossing the entrance aperture of the optic ends up on the receiver. These calculations allow us to obtain the spectral irradiance on the cell for the concentrated sunlight.

A concentrator has a geometrical concentration C defined as $C = A_E/A_C$, where A_E is area of the concentrator entrance aperture, and A_C the solar cell area (receiver area). Under concentration, the spectral irradiance on the solar cell is then $I_{DC}(\lambda) = C\eta_O(\lambda)I_{DS}(\lambda)$, and the current generated by the solar cell is:

$$I_{sc}^{conc} = \min\left(\frac{C A_C}{h_C} \int \lambda \eta_O(\lambda) I_{DS}(\lambda) Q_{Ej}(\lambda) d\lambda\right), \tag{5.38}$$

where $Q_{Ej}(\lambda)$ is either $Q_{ET}(\lambda)$, $Q_{EM}(\lambda)$ or $Q_{EB}(\lambda)$.

If the concentrator was ideal (no losses), we would have $\eta_O(\lambda) = 1$ for all wavelengths, and the current produced by the solar cell would be $C I_{sc}^{1sun}$. The polychromatic optical efficiency is then defined as [12]

$$\eta_{opt,polychrom} = \frac{I_{sc}^{conc}}{C I_{sc}^{1sun}} = \frac{\min\left(I_{sc,top}^{conc}, I_{sc,middle}^{conc}, I_{sc,bottom}^{conc}\right)}{C \min\left(I_{sc,top}^{1sun}, I_{sc,middle}^{1sun}, I_{sc,bottom}^{1sun}\right)}. \tag{5.39}$$

From Equations (5.37) and (5.38) we get

$$\eta_{opt,polychrom} = \frac{\min\left(\int \lambda \eta_O(\lambda) I_{DS}(\lambda) Q_{Ej}(\lambda) d\lambda\right)}{\min\left(\int \lambda I_{DS}(\lambda) Q_{Ek}(\lambda) d\lambda\right)}, \tag{5.40}$$

which is the ratio of the current per unit cell area generated by a solar cell under a real concentrator with spectral optical efficiency $\eta_O(\lambda)$ and the current per unit cell area generated by a solar cell under an ideal concentrator with spectral optical efficiency $\eta_O(\lambda) = 1$ for all wavelengths.

An approximation may be made considering that function $\eta_O(\lambda)$ has a constant value for the wavelengths of the top cell $\eta_O(\lambda) = \eta_T$, a different constant value for the middle cell $\eta_O(\lambda) = \eta_M$ and yet another constant value for the wavelengths of the bottom cell $\eta_O(\lambda) = \eta_B$. The current per unit area of each junction may be calculated by:

$$I_k = \frac{1}{h_C} \int \lambda I_{DS}(\lambda) Q_{Ek}(\lambda) d\lambda, \tag{5.41}$$

with $k = T$, $k = M$, or $k = B$ for the top, middle or bottom junctions, respectively. The current generated by a solar cell under a real concentrator is then, from Equation (5.38):

$$I_{sc}^{conc} = C A_C \min\left(\eta_T I_T, \eta_M I_M, \eta_B I_B\right). \tag{5.42}$$

If Φ is the flux captured by the entrance aperture of the concentrator, which has an area A_E, we can write $C\eta_k = A_E\eta_k/A_C = (A_E/\Phi)(\Phi\eta_k/A_C) = E_{Ck}/E_A$, where E_{Ck} is the irradiance on the solar cell for the wavelengths of junction k (top, middle, or bottom) and E_A is the irradiance at the entrance aperture of the optic.

Generally, the bottom cell generates more current that the top or middle cells, so the minimum current is either the top or middle current, and we may write:

$$I_{sc}^{conc} = A_E \min\left(\eta_T I_T, \eta_M I_M\right), \tag{5.43}$$

where $A_E = CA_C$ and A_E is area of the concentrator entrance aperture The efficiency is in this case obtained from Equation (5.40) as:

$$\eta_{opt,polychrom} = \frac{\min\left(\eta_T I_T, \eta_M I_M\right)}{\min\left(I_T, I_M\right)}. \tag{5.44}$$

The value of η_T may be estimated ray tracing the optic for different wavelenghts absorbed by the top cell and averaging the transmittance. The same for the value of η_M of the middle cell.

Different values of η_T and η_M can be obtained for different incidence angles α of sunlight on the optic (Fig. 5.2b). From Equation (5.43) we can then calculate the current $I_{sc}^{conc}(\alpha)$ as a function of the incidence angle of sunlight and, therefore, the concentrator acceptance angle based on the output current.

ACKNOWLEDGMENTS

The photos in this chapter are courtesy of Light Prescriptions Innovators (LPI) and the Technical University of Madrid (Universidad Politécnica de Madrid—UPM).

REFERENCES

1. R. Winston, J.C. Miñano, and P. Benítez (contributions by N. Shatz and J.C. Bortz) *Nonimaging Optics*, Elsevier Academic Press, Amsterdam (2005).
2. J. Chaves, *Introduction to Nonimaging Optics*, CRC Press/Taylor & Francis Group, Boca Raton, FL (2008).
3. R. Leutz and A. Suzuki, *Nonimaging Fresnel Lenses: Design and Performance of Solar Concentrators*, Springer, Berlin (2001).
4. W.T. Welford and R. Winston, *The Optics of Nonimaging Concentrators—Light and Solar Energy*, Academic Press, New York (1978).
5. W.T. Welford and R. Winston, *High Collection Nonimaging Optics*, Academic Press, San Diego, CA (1989).
6. J.J. O'Gallagher, *Nonimaging Optics in Solar Energy*, Morgan & Claypool Publishers, San Francisco (2008).
7. J.C. Miñano and J.C. González, New method of design of nonimaging concentrators, *Applied Optics* **31**, 3051–3060 (1992).
8. J.C. Minano, et al., High efficiency non-imaging optics, U.S. Patent 6639733 (2003).
9. P. Benítez, J.C. Miñano, J. Blen, R. Mohedano, J. Chaves, O. Dross, M. Hernández, and W. Falicoff, Simultaneous multiple surface optical design method in three dimensions, *Optical Engineering* **43**(7), 1489–1502 (2004).
10. P. Benitez, et al., Three-dimensional simultaneous multiple-surface method and free-form illumination-optics designed therefrom, U.S. Patent 7460985 (2008).
11. X. Ning, R. Winston, and J. O'Gallagher, Dielectric totally internally reflecting concentrators, *Applied Optics* **26**, 300 (1987).
12. P. Benitez et al., High performance Fresnel-based photovoltaic concentrator, *Optics Express* **18** (S1), 25 (2010).
13. P. Benítez, R. Mohedano, and J.C. Miñano, DSMTS: a novel linear PV concentrator, *Proc. 26 IEEE Photovoltaic Specialists Conference*, Anaheim, California (1997), p. 1145–1148.
14. P. Benítez, R. Mohedano, and J.C. Miñano, Manufacturing tolerances for nonimaging concentrators, *Nonimaging Optics: Maximum Efficiency Light Transfer IV, Proc. SPIE 3139* (1997), p. 98–109.
15. R.P. Friedman, J.M. Gordon, and H. Ries, Compact high-flux two-stage solar collectors based on tailored edge-ray concentrators, *Solar Energy* **56**, 607 (1996).
16. M. Victoria, C. Domínguez, I. Antón, and G. Sala, High concentration reflexive system with fluid dielectric, *Proceeding Twenty Third EPVSEC* (2008), p.132–136.
17. J.C. Miñano, J.C. González, and P. Benítez, RXI: a high-gain, compact, nonimaging concentrator, *Applied Optics* **34**(34), 7850–7856 (1995).

18. J.L. Alvarez, M. Hernández, P. Benítez, and J.C. Miñano, RXI concentrator for 1000X photovoltaic energy conversion, *SPIE Intern. Symposium of Optical Science, Engineering and Instrumentation*, Denver, CO (1999), p. 30–37.
19. A.L. Luque and V.M. Andreev, *Concentrator Photovoltaics*, Springer-Verlag, Berlin, Heidelberg (2010), chapter 6, Concentrator optics.
20. J.M. Gordon and D. Feuermann, Optical performance at the thermodynamic limit with tailored imaging designs, *Applied Optics* **44**(12), 2327–2331 (2005).
21. J.C. Miñano, P. Benítez, and J.C. González, RX: a nonimaging concentrator, *Applied Optics* **34**(13), 2226–2235 (1995).
22. R. Winston and J.M. Gordon, Planar concentrators near the étendue limit, *Optics Letters* **30**(19), 2617–2619 (2005).
23. S. Horne, et al., A solid 500 sun compound concentrator PV design, *IEEE 4th World Conference on Photovoltaic Energy Conversion*, Vols. 1 and 2 (2006), p. 694–697.
24. G.D. Conley, SolFocus Toward $1/W, NREL Growth Forum (November 8, 2005).
25. A. Plesniak, et al., Demonstration of high performance concentrating photovoltaic module designs for utility scale power generation, *Proc. ICSC-5*, Palm Desert, CA (2008).
26. P. Benitez, et al., Optical concentrator, especially for solar photovoltaics, U.S. Patent Publication 20080223443 (2008).
27. H. Ries, J.M. Gordon, and M. Laxen, High-flux photovoltaic solar concentrators with Kaleidoscope based optical designs, *Solar Energy* **60**(1), 11–16 (1997).
28. J.J. O'Ghallagher and R. Winston, Nonimaigng solar concentrator with near-uniform irradiance for photovoltaic arrays, *Nonimaging Optics: Maximum Efficiency Light Transfer VI, Proc. SPIE 4446*, (2001), p. 60–64.
29. D.G. Jenkins, High-uniformity solar concentrators for photovoltaic systems, *Nonimaging Optics: Maximum Efficiency Light Transfer VI, Proc. SPIE 4446* (2001), p.52–59.
30. W. Cassarly, *Nonimaging Optics: Concentration and Illumination, Handbook of Optics*, 2nd ed., McGraw-Hill, New York (2001).
31. L.W. James, Use of imaging refractive secondaries in photovoltaic concentrators, SAND89-7029, Albuquerque, NM (1989).
32. P. Benitez et al., Multi-junction solar cells with a homogenizer system and coupled non-imaging light concentrator, U.S. Patent Publication 20090071467 (2009).
33. J.C. Miñano, et al., Free-form integrator array optics, *Nonimaging Optics and Efficient Illumination Systems II, Proceedings of the SPIE*, Volume 5942 (2005), p. 114–125.
34. J.C. Miñano et al., Free-form lenticular optical elements and their application to condensers and headlamps, U.S. Patent Publication 20080316761 (2008).
35. M. Hernández, A. Cvetkovic, P. Benítez, and J.C. Miñano, High-performance Köhler concentrators with uniform irradiance on solar cell, Invited paper, Nonimaging Optics and Efficient Illumination Systems V, SPIE Vol. 7059 (2008).
36. P. Benitez, J. Wright, et al., High-concentration mirror-based Köhler integrating system for tandem solar cells, *4th World Conference on Photovoltaic Energy Conversion*, Hawaii (2006).
37. M. Hernandez et al., Sunlight spectrum on cell through very high concentration optics, *3th World Conference on Photovoltaic Energy Conversion*, Osaka, Volume 1 (2003), p. 889–891.
38. P. Zamora, et al., Advanced PV concentrators, *34th IEEE PVSC*, (2009).
39. A. Cvetkovic et al., Characterization of Fresnel-Köeler concentrator, *25th European Photovoltaic Solar Energy Conference and Exibition, Valencia, Spain, Proc 1BO.7.6* (August 2010).
40. P. Benitez et al., Köhler concentrator, U.S. Patent Publication 20100123954 (2010).
41. A. Cvetkovic, *Free-form optical systems for nonimaging applications*, PhD thesis, Universidad Politécnica de Madrid, (2009) (http://oa.upm.es/1782/).

CHAPTER 6

LIGHTPIPE DESIGN

William J. Cassarly

6.1 BACKGROUND AND TERMINOLOGY

6.1.1 What is a Lightpipe

Lightpipes are a key optical technology used to transport light from source to target using multiple reflections. Like the water pipes in a building, lightpipes are typically hidden from direct view. They are used in a wide variety of applications, including instrument lighting, automotive dashboards, pool and spa lighting, liquid crystal backlights, and projection systems. Some illustrative examples are shown in Figure 6.1.

Lightpipes are commonly solid entities that use total internal reflection (TIR) at dielectric interfaces (e.g., PMMA to air) to propagate light from source to target. However, a lightpipe can be a hollow reflective structure (e.g., Reference 1) or a hollow lightpipe that uses TIR (e.g., References 2 and 3).

When the amount of delivered light is small, the lightpipe performance criteria may primarily be the ability to route the light from the source to target. Especially in high flux applications, key performance criteria include the efficiency of coupling light from source to target and the distribution of the light exiting the lightpipe.

Lightpipes can provide a uniform light distribution at the lightpipe output. With the proper geometry and input distribution, the output can be so uniform that providing uniformity becomes the primary purpose of the lightpipe. This type of lightpipe is often called a mixing rod or mixer. Square, rectangular, and hexagonal cross section-shaped mixing rods are common.

Lightpipes are also used as angle-to-area converters to concentrate or collimate light. A slow change in the area of the lightpipe along its length is usually sufficient to provide the concentration/collimation. As the required length becomes small, special profiles are required to maximize the concentration/collimation of the lightpipe. Examples from the nonimaging optic literature include the dielectric total internal reflecting concentrator (DTIRC) [4], θ_1/θ_2 converter (e.g., References 5, 6, and 66), and the dielectric compound parabolic concentrator (DCPC) [7].

Lightpipes are used to distribute source flux over a large area for use as backlights. One approach couples light into the edge of the lightpipe and controls the flux that escapes out of the sides of the lightpipe. Because a thin lightpipe can

Illumination Engineering: Design with Nonimaging Optics, First Edition. R. John Koshel.
© 2013 the Institute of Electrical and Electronics Engineers. Published 2013 by John Wiley & Sons, Inc.

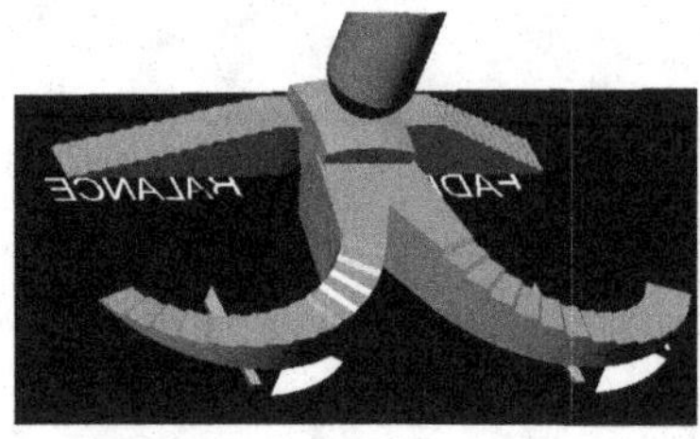

Figure 6.1 Multi-segment lightpipe, LCD backlight, and tapered rippled mixer.

uniformly illuminate a large area, this approach minimizes the use of available package space. The lightpipe is commonly placed close to an appliqué in order to backlight the appliqué in signs and instrument lighting. They are also used to backlight large-area spatial light modulators, such as liquid crystal displays.

Lightpipes are used to distribute source flux over a large area for use as luminaires. Lightpipe luminaires can be hollow or solid, with solid rod versions sometimes used to replace conventional linear fluorescent lamps. Large hollow lightpipes can also be used to create luminaires, such as the system used to illuminate the Space Hall of the Smithsonian's National Air and Space Museum.

While some applications transport the light from one source to one output area, lightpipes are also used to split the light from the source and illuminate multiple separated output locations. The reverse is also true, where light from multiple sources is combined to create a single output.

Many different terms are either synonyms for lightpipe or represent special categories of lightpipes. Example terms include light guides, single mode fibers, graded index fibers, step-index fibers, waveguides, large core plastic optical fibers, mixing rods, mixers, homogenizers, and luminaires.

6.1.2 Lightpipe History

The concept of a lightpipe has been around for a long time. One of the best-known early examples of lightpiping was performed by John Tyndall in 1870. He used a jet of water flowing from one container to another and a beam of light to demonstrate that light used internal reflection to follow a specific path. As the water exited the first container, it followed a curved path into the second container. Tyndall directed a beam of sunlight at the water so that the audience could see that the light followed a zigzag path inside the curved path of the water. Gelatin has also been used to illustrate lightpipe principles [63].

In 1880, William Wheeling patented a method of light transfer called "piping light." Wheeling's idea was to use mirrored pipes branching off from a single source of illumination to light many different rooms. This idea is similar to the way that water, through plumbing, is carried throughout buildings. The use of mirrored pipes limited the efficiency of early systems, but large core plastic optical fibers can now make the concept practical. Similar concepts are used to couple sunlight into the interior rooms of large buildings.

Even earlier reports of light piping were provided by Daniel Colladon [8]. He described a "light fountain" in an article titled "On the Reflections of a Ray of Light

Inside a Parabolic Liquid Stream." Jeff Hecht's *The Story of Fiber Optics* provides more about the history of light piping.

6.2 LIGHTPIPE SYSTEM ELEMENTS

Systems that employ lightpipes usually have three elements: a source, the lightpipe distribution optics, and the delivery optics (Fig. 6.2).

6.2.1 Source/Coupling

Light-emitting diodes (LEDs) are currently the most common light source designed with lightpipes. LEDs have replaced the use of incandescent lamps in most applications. Some lightpipe systems use higher power sources, such as discharge arc sources. Coherent, incoherent, and partially coherent sources are all used with lightpipes.

In order to couple light into the lightpipe, the source can simply be placed close to the lightpipe. Other optical elements, such as reflectors or lenses, are added to increase the coupling optic collection efficiency.

6.2.2 Distribution/Transport

For short distances, both acrylic (e.g., PMMA) and polycarbonate lightpipes are used. For longer distances, PMMA and glass are most common because they can have low absorption with minimal volumetric scatter over a wide range of wavelengths. The material choice is often driven by the environmental and cost factors of the application. For example, high temperature applications tend to use glass or silicone lightpipes, while cost-sensitive applications tend to use PMMA. In some cases, a high temperature lightpipe (e.g., a straight glass rod) is used near the source, and a lower temperature lightpipe is used to distribute the light to remote locations.

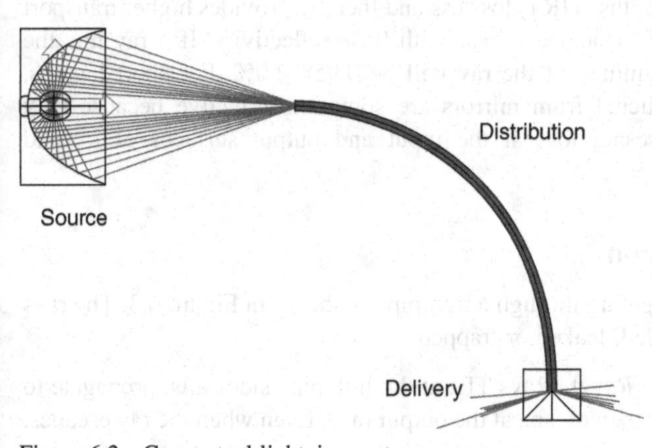

Figure 6.2 Conceptual lightpipe system.

6.2.3 Delivery/Output

The light exiting the end of the lightpipe can be used directly, or a delivery optic can be attached. The simplest and most common delivery optic is a thin diffuser that is added to ensure angular uniformity of the exiting light. The diffuser can be a separate optic element or built into the lightpipe by adding surface texturing on the lightpipe output face or by adding volumetric scattering along a length of the lightpipe.

Light can be delivered from the input end to the output end of a lightpipe. This is usually called an end light system. Light can also be extracted along the length of the lightpipe, and this is sometimes called a side light system. Backlights are a type of side light system; however, that is not a common use of the terminology.

6.3 LIGHTPIPE RAY TRACING

Most of the lightpipe examples shown in this chapter use the concept of total internal reflection (TIR).

6.3.1 TIR

When a ray hits a dielectric interface, the ray splits into a transmitted and reflected component. The percent reflectance and transmittance depends upon the index of refraction on the two sides of the surface and the angle of incidence. For a smooth surface, the ratio of transmitted and reflected energy can be computed using the standard Fresnel equations. When the angle of incidence, measured relative to the surface normal, is greater than the critical angle, the ray is completely reflected, and this is called TIR. Quantitatively, the critical angle is $\sin^{-1}(n_{out}/n_{in})$, where n_{out} is the index of refraction outside of the lightpipe, and n_{in} is the index of refraction inside the lightpipe.

When a ray experiences multiple reflections, which is typical of propagation through long lightpipe lengths, TIR is lossless and thereby provides higher transport efficiency than mirrors. Consider a mirror with 95% reflectivity. If a ray hits the mirror 10 times, the magnitude of the ray will be 0.95^{10}~60%. For short lengths, hollow lightpipes constructed from mirrors are sometime effective because they do not introduce the Fresnel loss at the input and output surfaces of a solid lightpipe.

6.3.2 Ray Propagation

A selection of rays propagating through a lightpipe is shown in Figure 6.3. The rays can be classified as coupled, leaked or trapped.

1. *Coupled Rays (Blue Rays).* Rays TIR at the lightpipe sidewalls, propagate to the output face, and then escape at the output face. Even when the ray escapes,

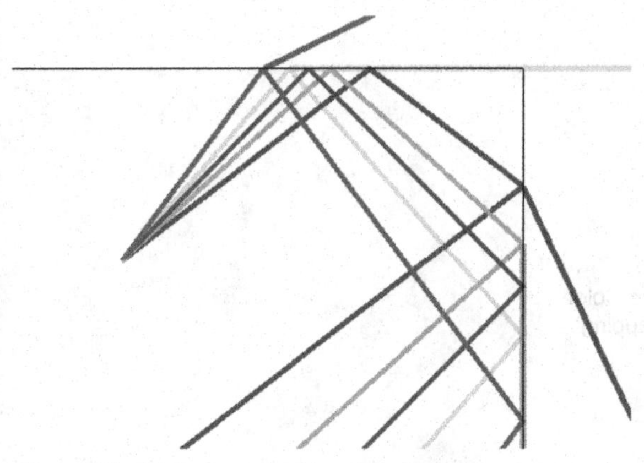

Figure 6.3 Representative rays in a lightpipe.

a small fraction of an unpolarized ray's flux is reflected. If it is critical to avoid this reflection, antireflection coatings can be used.

2. *Leakage Rays (Magenta Rays).* Some of the ray's energy leaks out of the lightpipe sidewalls. In most cases where leakage occurs, a fraction of the ray's energy will still propagate to the output face of the lightpipe.
3. *Trapped Rays (Red Rays).* Rays propagate to the output of the lightpipe, but cannot escape because the ray TIRs. Even when TIR does not occur, a large fraction of the ray's energy can still be reflected.

The boundary between the three cases is shown with the yellow and green rays in Figure 6.3. The yellow ray hits the lightpipe side wall near the critical angle. The green rays hit the lightpipe output face near the critical angle.

Rays that would normally be coupled can change to rays which leak or are trapped as a result of volume scattering, surface imperfections, fluorescence, and changes in the lightpipe geometry. Flux confinement diagrams [9], which are unit sphere constructions based upon an index of refraction of the lightpipe and its geometry, have been introduced as a means to help quantify light propagation in lightpipes.

6.4 CHARTING

LightTools® [10] was used to analyze, design, and optimize all the lightpipe examples shown in this chapter. Simulation results are presented using either false color raster charts or a pseudo true color output (e.g., RGB). For simplicity, chart labels do not include the "false color" designation, but the relative color mapping used in the illuminance charts is shown in Figure 6.4. For clarity, charts that use pseudo true color include "RGB" in the chart name (e.g., RGB illuminance).

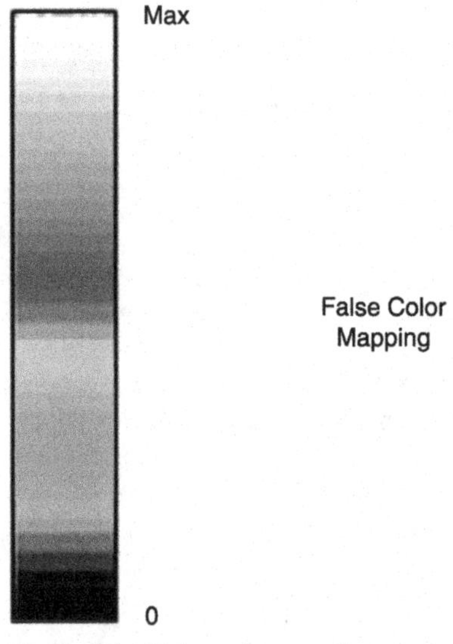

Figure 6.4 False color mapping used with illuminance, luminance, and intensity charts.

6.5 BENDs

When transporting light from one location to another, leakage out of a lightpipe is generally undesired because it reduces the efficiency of the optical system. In some instrument lighting applications, the leaked light may be a small fraction of the total source light, but it can produce enough "glare" that it must be blocked. In some cases, mechanical blockage of the light must be used, but sometimes the bend can simply be oriented so that the leakage is away from the viewer.

A long slowly bent lightpipe will transport light similar to a straight lightpipe. As the bend becomes tighter, leakage can occur. Sometimes, bundles of small-clad fibers are used so that even though the bundle may have a tight bend, the light is transported with high efficiency because the individual fibers have a loose bend.

6.5.1 Bent Lightpipe: Circular Bend

In this section, we want to examine some of the tradeoffs found with bent lightpipes.

6.5.1.1 Setup and Background A slice through the plane of symmetry of a lightpipe with a 90° circular bend is shown in Figure 6.5. This plane of symmetry can be called the *Principal section* [11]. For the circular bend, the lightpipe outer radius is r_2 and the inner radius is r_1. r_1 is typically called the *bend radius*, since r_1 is the radius of a drum that the lightpipe could be wrapped around. The *outer to inner bend ratio* is

$$m = \frac{r_2}{r_1} = \frac{r_o + h}{r_o - h}, \tag{6.1}$$

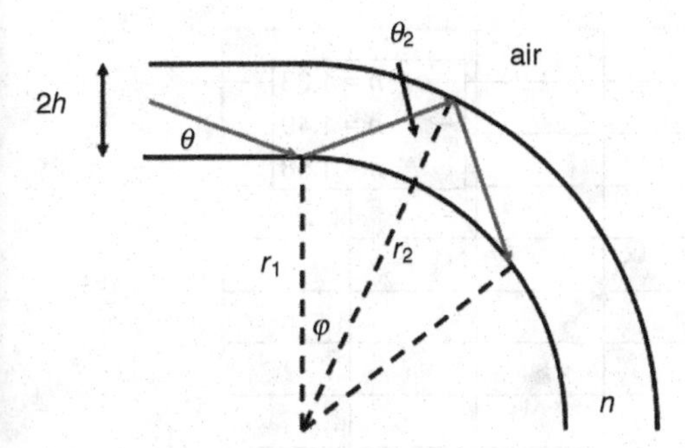

Figure 6.5 Lightpipe with circular bend.

where $r_2 - r_1 = 2h$, $(r_2 - r_1)/2 = r_0$, and r_0 is the *center line radius*. The *bend radius to thickness ratio* is

$$r_1/2h = 1/(m-1). \tag{6.2}$$

Both m and $r_1/2h$ sometimes called the bend ratio (e.g., Reference 67). To avoid confusion, we will refer to m as the *bend index*.

As depicted in Figure 6.5, the angle of the propagating ray, θ, is measured inside of the lightpipe. If the ray refracts at a flat surface before entering the material of the bend, Snell's law ($\sin\theta_{air} = n\sin\theta$, where n is the index of refraction of the lightpipe) can be used to find the angle in air.

As the ray propagates around the circular bend, the skew invariant (e.g., conservation of angular momentum) requires that

$$r_1 \sin(90-\theta) = r_1 \cos\theta = r_2 \sin\theta_2 = s, \tag{6.3}$$

where s is the closest distance that the ray could approach the axis or rotation. To ensure that TIR occurs at the outer surface of a lightpipe, we need

$$\sin\theta_2 > 1/n, \tag{6.4}$$

where θ_2 is the angle of incidence at the outer surface. Rays that do not TIR will split into portion that reflects and another portion that transmits. The portion that transmits and therefore leaks out of the bend is called the *bend leakage*. Bend leakage reduces the bend's light transfer efficiency and sometimes produces unwanted glare that must be blocked.

The ray hits the inner bend surface with angle of incidence equal to $90 - \theta = \theta_2 + \phi$. Since this angle is larger than θ_2, a ray that TIRs at the outer surface and hits the inner surface will TIR at the inner surface.

6.5.2 Bend Index for No Leakage

We can combine Equations (6.3) and (6.4) to provide the no leak relationship

$$\frac{r_2}{r_1} = m < n\cos\theta_{\text{max\_in}}, \tag{6.5}$$

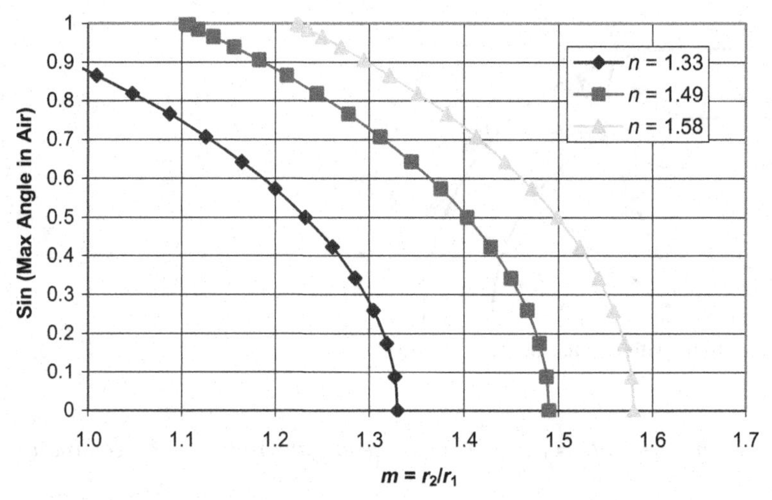

Figure 6.6 No leakage for circular bends (Eq. 6.6) using three common values for n.

where θ_{max_in} is the largest angle inside of the start of the bend that will TIR. Using $\sin\theta_{max\_air} = n\sin\theta_{max\_in}$, we can rewrite the no leakage of Equation (6.5) in the form

$$m < \sqrt{n^2 - \sin^2\theta_{max\_air}}. \tag{6.6}$$

For comparison, the relationship for the cladding index where leakage occurs in a straight stepped index fiber is

$$n_{clad} < \sqrt{n_{core}^2 - \sin^2\theta_{max\_air}} \tag{6.7}$$

where n_{core} is the core index and n_{clad} is the cladding index. Comparing Equations (6.6) with Equation (6.7) shows that m is directly analogous to the clad index. To highlight this relationship and also avoid confusion with different definitions of bend ratio; we call m the *bend index*.

The relationship between bend index and $\sin\theta_{max\_air}$ defined by Equation (6.6) is plotted in Figure 6.6 for three typical indices of refraction ($n = 1.33$ for Teflon, $n \sim 1.49$ for PMMA, and $n \sim 1.58$ for polycarbonate). Tight bends with $m > n$ can produce leakage from the bend surface and looser bends with $m < \sqrt{n^2 - 1}$ avoid leakage from the bend surface.

6.5.3 Reflection at the Output Face

In many practical situations, leakage at the outer surface of the bend is not the only issue to consider. Typically, there is an air interface at the output face of the bend. After propagating around the bend, the angle of incidence at the output face of the bend can be larger than the angle at the input face. This means it is possible for the ray to propagate to the output face, but then TIR at the output face. If the ray experiences TIR at the output face, or partially reflects because of Fresnel losses, then the

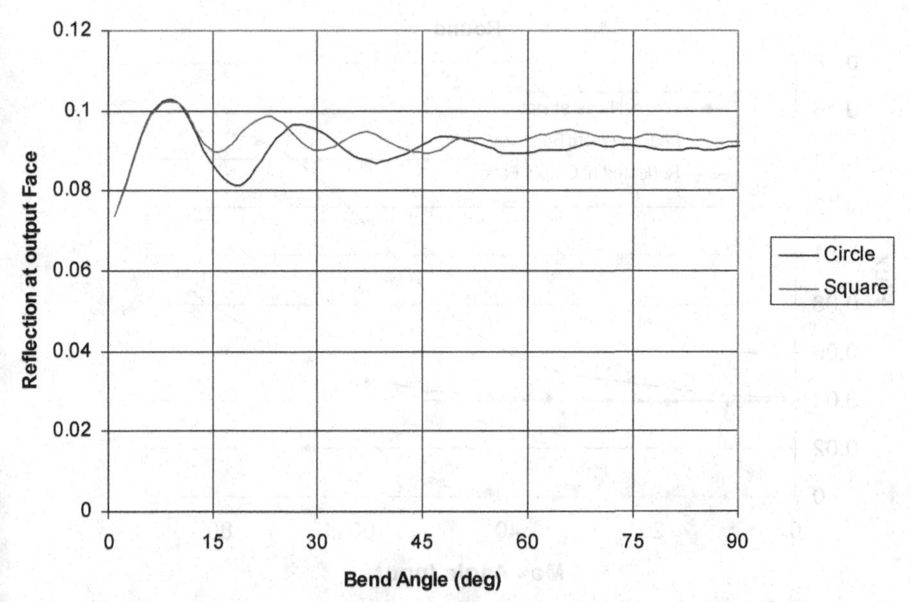

Figure 6.7 First hit reflection at the bend output face as a function of bend arc angle. This cases uses $m = 1.1146$, $n = 1.49378$.

reflected ray will return to the input of the bend since the reflected ray will still satisfy the no-leak criteria.

A 90°-arc angle is depicted in Figure 6.5, but other arc angles can, of course, be used. The amount of light which is reflected from the output face the first time that a ray hits the output face depends upon the arc angle of the bend, as well as the angle and position of the ray at the input face. This effect was analyzed using LightTools with a clipped Lambertian source, with a maximum angle of 84° in air, $n = 1.49378$, an air to n interface at both the input and output faces of the bend, $m = 1.1146$, and the source sized to match the size of the input face. With a 0° arc angle, the input and output faces of the bend are parallel. As the arc angle increases, the fraction of light reflected at the output face will first increase, then decrease, and converge to a relatively constant value as the arc angle gets large. Results for both a circular and a square cross-section lightpipe are shown in Figure 6.7. The ~2% difference between the 0°-arc angle and the 90°-arc angle has been observed experimentally (e.g., Reference 12). Please note that the ~7.4% reflection is higher than the ~3.9% found for normal incidence because Fresnel reflections increase with angle of incidence, and the source in this case is +/−84° in air.

6.5.4 Reflected Flux for a Specific Bend

Now, let us consider a specific bend configuration and examine its leakage and first reflection at the output surface for different θ_{max_air}. Simulation results for the case of $m = 1.408$, and $n = 1.49378$ are shown in Figure 6.8. The no leakage condition, $m < \sqrt{n^2 - \sin^2 \theta_{max\_air}}$, is satisfied with $\theta_{max\_air} < 30°$. Even for θ_{max_air} larger than 30,

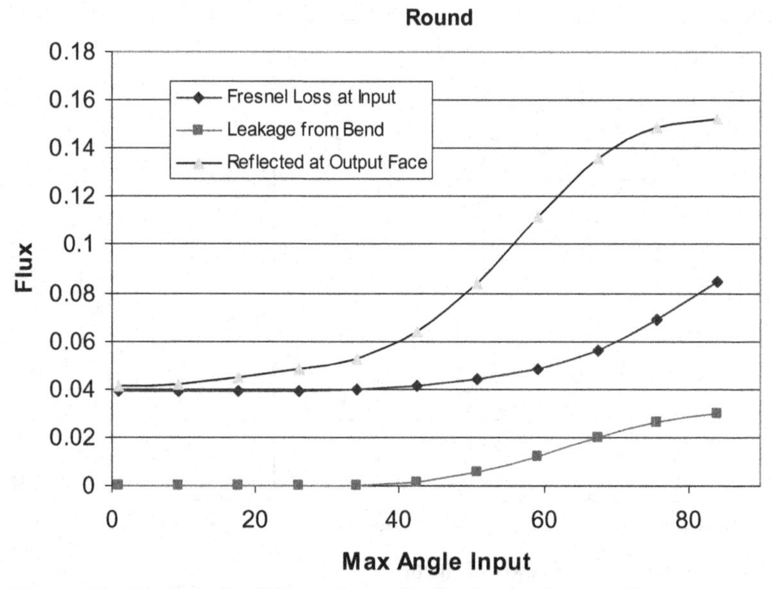

Figure 6.8 Leakage for different input distributions and a specific m.

the loss from reflections at the output face can be larger than the leakage out the bend surface.

6.5.5 Loss Because of an Increase in NA

Now, we will consider a specific θ_{max_air} and examine how the flux coupled to the output face varies for different bend ratios. In this case, we want to include the fact that the angular distribution of the flux propagating around the bend increases, since that increase can be critical if the flux is going to be transported using a clad fiber. We would have an additional loss if the bend itself is clad with a lower index material than air, but even if the bend is immersed in air, very small m are required to avoid losses. Figure 6.9 shows efficiency results with four different clipped Lambertian sources (5, 15, 30, and 45°), where $m < \sqrt{n^2 - \sin^2 \theta_{max\_air}}$. Efficiency is flux within the clip angle divided by the total flux exiting the bent lightpipe. Even though no leakage occurs, the efficiency drops because the angular distribution widens compared with the original distribution.

Depending upon the application, the increase in NA caused by the bend can be a bigger effect than "leakage" out of the bend. One example occurs when a clad fiber is used to distribute the light after the bend. The increase in NA caused by the bend may produce a significant reduction in the amount of light propagated by the clad fiber. Another example occurs when considering uniformity. With a bent lightpipe, the change in the angular distribution NA can result in a corresponding change in the spatial uniformity, even though there is no leakage. In contrast, if the input face of a straight lightpipe is uniformly filled and there is no leakage, then the output is also uniform.

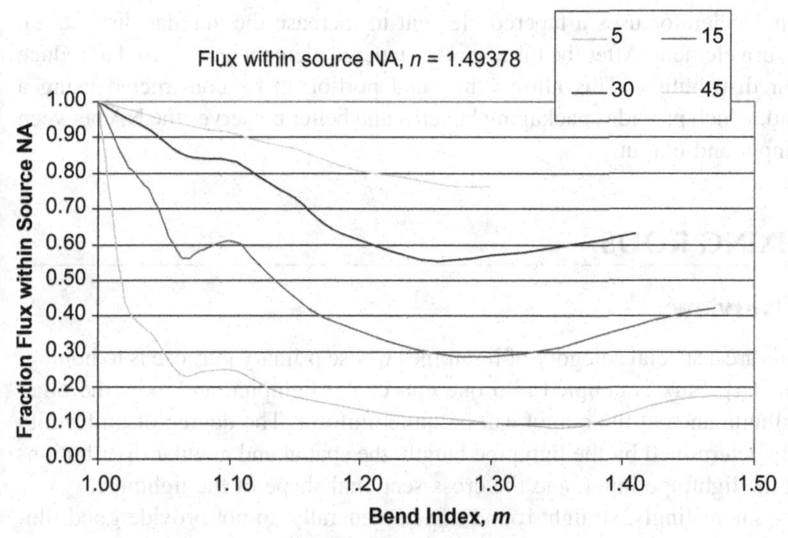

Figure 6.9 NA preservation for circular bend.

It is also worth noting that the amount of leakage varies as a function of position at the input lightpipe surface. The total leakage is roughly the same if source uniformly fills the input face of a circular or a rectangular lightpipe; however, if a small source is placed in the middle of the input face, then the square lightpipe leaks less than the circular lightpipe (e.g., Reference 11; Figure 6.9 is for a small source at center input face).

6.5.6 Other Bends

A bent lightpipe does not need to follow a circular arc. The arc can follow other curves. Elliptical arc paths can sometimes be used to make modest improvements in efficiency and spatial uniformity.

When the primary design goal of a bend is to avoid leakage out of the TIR surface, the equiangular spiral can be used to specify the surface profile. This concept has been used with waveguides [13]. Equiangular spirals have also been applied to no-loss lightpipes [14], albeit with geometries where the cross-sectional area varies along the bend path. For modest arc angles, the equiangular path is nearly an ellipse.

With rectangular lightpipe cross sections, an "angle rotator" [15, 61] can be used to rotate the direction of the radiation without changing the extents of its angular aperture. This concept is ideal for rays in the principal plane that are within $\pm\theta$ of the input surface normal. Except with $\theta = 90°$, the shape of the preserved angular aperture is not rotationally symmetric (e.g., Reference 62). The preserved angular aperture is $|M/N| < \tan\theta$ and $|L/N| < \tan\theta$, where L, M, and N are the direction cosines of rays relative to the exit faces of the angle rotator.

When the angular distribution entering the lightpipe is less than a hemisphere, an "accelerator" [16] can be used to improve uniformity and efficiency. The simplest form of an accelerator uses a tapered element to increase the angular distribution entering a turn element. After the turn, another tapered element can be used to reduce the angular distribution. This allows the bend portion to be constructed using a tighter bend, which provides packaging benefits and better preserves the NA between the bend input and output.

6.6 MIXING RODS

6.6.1 Overview

Mixing rods are a special category of lightpipes whose primary purpose is to homogenize light [64]. Flux is coupled into one end of the lightpipe and exits the other end. The illuminance at the output can be quite uniform. The degree of uniformity is primarily determined by the lightpipe length, the spatial and angular distributions of light at the lightpipe input, and the cross-sectional shape of the lightpipe.

Quite interestingly, straight round mixers generally do not provide good illuminance uniformity; whereas square and hexagonal shaped mixers do provide good illuminance uniformity. An overview example, where a source is coupled into a mixing rod using an elliptical reflector, is shown in Figure 6.10.

Many applications use mixing rods to provide uniform distributions. Some examples include down lighting, theater lighting, solar [17, 18, 69], lithography [19, 20], treatment of skin lesions [21, 22], laser beam shaping [23, 24], laser scanning [25], fiber-to-fiber couplers [26], fiber bundles with discharge lamps [18], and displays [27, 28]. There have been some recent articles that help to quantify the performance of straight mixers [29–31].

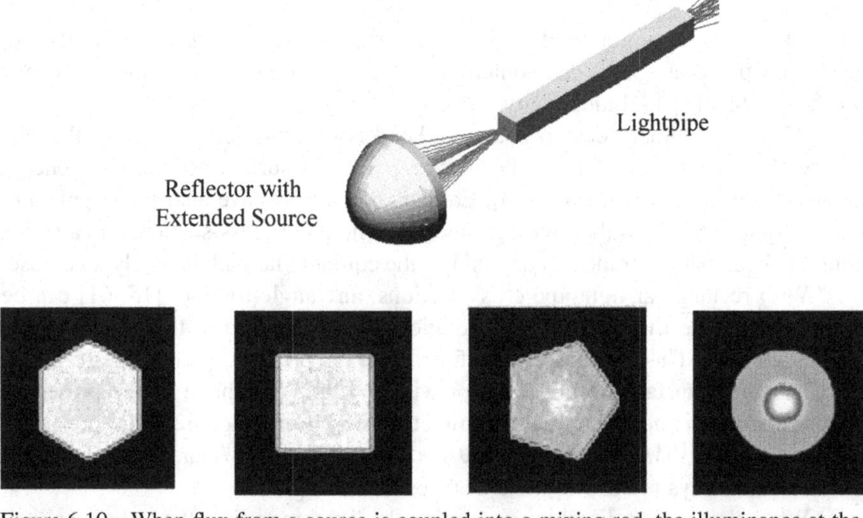

Figure 6.10 When flux from a source is coupled into a mixing rod, the illuminance at the output end of the mixing rod is significantly affected by the shape of the lightpipe.

A mixer can be used in ways that are unrelated to providing uniformity at the mixer output. Examples range from kaleidoscopes to Talbot plane coupling to optical computing (e.g., References 32–34).

6.6.2 Why Some Shapes Provide Uniformity

Square mixers are used to provide extremely uniform illuminance distributions. To understand how this works, consider the illuminance distribution that would occur at the output plane of the lightpipe if the lightpipe sidewalls are removed. That no-lightpipe distribution can be decomposed into N subregions. The lightpipe sidewalls cause the N subregions to be superimposed. This is illustrated in Figure 6.11 using a Williamson-type construction [35, 36], which is also called a tunnel diagram.

In the case of a square mixer, each point in the mixer output face receives N contributions, and each point averages those contributions. If the N contributions were statistically independent, then the central limit theorem would predict uniformity with a standard deviation that is proportional to $1/\sqrt{N}$. However, many source distributions used with mixing rods are symmetric about the optical axis. As a result, the subregion distributions with a positive slope and negative slope are superimposed and cancel each other, which further enhances the uniformity compared with a central limit theorem prediction.

Figure 6.12 shows the results of a Monte Carlo simulation where a source with a Gaussian illuminance distribution and a Gaussian intensity distribution is

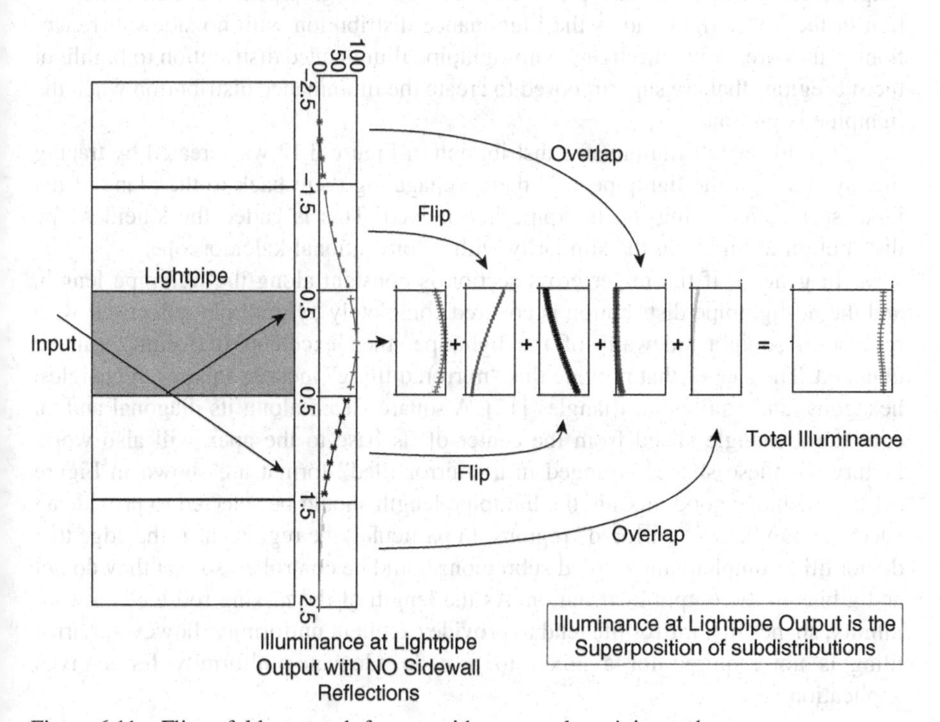

Figure 6.11 Flip-n-fold approach for use with rectangular mixing rods.

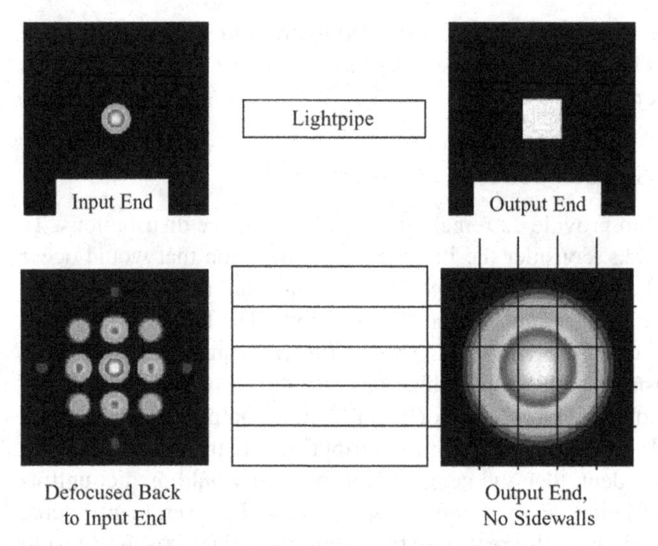

Figure 6.12 Mixer example showing the illuminance at the input and output of a mixer (upper images), as well as the no-mixer distribution (lower right), and the kaleidoscope distribution (lower left).

propagated through a square lightpipe. The illuminance at the lightpipe input and output faces is shown in the top of the figure. The no-lightpipe illuminance distribution in the lower right shows the illuminance distribution with no sidewall reflections. Lines are drawn through the no-lightpipe illuminance distribution to highlight the subregions that are superimposed to create the illuminance distribution when the lightpipe is present.

The lower left illuminance distribution in Figure 6.12 was created by tracing the rays through the lightpipe, and then propagating them back to the plane of the input surface assuming the lightpipe is removed. This is called the kaleidoscope distribution to highlight the similarity with a conventional kaleidoscope.

In general, if the mixer cross section is constant along the lightpipe length, and the no-lightpipe distribution is covered completely by multiple reflections with respect to straight sidewalls of the lightpipe, then excellent uniformity can be obtained. The shapes that provide this "mirrored tiling" include squares, rectangles, hexagons, and equilateral triangles [17]. A square sliced along its diagonal and an equilateral triangle sliced from the center of its base to the apex will also work. Pictures of these shapes arranged in a "mirror tiled" format are shown in Figure 6.13. To achieve good mixing, the lightpipe length should be selected to provide an adequate number of "mirrored" regions. In particular, the regions near the edge that do not fill a complete mirror tiled subregion should be controlled so that they do not add a bias to the output distribution. As the length of the mixing rod tends toward infinity, shapes that mirror tile tend to provide excellent uniformity; however, mirror tiling is not required for a mixer to provide adequate uniformity for a given application.

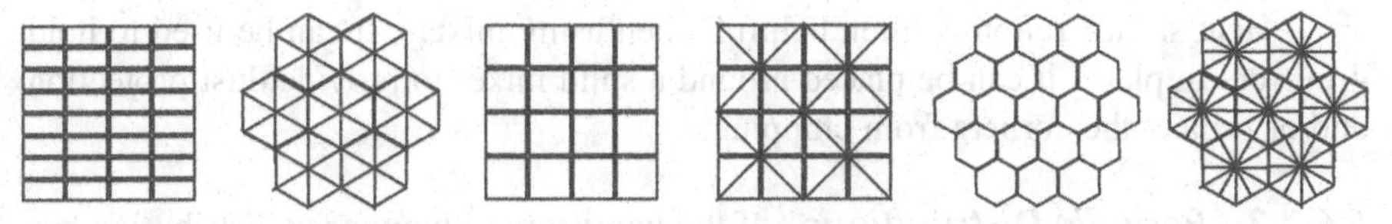

Figure 6.13 Mirror tiled shapes that can ensure uniformity with adequate length.

Round lightpipes can provide annular averaging of the angular distribution of the flux that enters the "input" end. Because of this, the illuminance distribution along the length of the lightpipe tends to be rotationally symmetric; however, the illuminance distribution tends to be peaked near the center of the lightpipe. The magnitude of this peak varies along the length of the lightpipe. If the distribution entering the lightpipe is uniform across the whole input face, and the angular distribution is a clipped Lambertian, then the round lightpipe maintains the uniformity that was present at the input end. However, if either the input spatial distribution or the input angular distributions are not uniform, then nonuniform illuminance distributions can occur along the lightpipe length. This means that a source can be coupled into a square lightpipe of adequate length to provide a uniform illuminance distribution at the lightpipe output face; however, if this uniform distribution is then coupled into a round lightpipe, then the illuminance at the output face of the round lightpipe may not be uniform.

6.6.3 Design Factors Influencing Uniformity

6.6.3.1 Length The length of the lightpipe that is required to obtain good uniformity will depend upon the details of the source intensity distribution. In general, overlapping a 9 × 9 array of "mirrored" regions provides adequate uniformity for most rotationally symmetric distributions. For an f/1 distribution with a hollow lightpipe, this means that the lightpipe aspect ratio (length/width) should be greater than 6:1. For a lightpipe made from acrylic, the aspect ratio needs to be more like 10:1 to compensate for the fact that the angular distribution inside of the mixer is lower than the angular distribution in air.

6.6.3.2 Solid versus Hollow Square cross section mixers are perhaps the simplest mixers to fabricate. Four first surfaces mirrors or a piece of glass or plastic polished on 6 sides is all that is required.

The design of the solid mixer should consider Fresnel surfaces losses at the input and output ends, material absorption, cleanliness of the sidewalls, and chipping of the corners. With high power lasers and high power Xenon lamps, a solid coupler may be able to handle the average power density, but the peak that occurs at the input end may introduce damage or simply cause the solid lightpipe to shatter.

The design of a hollow mixer should consider dust on the inside surfaces, coating losses, coating angular variations, coating color variations, and chips in the mirrors where the sidewalls join together.

Heat shrink Teflon is often helpful when using mixers. It can be used to hold 4 mirrors in place. It can be placed around a solid mixer to provide dust protection and/or protect the corners from chipping.

6.6.3.3 Periodic Distributions If the unmirrored illuminance distribution has a sinusoidal distribution with certain periods that matches the width of the lightpipe, then the peaks of the sinusoid can overlap. A Fourier analysis of the lightpipe illuminance distribution can help to uncover potential issues [35].

6.6.3.4 Coherence Mixing rods are sometimes used with coherent sources. If the coherence length of the laser is not small compared with the path length for overlapping distributions, then speckle effects must be considered (e.g., References 19 and 23). Speckle effects can be mitigated if the lightpipe width is made large compared with the size of the focused laser beam used to illuminate the input end of the lightpipe (e.g., References 19 and 37). A negative lens for a laser illuminated lightpipe is described by Reference 24; however, this means that the étendue of the source is less than the étendue of the lightpipe output.

Speckle can be controlled in lightpipes by adding a time-varying component to the input distribution. Some methods include a rotating diffuser at the lightpipe input end and moving the centroid of the distribution at the input end. In principle, the diffuser angle can be small so that the change in étendue produced by the diffuser can also be small.

Coherent interference can also be used advantageously, such as in the use of mixing rods of specific lengths that provide sub-Talbot plane reimaging. This was successfully applied to an external cavity laser [34].

6.6.3.5 Angular Uniformity The angular distribution exiting a mixer depends upon both the spatial and angular distribution entering the mixer. Consider a small 30° clipped Lambertian source placed in the center of a 100-mm long, 10-mm × 10-mm square hollow mixer. Because the source intensity distribution is symmetric, the output intensity is also symmetric. The illuminance and intensity of the light exiting the square mixer is shown in Figure 6.14, where the intensity plot extents are +/–40°. Now, shift the center of the mixer so the source is in the corner of the

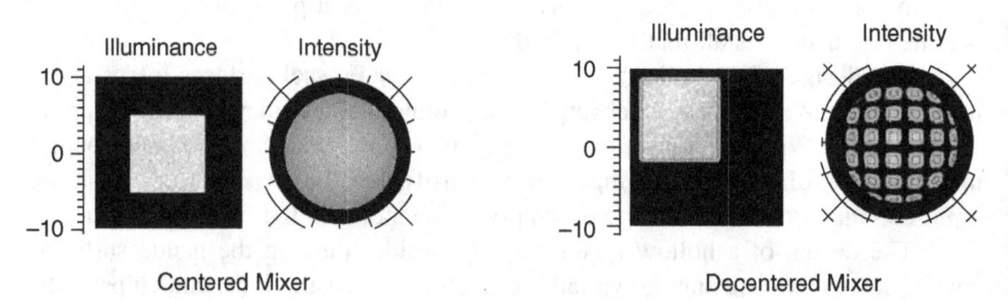

Figure 6.14 Illuminance and intensity exiting a square mixer with a 30° clipped Lambertian source placed at the center (left) and corner (right) of the mixer input.

mixer. The angular distribution now has peaked regions surrounded by regions of 0 intensity. At this specific length, there is also a slight corner to corner variation in the illuminance at the exit face of the mixer. This type of slight variation is typical tolerance to consider when using mixers. This tolerance should be considered with a decentered source and also a source that enters the mixer with a slight tilt angle.

The variation in the angular distribution can often be minimized by adding a low angle diffuser at the output end of the mixer (e.g., Reference 38). Nominally, the full-width half-max of the diffuser is one to three times the angle subtended by the mixer input when viewed from the mixer output. By making the mixer longer, the subtended angle is reduced and the increase in Etendue caused by the diffuser can be reduced. Lens arrays and lightpipes can also be combined to provide control over both the output illuminance and output intensity [68].

An optical system that reimages the output face of a mixer often assumes a "pupil" that is conjugate to the input end of the mixer. When the pupil reimages the input face of the mixer, the illuminance in the pupil typically shows more structure than the intensity distribution exiting the mixer.

When a system requires a uniform intensity distribution, as compared with a uniform illuminance distribution, the face of the mixer can be imaged into the far field. In order to minimize mixer length, the angular distribution of the light entering the mixer is typically large, and the use of aspheric lenses is common.

6.6.3.6 Circular Mixer with Ripples A circular mixer does not provide the exceptional spatial uniformity obtained with a square mixer; however, similar performance has been achieved by adding ripples around the perimeter of a round mixer [39]. To illustrate, consider the source ellipse mixer system shown in Figure 6.15. The illuminance at the input and output end of a 10:1 aspect ratio circular mixer is shown. The resulting illuminance is not very uniform. The circular mixer is then

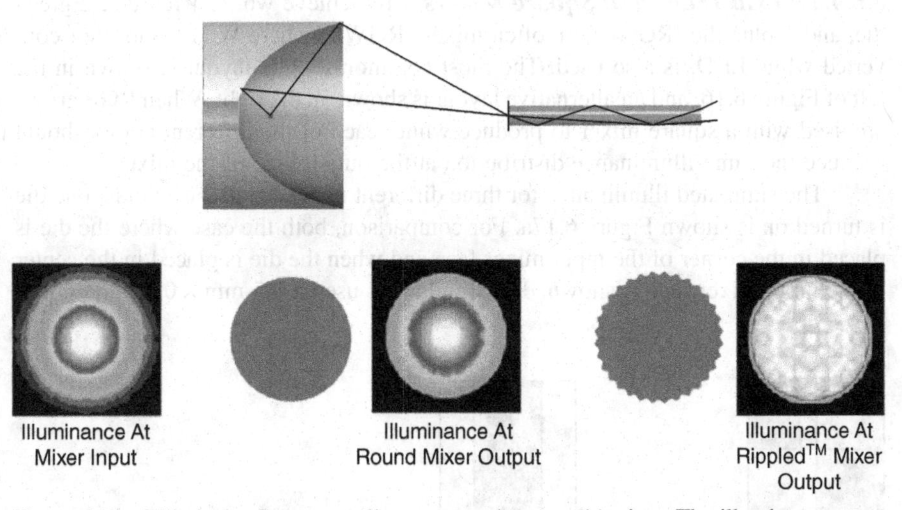

Figure 6.15 Elliptical reflector coupling a source into a solid mixer. The illuminance at the mixer input and output is shown for a circular mixer and a circular mixer with ripples.

replaced with a constant cross-section mixer that has ripples around the perimeter. The resulting illuminance is very uniform.

The simulation result for the rod with ripples in Figure 6.15 uses a sinusoidal ripple profile with 128 peaks and valleys going around the circumference of the mixer. The height of the sinusoid was adjusted to provide an angle of ~30° between the peaks and valleys. To make the profile more observable, 32 peaks and valleys are depicted in the mixer end view image shown in Figure 6.15. A significant number of ripples is typically used because good mixing does not depend upon the exact number of ripples and the ripple size becomes small compared with the mixer.

There are two key aspects to the ripple profile: rounded ripples and an extruded ripple profile. The rounded profile introduces a range of slope perturbations. Sinusoidal, multi-segment faceted profiles, and randomized facet angles are some of the many ways to provide a range of slope perturbations. V-groove profiles will work in some cases; however, the V-groove angle required to achieve good uniformity can be sensitive to the light distribution at the mixer input. The use of an extruded ripple profile helps to ensure that the numerical aperture of the input flux distribution is not increased. An extruded ripple profile also tends to make fabrication simpler.

6.6.4 RGB LEDs

LEDs are used in many different applications. The need for uniformity with LEDs has raised interest in the use of mixers with LEDs. Of particular interest is the use of mixers to combine the output of different colored LEDs, such as red (R), green (G), and blue (B) to produce white. White is achieved by adjusting the relative output of the individual die. The ratio of the output of the different colored LEDs can be changed to adjust the color temperature of the output. Much of this section follows the treatment found in Reference 40.

6.6.4.1 RGB LEDs with Square Mixers To achieve white, 1 red die, 2 green die, and 1 blue die (RGGB) are often mixed. RGWB, where W is a phosphor converted white LED, is also used. The most common RGGB layout is shown in the left of Figure 6.16, and an alternative layout is shown on the right. When RGB arrays are used with a square mixer to produce white, each of the different colors should produce the same illuminance distribution at the output face of the mixer.

The simulated illuminance for three different mixer lengths when just one die is turned on is shown Figure 6.17a. For comparison, both the case where the die is placed in the corner of the input mixer face and when the die is placed in the center of the input mixer face is shown. The simulations used a 0.5-mm × 0.5-mm emitter

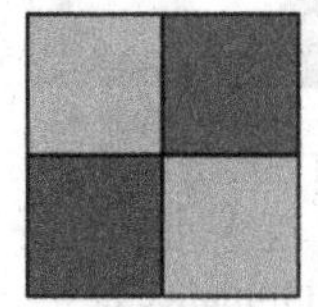

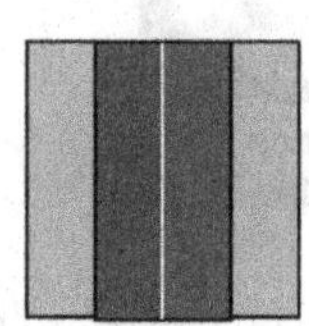

Figure 6.16 Two rossible RGGB layouts.

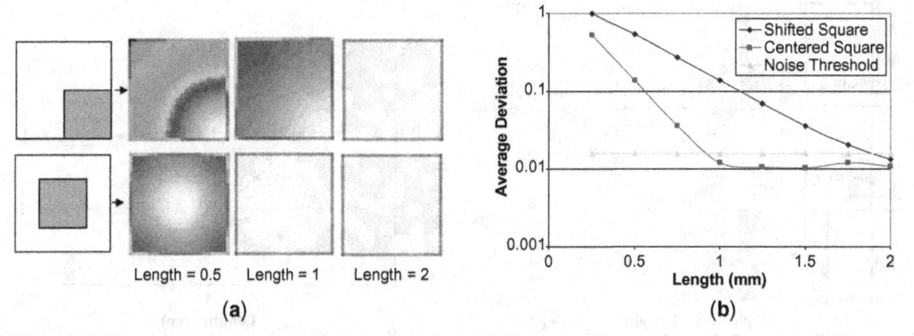

Figure 6.17 (a) Illuminance at mixer lengths of 0.5, 1, and 2 mm for a 0.5-mm × 0.5-mm emitter placed in the corner (shifted square) and in the center (centered square). (b) The relative deviation as a function of mixer length.

with constant luminance into a hemisphere. The 1-mm × 1-mm mixer is hollow with 100% reflectivity. The simulation results are obtained using 400,000 rays, and the illuminance is computed using a 10×10 array of bins. As Figure 6.17 shows, a longer mixer is required for the die placed in the corner of the input mixer face than for the case where the die is placed in the center of the mixer. The normalized standard deviation of the illuminance for the 10×10 bins is plotted in Figure 6.17b. Also included in the plot of Figure 6.17b is the noise threshold computed using $\sqrt{1/n}$, where n is the average number of rays in each bin.

One of the key issues to note in Figure 6.17a is that the nonuniformity of the illuminance is dominated by a graded variation from corner to corner. Because of symmetry with the two green emitters, the required mixer length is therefore dominated by the length required to achieve uniformity with the R and B die. This 2×2 emitter layout is natural, considering that the common emitter size is square. Square die are common, but the die could be rectangular and a 1×4 layout used to provide similar source output. Consider the cases of a 0.25-mm × 1-mm emitter as shown in Figure 6.18a. Like in Figure 6.17a, the illuminance at three different mixer lengths is shown, and the corresponding normalized standard deviation of the illuminance is plotted in Figure 6.18b. These simulation results show that if the GRBG layout shown on the right of Figure 6.16 is used, then the uniformity is superior because the Rect_Shift case is better than the Rect_Edge case, and the use of two green emitters allows much of the nonuniformity of the Rect_Edge case to be cancelled. For a desired level of uniformity, these simulations show that a 4×1 layout requires ~0.3-mm shorter length with a 1-mm × 1-mm mixer than the 2×2 layout.

6.6.4.2 *RGB LEDs with Circular Mixers* Like the case of RGB LEDs used with square mixers, excellent uniformity can be achieve through the use of ripples on a round mixer. Figure 6.19 shows an RGGB LED array placed at the input face of a straight round dielectric mixer. The rod diameter is 3 mm, and the length is 9 mm, and the index of refraction is 1.5. The illuminance of the individual emitters is also shown in Figure 6.19. There is a noticeable variation in the color, especially

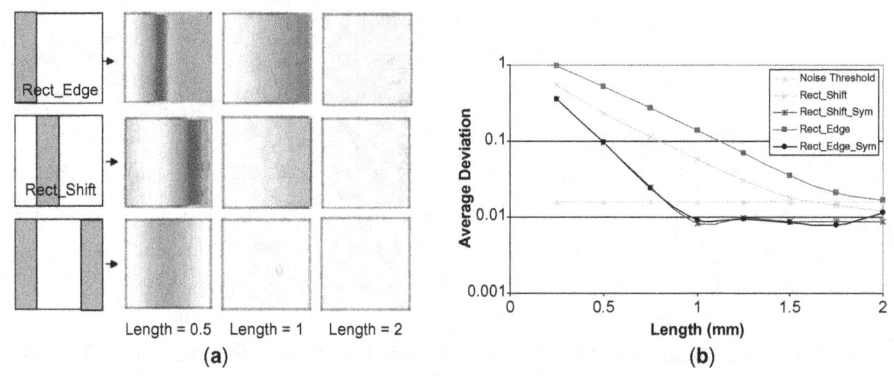

Figure 6.18 (a) Illuminance at mixer lengths of 0.5, 1, and 2 mm for a 0.25-mm × 1-mm emitter placed at the left edge (Rect_Edge), slight shifted from the center (Rect_Shift) and two emitters at the two edges. (b) The relative deviation as a function of mixer length.

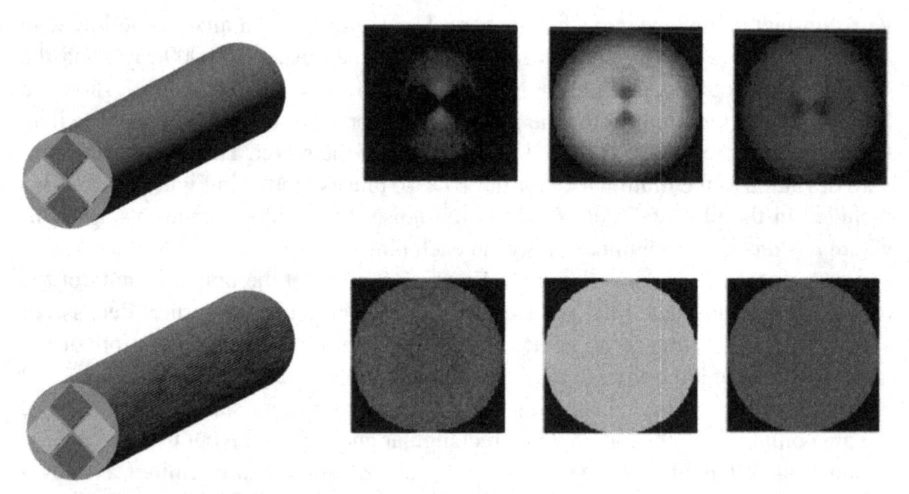

Figure 6.19 RGGB LED array with a round mixer (upper) and a round mixer that has ripples applied around the perimeter (lower). The illuminance at the output of the rod for each of the emitters is shown on the right.

near the center of the mixer, for the case of a smooth mixer. In contrast, a mixer with ripples applied around the circumference is also shown in Figure 6.19. In this case, the uniformity at the output is significantly better.

6.6.5 Tapered Mixers

Tapered mixers are used in some applications because they can shorten the required mixer length compared with a straight mixer. Tapered mixers have been used with LCD projector systems, multimode lasers (e.g., Reference 25), and solar furnaces [17].

The change in area of a tapered mixer means the mixer also provides an angle-to-area conversion. In the ideal case, the projected solid angle (PSA) of the collected light times the area of the input is equal to the PSA of the output light times the area of the output.

Consider the situation where the optics that couple the light into the mixer are not predetermined. Better mixing is obtained by starting with high angles at the lightpipe input and tapering to lower angles than if the input distribution starts with the lower angles [18]. A Williamson construction of a tapered mixer can be used to show that the tapered pipe has more virtual images than the straight tapered case.

6.6.5.1 Length With an infinite length, a straight tapered mixer can provide an ideal angle-to-area transformation and perfect mixing. This is sometimes called the adiabatic theorem [41]. If the mixer or the light source distribution is asymmetric, the adiabatic theorem does not mean that the angular distribution for an infinite length will become rotationally symmetric.

Nonimaging optics (e.g., Reference 42) provides design approaches for "ideal" angle-to-area converters with reasonable lengths (e.g., CPCs, DCPC, theta1-theta2 converters, and DTIRC). Mixers can take advantage of these design forms; however, ideal angle-to-area converters often assume that the source has constant luminance. When a source that does not have constant luminance is used with a device designed to produce ideal angle-to-area transformation with a constant luminance, the result is typically a nonuniformity in the exiting angular or spatial distribution (or both).

Often, short nonimaging optic angle-to-area converters can be used as starting points for a mixer design. In some cases, the profile versus length of an ideal angle-to-area converted can be adjusted to enhance uniformity. More typically, the length will also need to be increased to provide acceptable spatial uniformity at the output face of the mixer.

Illumination optimization is often effective for finding the optimum coupler profile that provides a desired compromise between uniformity and angle-to-area conversion. Choosing effective variables to minimize optimization time can be important. Optimization of lightpipes is an active area of research (e.g., References 43 and 67). A general trend is that the concavity of the lightpipe shape can be used to optimize lightpipe output uniformity [43]. If the output of a mixer of a given length is dipped in the center, then the mixer profile will generally need to be made more concave to reduce the peak and provide better uniformity. Likewise, if the output of the mixer is dipped in the center, then the profile versus length will generally need to be made more convex. In many cases, these types of subtle shape changes can mean that there will be tight fabrication tolerances.

6.6.5.2 Straight Taper Plus Lens Just for fabrication simplicity, the most common tapered mixer approach is a straight tapered mixer whose length is long enough to provide adequate uniformity and adequate angle-to-area conversion. Sometimes, this results in a mixer which provides acceptable uniformity but less than ideal angle-to-area conversion. A lens added near the output face of the mixer can improve the angle-to-area conversion; however, the impact of the lens becomes less significant as the mixer becomes long enough to provide adequate uniformity.

To illustrate the basic tradeoffs, we examined a 2 × 2 RGGB LED with a 2.2-mm × 2.2-mm hollow mixer that tapers to a 6.6-mm × 6.6-mm output. A side-view of the mixer is shown in Figure 6.20. The luminance of the 1 mm × 1 mm LEDs is Lambertian, and there is a 0.1-mm gap between the die. Ideally, 100% of the light could be contained within a 19.5° half-cone angle at the mixer output face. We simulated the output for different lengths, and report the angle within which 90, 95 and 99% of the light is collected. This is shown on the right of Figure 6.20. The results show that it takes about a 50-mm coupler to achieve nearly ideal angle-to-area conversion. Also shown in Figure 6.20 is the 95% half-cone angle for the case where a thin ideal Fresnel lens is placed at the output face of the coupler. In terms of angle-to-area conversion, the impact of the lens is significant, especially at shorter coupler lengths.

Of course, the angle-to-area results in Figure 6.20 are only part of the story. The spatial uniformity at the mixer output must also be considered. In Figure 6.21, we show the illuminance at the output face with just one of the die turned on for the four different lengths. The location of the die that is emitting is shown using a dashed line in the length = 5 mm result. Two million rays with 20 × 20 bins (1.4% noise) were used in the simulations.

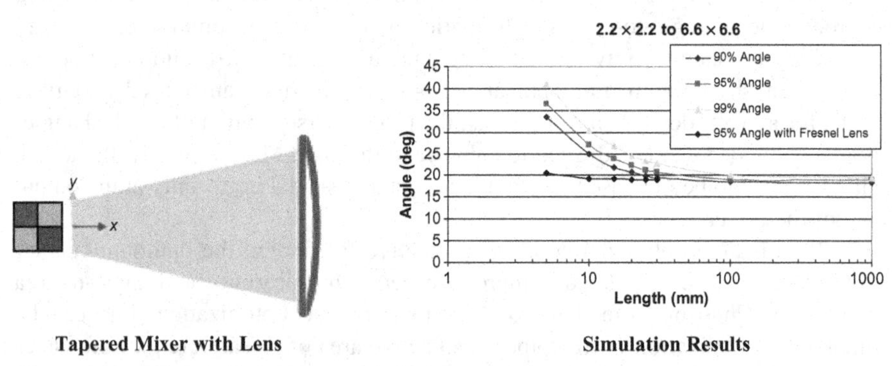

Figure 6.20 Mixer with a lens and the collection angle of the light exiting the lens.

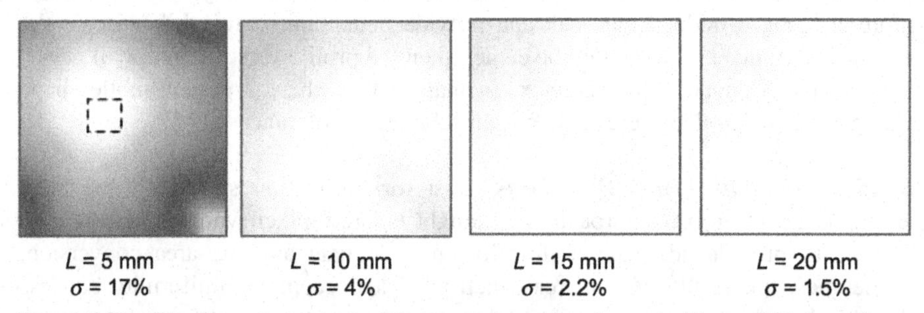

Figure 6.21 Illuminance at output of the tapered coupler shown in Figure 6.20. Lengths of 5, 10, 15, and 20 mm are shown, along with the corresponding relative standard deviation of the illuminance (σ).

These results illustrate that the use of a lens with a tapered coupler does reduce the length of an angle-to-area converter; the length required for uniformity is often long enough that the impact of the lens is modest.

To quantify color performance, the Cx,Cy chromaticity of the LEDs is R = (0.7,0.3), G = (0.181,0.726), and B = (0.153, 0.027). The flux of the LEDs is adjusted so that the average chromaticity of all the emitters is (0.331, 0.331). We compute the color difference for each bin in the mixer output face using

$$|u'v'|_i = \sqrt{(u_i' - u_{\text{avg}}')^2 + (v_i' - v_{\text{avg}}')^2}, \tag{6.8}$$

where u′,v′ is the chromaticity of the ith bin and u'_{avg} and v'_{avg} is the average of all the bins. When using Monte Carlo simulations, smoothing of simulation data can be beneficial in order to reduce the required number of traced rays (e.g., Reference 44).

We computed the mean, $<|u'v'|>$, and standard deviation, $\sigma_{u'v'}$, of the color differences among the bins. In many cases, $<|u'v'|> + 2,\sigma_{u'v'}$ is a reasonable performance criteria and is similar to the maximum $|u'v'|$ of all the bins. Typically, when $<|u'v'|> + 2\sigma_{u'v'}$ is >0.006, the color differences are observable. $<|u'v'|> + 2\sigma_{u'v'} < 0.003$ is a reasonable criteria for color uniformity in some applications. $<|u'v'|> + 2\sigma_{u'v'}$ for lengths of 5 mm is ~0.03, 10 mm is ~0.0071, 15 mm is ~0.0048, and 20 mm is ~0.0030.

6.6.5.3 *Angular Uniformity*

Some applications require both angular and spatial uniformity. Examples include virtual displays, lithography, and general lighting applications.

In virtual displays, a mixer with a rectangular output face is sometimes imaged onto a small spatial light modulator (SLM). Optics are placed between the SLM and the viewer so that the viewer sees a magnified image of the SLM. In this case, the design for uniformity needs to include the fact that the eye can move around, and for a specific eye position, the eye will only collect a small portion of the light which illuminates the SLM.

An example system is shown in Figure 6.22 (similar to Reference 28). An RGGB LED array is coupled into a 2.4-mm × 2.4-mm to 9.3-mm × 5.6-mm solid mixer with 52-mm length. Relay optics image the mixer output face onto the SLM, and the pupil of these relay optics is imaged near the pupil of the viewer's eye. In

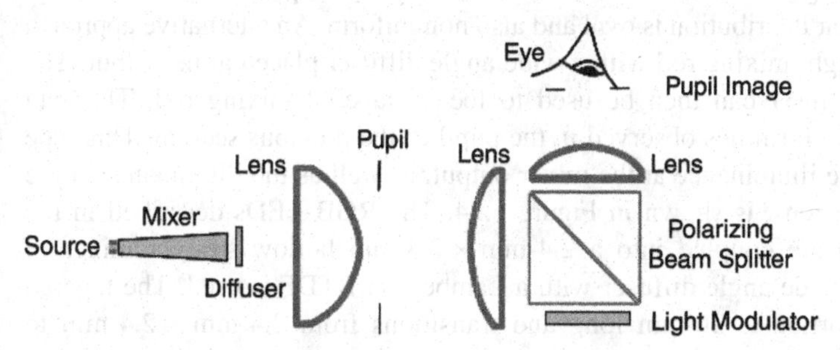

Figure 6.22 Virtual display system using a tapered mixer.

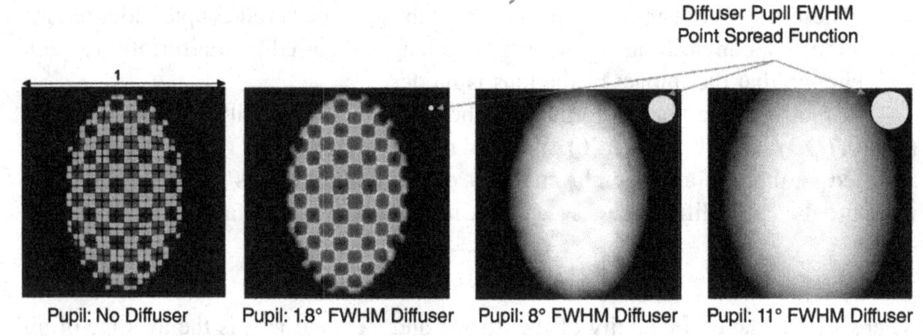

Figure 6.23 RGB Illuminance at the pupil.

our simulations, the Cx,Cy chromaticity of the LEDs is R = (0.7,0.3), G = (0.181,0.726), and B = (0.153, 0.027). The output of the LEDs is adjusted so that the average chromaticity of all the emitters is (0.331, 0.331).

We assume ideal imaging optics to examine the illuminance in the pupil. RGB illuminance plots of the pupil distribution are shown in Figure 6.23. The pupil distribution is oval because the mixer has a square to rectangular shape. The pupil shows a colored kaleidoscope pattern resulting from the RGGB LED array. This type of spatial structure can result in color and luminance variations as the viewer's eye is shifted. To minimize this eye position dependence, a diffuser is added near the output face of the mixer. A Gaussian diffuser with BSDF $\propto e^{-0.5*(0.6*\theta/\mathrm{FWHM})^2}$ is used to model the diffuser. Diffuser full-width half-maximums (FWHM) of 1.8°, 8°, and 11° were simulated. The RGB illuminance distributions show that with FWHM = 11°, the color uniformity in the pupil is good. At FWHM = 1.8°, the kaleidoscope pattern is still readily observed. To understand why 1.8° and 11° provide such different results, we included a small circle to depict the diffuser FWHM in the pupil for the different cases. The diffuser FWHM required to achieve "good" uniformity is about three times the imaged area of the input face of the tapered mixer. Volume scattering within the bulk of the dielectric mixer can also be used instead of adding the diffuser to the output end of the mixer (e.g., Reference 45).

6.6.5.4 Straight + Diffuser + Taper As the previous section illustrates, with a square to rectangular mixer, one can achieve a uniform spatial distribution, but the exiting angular distribution is oval and also nonuniform. An alternative approach uses a short straight mixing rod with a wide angle diffuser placed at its output. The output of the diffuser can then be used to feed a tapered mixing rod. This can remove the nonuniformities observed in the pupil of the previous section. This type of system and the illuminance at the mixer output, as well as the illuminance at the pupil of an ideal lens, is shown in Figure 6.24. The RGB LEDs described in the previous section are coupled into a 2.4-mm × 2.4-mm hollow straight mixer of length 3 mm. A wide angle diffuser with a Lambertian BTDF is used. The tapered hollow mixer portion is 26-mm long and transitions from 2.4 mm × 2.4 mm to 5.6 mm × 9.3 mm.

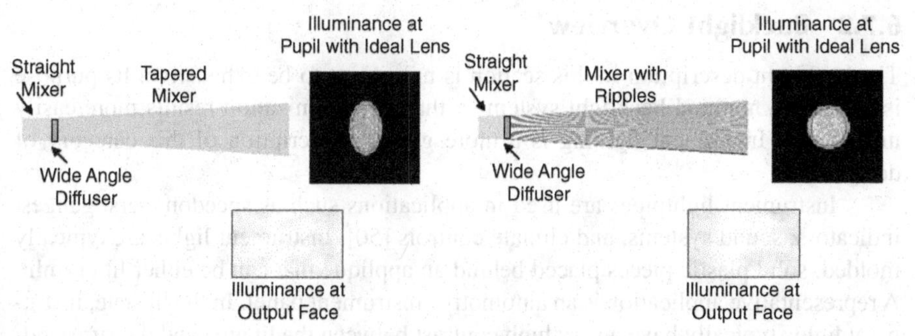

Figure 6.24 Short straight mixer with a wide angle diffuser and a tapered mixer. The addition of ripples to the perimeter of the tapered mixer circularizes the pupil.

In some cases, the optical system that will collect this light may require a round pupil, and in others, the oval is rotated 90° from the desired direction. One way to circularize the pupil without a significant increase in the étendue is to replace the smooth tapered mixer with a mixer that uses ripples. We used ripples that have a 30° peak to valley angle, and the height of the ripples is adjusted along the length to keep that angle constant. Simulation results are shown in Figure 6.24. The illuminance output at the end of the mixer is still uniform, but the pupil shape is now circular.

This type of short straight mixer plus diffuser plus tapered mixer is somewhat idealized, but it serves to illustrate the use of short straight mixers with RGB LEDs and the use of a mixer with ripples to circularize the angular distribution exiting the mixer. Anamorphic transformations are also used, for example, in the areas of solar concentration [46, 47] and laser diodes arrays [48].

6.7 BACKLIGHTS

6.7.1 Introduction

Display devices, such as backlights and keypads, require uniform spatial luminance over a specified area and for a range of view angles. Historically, display devices have been designed through an iterative process using hardware prototypes. This design process is effective, but the number of iterations is limited by the time and cost to make the prototypes. In recent years, virtual prototyping using illumination software modeling tools has replaced many of the hardware prototypes. Typically, the designer specifies the design parameters, builds the software model, predicts the performance using a Monte Carlo simulation, and uses the performance results to repeat this process until an acceptable design is obtained. Illumination optimization provides the ability to automate this design process while also providing improved performance. Much of the treatment in this section follows [49].

6.7.2 Backlight Overview

The backlight description in this section is not meant to be exhaustive. Its purpose is to show a nominal backlight system so that the optimization results more easily understood. Instrument lighting is a more general description of this category of designs.

Instrument lightpipes are used in applications such as speedometers, gauges, indicators, sound systems, and climate controls [50]. Instrument lights are typically molded, solid plastic pieces placed behind an appliqué that can be either lit or unlit. A representative application is an automotive instrument panel. In the lit state, instrument lights typically have a very high contrast between the lit area and the surrounding area. The sources in an instrument light can be used to illuminate a number of spatially separated areas or just one large area.

Backlights represent a large class of instrument lights and take on many forms. The most common backlights are composed of a source coupled into the edge of a plastic lightpipe, where a spatially varying extraction pattern couples flux out of the lightpipe to produce a luminance distribution that is uniform over an area and for a range of view angles. A spatial light modulator, typically a liquid crystal based device, is then placed over the backlight. The extraction pattern and the power within the light-guide are balanced to provide the desired output luminance distribution. This balance is depicted conceptually in Figure 6.25a. Determining the extraction pattern for a given backlight geometry is greatly facilitated through the use of illumination optimization.

The typical elements of an edge lit backlight are depicted in Figure 6.25b. Many backlights include brightness-enhancing films (BEFs) to recycle one of the polarization states or to selectively control the angular luminance distribution. A diffuser is often included with the BEFs to enhance fine scale uniformity. A reflective frame and backplane are often included to minimize lost light.

Typically, the sources used in backlights are either small fluorescent lamps (e.g., CCFLs) or LEDs. With CCFLs, the extraction patterns typically vary macroscopically in one dimension. With LEDs, the required pattern often varies in two dimensions, which often makes them more difficult to design. LEDs without domed lenses are typical.

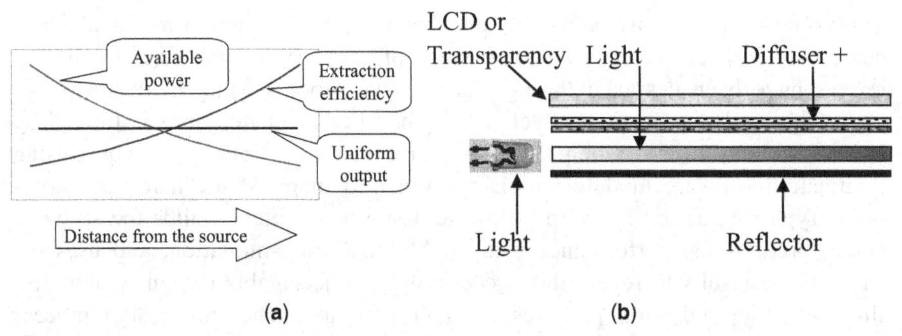

Figure 6.25 (a) Conceptual diagram showing how extraction efficiency varies with the available power to provide uniform output. (b) Schematic of a typical backlight.

Extractors are placed on at least one surface of the lightpipe to couple the light out of the lightpipe. There are many different extractor approaches. Some of the common approaches are paint, etched patterns, diffractive structures, and 3D structures that are larger than the wavelength of the light. The 3D structures, which we also call textures, can include many different shapes. Some common 3D textures include lenses, prisms, pyramids, cones, and hemispheres. The surface properties of the 3D textures can range from nearly specular to nearly diffuse.

6.7.3 Optimization

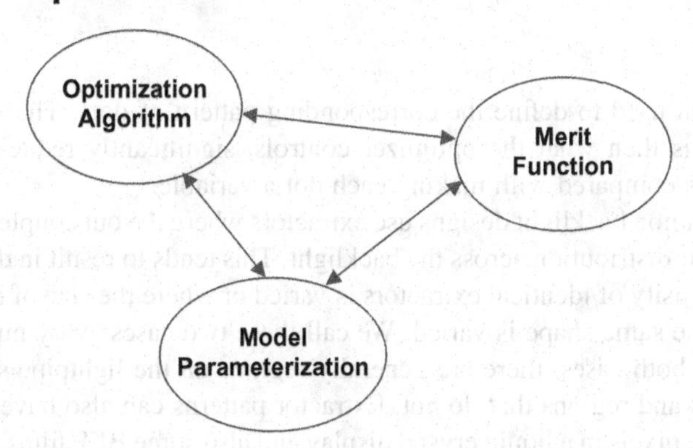

Optimization is the ability to automatically refine the performance of a system based upon a specified performance criterion. Classical optimization approaches have three main elements: parameterization, merit function, and the optimization algorithm. These elements are depicted in the insert on the right. The user interface ties these three elements together. Optimization has been applied successfully to imaging systems for many years [51]. The use of optimization in commercial illumination software packages is still relatively new, but its use is growing quickly.

The optimization of the density patterns for backlights requires these three optimization elements. The optimization setup needs to define what is allowed to vary (the parameterization), a means to evaluate the output (the merit function) and an algorithm to vary the parameterization so that the merit function is minimized. Backlights have the special attribute that the extractor density is the main parameter to vary and there is usually a 1 : 1 correlation between the density in one region of the backlight and the output of the backlight in that region. To optimize a backlight, the optimization algorithm can reduce the density where the output is too high and increase the density where the output is too low. Performing these adjustments iteratively can provide convergence to a pattern that minimizes the merit function in a small number of iterations.

6.7.4 Parameterization

If size and location of each dot in a backlight is added as an optimization variable, the number of variables to optimize could be very large. To reduce the number of variables, the extractor density can be defined as a 2D grid of values. The values in

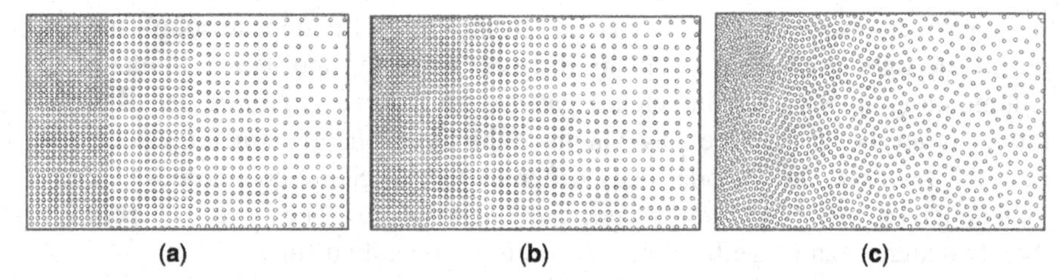

Figure 6.26 Vary number examples to control extraction efficiency: (a) constant spacing within discrete regions, (b) smooth gradient in spacing, and (c) smooth gradient with sinusoidal shift.

the 2D grid are then used to define the corresponding pattern of dots. The grid of extractor densities is then what the optimizer controls, significantly reducing the number of variables compared with making each dot a variable.

The most common backlight designs use extractors where the out coupled light has the same angular distribution across the backlight. This tends to result in designs where the spatial density of identical extractors is varied or where the size of extractors with roughly the same shape is varied. We call these two cases "vary number" and "vary size." In both cases, there are generally regions on the lightpipe surface that have extractors and regions that do not. Extractor patterns can also have a correlation between the pixels in a liquid crystal display and also some BEF films. These correlations can produce Moiré effects, which are generally undesired in backlights. In many cases, a diffuser is used so that the substructure of the extractor pattern is not visible. If the diffuser scatter angle does not vary spatially, the region with the largest spacing between extractors often dominates the required diffuser scatter angle, with large gaps needing larger diffuser angles than small gaps.

6.7.4.1 Vary Number When creating a spatially varying density pattern, it is important to create a smooth variation in the density of extractors. Regions of uniformly spaced extractors can produce a mosaic effect, as can be seen by examining Figure 6.26a,b. These two patterns have the same average density, but Figure 6.26b is superior because the gradient in the density is smooth. Even with Figure 6.26b, the dot pattern can have some residual artifacts because of the underlying rectilinear dot placement. Figure 6.26c shows a simple sinusoidal shift that removes most of the artifacts. Some of the other possible approaches are based on dither, molecular dynamics [52], and polyominoes [53].

6.7.4.2 Vary Size It is also important to create a smooth variation in the density of extractors when varying the size. In many applications, a hexagonal pattern is sufficient, but regions of uniformly sized extractors can produce a Mosaic effect, which can be seen by examining Figure 6.27a,b. These two patterns have the same average density and the same hexagonal placement, but Figure 6.27b is superior because the gradient in the density is smooth. For comparison, Figure 6.27c shows the same density using a rectilinear placement. Dither can also be used with vary size, both in terms of dithering the size and dithering the placement.

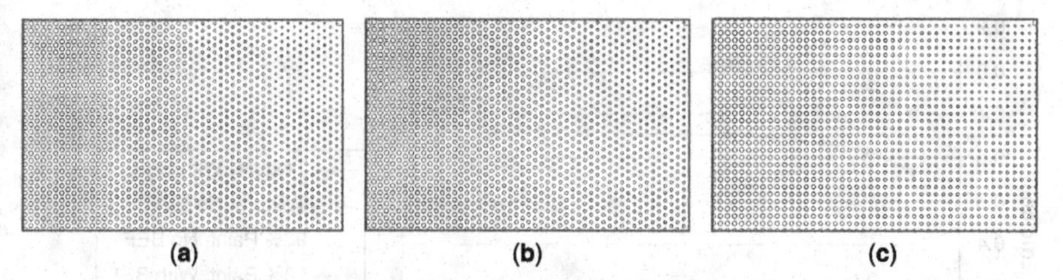

Figure 6.27 Vary size examples to control extraction efficiency: (a) hexagonal spacing where size changes discretely between regions, (b) hexagonal spacing where size varies smoothly, and (c) rectilinear spacing where size varies smoothly.

6.7.5 Peak Density

A backlight extractor pattern can be characterized by the region that has the highest density. In most cases, the higher the peak density, the higher the efficiency. Typically, the efficiency versus peak density has a strong dependence for low peak densities, but then plateaus and has only a second-order effect as the peak density is increased further. Large peak densities in the extractor pattern generally introduce a larger difference between the peak density and the minimum density compared with smaller peak densities. Large peak densities also require more care during manufacturing to ensure that dots do not overlap.

To illustrate the typical efficiency versus peak density relationship, a 1.2-mm × 42-mm × 32-mm lightpipe with an output region of 39 mm × 29.25 mm was constructed. Two LEDs were placed 30 mm apart on the 42-mm edge of the lightpipe. The lightpipe was surrounded by a frame with 95% reflectivity and a 98% reflectivity mirror below the lightpipe. A 20° Gaussian diffuser was placed over the lightpipe. Four different cases were considered. Two used 98% reflective paint dots, and two used nearly hemispherical textures. For each extractor type, the case of using and not using two crossed BEFII films was evaluated. The efficiency versus peak density results are shown in Figure 6.28. The addition of BEFII films decreases the total efficiency, but increases the on-axis luminance.

6.7.6 Merit Function

The merit function is the numeric feedback that tells the optimization algorithm whether or not the system is improving as the optimization procedure progresses. There are several possible ways of constructing a merit function, but perhaps the most common involves the sum of the squares of the differences between a set of figures of merit (merit function items) and their associated target values. The basic relationship is

$$\text{Merit Function} = \text{MF} = \sum w_i^2 (V_i - T_i)^2, \tag{6.9}$$

where w_i is the weight of ith merit function item, V_i is the current value of the ith merit function item, and T_i is the target value for the ith merit function item. For the

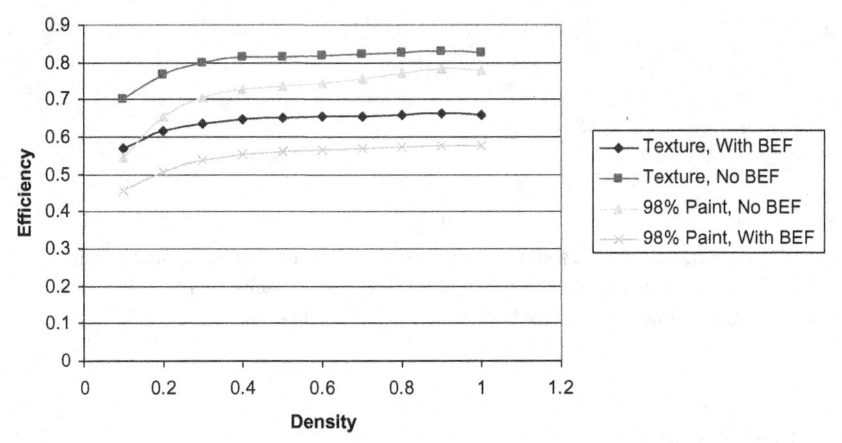

Figure 6.28 Backlight efficiency as a function of peak density for four cases.

type of merit function construction described by Equation (6.9), the system is at its optimal solution when the value of the merit function is equal to zero. It is not required that the system is actually able to go all the way to $MF = 0$ for the system to optimize.

When Monte Carlo simulations are used, the results include statistical errors that depend upon the number of traced rays. That noise contribution should be considered when optimizing so that the optimization algorithm does not spend lots of time trying to remove artifacts in the output that are primarily noise. The merit function relationship in Equation (6.9) will have a noise component when Monte Carlo simulations are used to evaluate the *MF*:

$$\mathrm{MF} = \sum w_i^2 (V_i - T_i)^2 + \mathrm{Noise}. \tag{6.10}$$

This noise contribution can be estimated, and the optimization terminated when the *MF* approaches the same value as the noise estimate. Alternatively, the number of traced rays can be increased to minimize the noise. It is also possible to use low-resolution output distributions with a small number of traced rays and then use a larger number of rays when higher resolution patterns are required.

6.7.7 Algorithm

Most backlights represent a very special type of illumination system in that there is a 1:1 spatial correlation between the extractor positions and the spatial luminance distribution. Increasing the extraction efficiency in a given region of the lightpipe can alter the distribution over the whole display, but there tends to be a direct correlation for the luminance directly above the extractor. This means that locally increasing the extraction efficiency will increase the local luminance, and locally decreasing the extraction efficiency will reduce the local luminance. Using this special correspondence can enhance the speed of convergence of the optimization

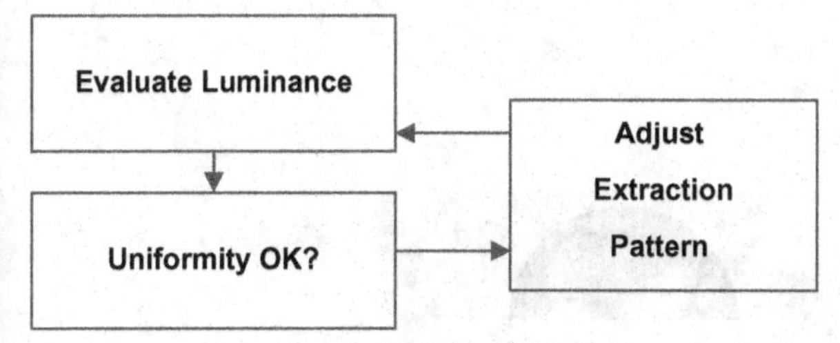

Figure 6.29 Backlight optimization flowchart.

algorithm. The iterative nature of the optimization is depicted in Figure 6.29 (e.g., References 54–57).

6.7.8 Examples

We'll now consider a number of examples. The purpose of these examples is to illustrate. Images of the output luminance and the extractor density will be shown. They all use the same color mapping, which is shown in Figure 6.4.

6.7.8.1 Peaked Target Distribution It is typical to characterize the output of a backlight in terms of its luminance in the center of the display. The luminance near the corners can be lower than the center. The corner luminance is typically measured 5% in from the edge using the ANSI specification (e.g., Reference 58).

The optimization target for a backlight can be specified by a distribution that varies slowly from the center to the edge. This can provide an increase in the luminance in the center of the display. This occurs for two main reasons. First, if the total lumens are the same for a constant distribution compared with a peaked distribution, the peaked distribution will be higher in the center. Second, to achieve a peaked distribution, the relative extractor density is normally increased in the center and reduced near the edges, compared with the extractor pattern for a constant luminance distribution. This tends to result in a slight increase in the total output flux, since losses near the edges can be smaller.

To provide an example, we used the 2 LED display geometry shown Figure 6.30. A diffuser and crossed BEF films were used with an extractor pattern of nearly hemispherical 3D textures.

The optimized density pattern and luminance output for a constant target distribution are shown in the left of Figure 6.31. The target was optimized for a distribution that smoothly varies from 1.125 in the center to 0.7875 at the corner (a corner-to-center ratio of 0.7) using a quadratic function to define the shape of the target going from center to edge. This peaked distribution was optimized with results shown in the right of Figure 6.31.

In absolute terms, the peaked target produced a 13.5% increase in the center luminance and a 2% increase in the total output. A 13.5% increase is close to the

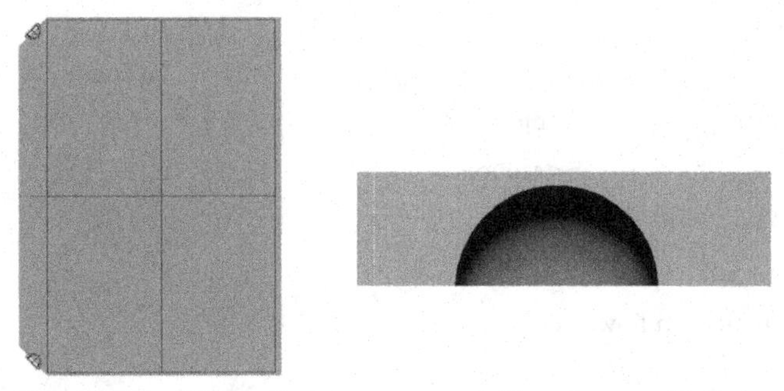

Figure 6.30 2 LED display geometry (left) and a magnified view of a hemispherical extractor (right).

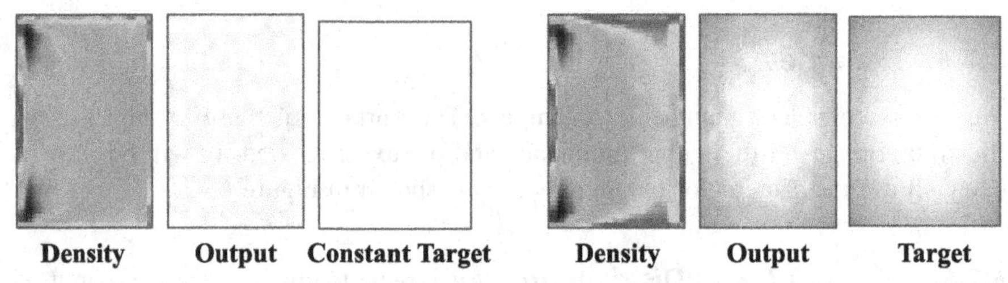

Figure 6.31 Density distribution for a constant target and a peaked target.

expected 12.5%. Notice that the extractor pattern for the peaked target shows an increase in the center density relative to the center density for the uniform target.

6.7.8.2 Border Extractors The output area of most LCD backlights is rectangular and is smaller than the size of the lightpipe used in the backlight. It is generally preferred to create an extractor pattern that is slightly larger than the desired output area. This avoids potential problems with dim regions near the edge when the display is viewed off-axis. It also tends to produce more uniform extractor density patterns because the edge region then has extractors on both sides of the edge bins.

To illustrate this relationship, a 1.2-mm × 42-mm × 32-mm lightpipe with an output region of 39 mm × 29.25 mm was constructed. Two LEDs were placed 30–mm apart on the 42–mm edge of the lightpipe. The lightpipe was surrounded by a frame with 95% reflectivity and a 98% reflectivity mirror below the lightpipe. A 20° Gaussian diffuser was placed over the lightpipe. Figure 6.32 shows the optimized density pattern with zero, one, and two extra sets of dots along the outside edge. While these extra dots reduce the increase in density observed along the edge in the zero extra dot case, they do introduce a slight reduction in the total efficiency of light into the desired target area since more light is now extracted outside of the desired area.

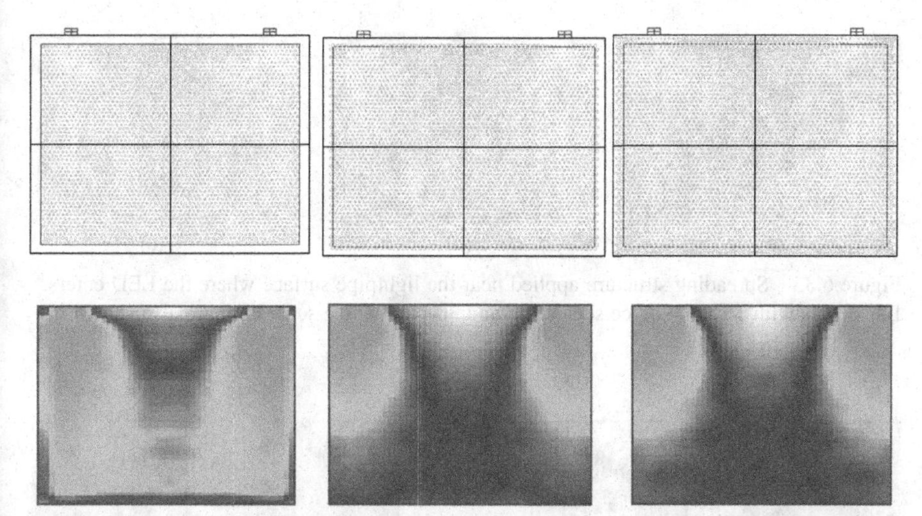

Figure 6.32 Optimized extractor pattern for zero, one, and two extra rows of extractors (going left to right, respectively).

The output efficiency of the zero, one, and two extra rows of dots is 74, 73.5, and 70.5%. These efficiency values depend upon the details of the display design and the relative size of the output area compared to the extra dots. A reflector could be placed around the edge of the output face of the backlight to "recycle" the light.

6.7.8.3 Input Surface Texturing Discrete LEDs placed along the edge of an edge lit display often results in optimized extractor density profiles that are very low near the LED compared with the peak in the extractor density. In some cases, the gradients in the density profile are very large, and the extractor profile can be overly sensitive to the placement of the LED and to fabrication of the extractor density profile. This often drives LED backlight to include a "mixing gap" between where the LED enters the lightpipe and where the desired output begins. One of the reasons why this occurs is that an angle of 90° in air results in ~42° inside of the lightpipe using Snell's law. It is often desired to avoid making this gap distance too large. Narrower cone angles require a larger mixing gap distance to achieve the same amount of spreading. One technique to increase the cone angle is obtained by adding a spreading structure on the face of the lightpipe where the LED is placed. To ensure that light still TIRs in one axis, the spreading structure can be asymmetric. An example with cylindrical lenses is shown in Figure 6.33a. A spreading structure of this type can be especially valuable when the LED is placed near the corners of the lightpipe.

To illustrate the impact of the spreading structure, a 2 LED backlight model was used with two different separations between the LED. Figure 6.34b shows a 15-mm gap between 2 LEDs, and Figure 6.34c shows a 30-mm gap. The optimized distributions for the 15-mm and 30-mm gaps with the spreading structure are shown

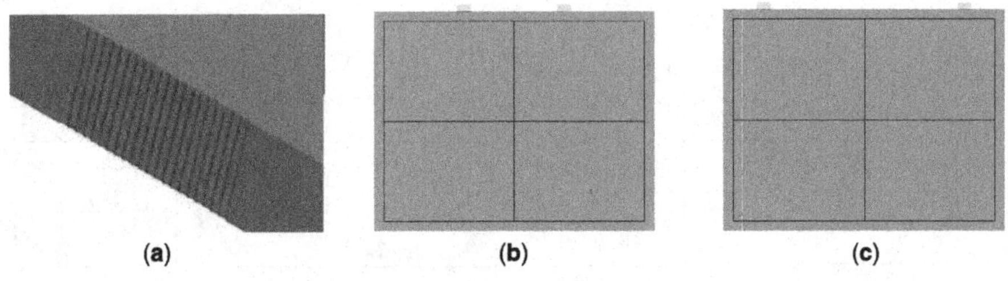

Figure 6.33 Spreading structure applied near the lightpipe surface where the LED enters. Backlights with 15-mm source separation and 30-mm source separation are also shown.

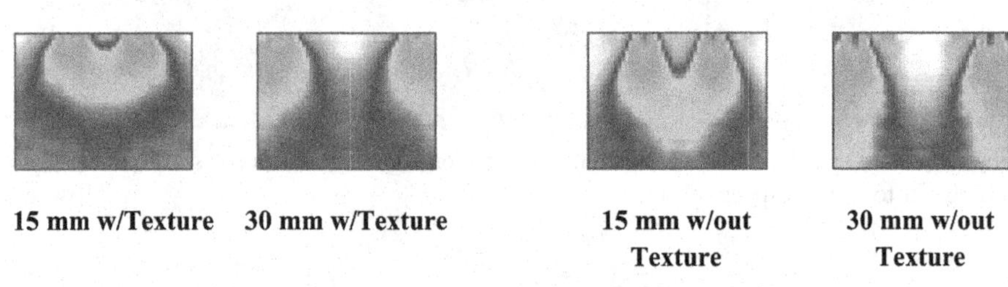

Figure 6.34 Optimized density pattern for 15- and 30-mm LED separations with the spreading structure (a and b) and without the spreading structure (c and d).

in Figure 6.34a,b. For comparison, the 15-mm and 30-mm gap cases without spreading structure are shown in Figure 6.34c,d. A significant difference in the optimized extractor pattern is evident by comparing the optimized density patterns.

6.7.8.4 Variable Depth Extractors One approach to fabricating arrays of extractors uses a tool tip whose depth varies from position to position. This is also a common approach used to make the mold for the micro lens arrays used in the cover plate for many lamps (e.g., look at most PAR38 lamps). For the example shown in Figure 6.35a, the shape of the extractor has a rounded tip and then a body. The depth of the tip is used to control the extraction efficiency of a given extractor dot. A single extractor is shown in Figure 6.35b, and a small region with multiple extractors is shown in Figure 6.35c.

To illustrate a design with the extractor of Figure 6.35b, the three-LED backlight shown in Figure 6.36a was optimized. The starting point was the uniform tool tip depth distribution shown in Figure 6.36b, which results in the output luminance shown in Figure 6.36c. After optimization, the tool tip depth distribution shown in Figure 6.36d provides the desired output luminance shown in Figure 6.36e.

6.7.8.5 Inverted 3D Texture Structure Many backlight extractor concepts place the texture on the surface opposite to the surface where the light exits toward the LCD. An alternative approaches use inverted structures that allow light to couple into structures that touch the top surface of the lightpipe and then use a TIR surface

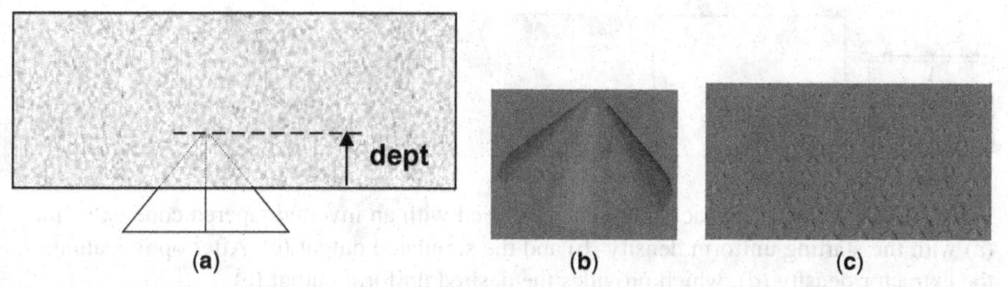

Figure 6.35 Illustration of how a tool tip depth is varied (a). The resulting 3D texture (b) and a region of an extractor pattern where the depth is varied.

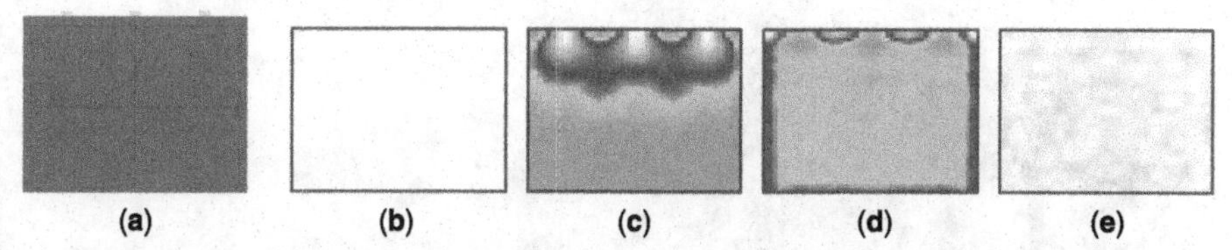

Figure 6.36 Three LED backlight geometry (a) with the starting uniform tool tip depth (b) and the simulated output (c). After optimization, the tool tip depth (d), which provides the desired uniform output (e).

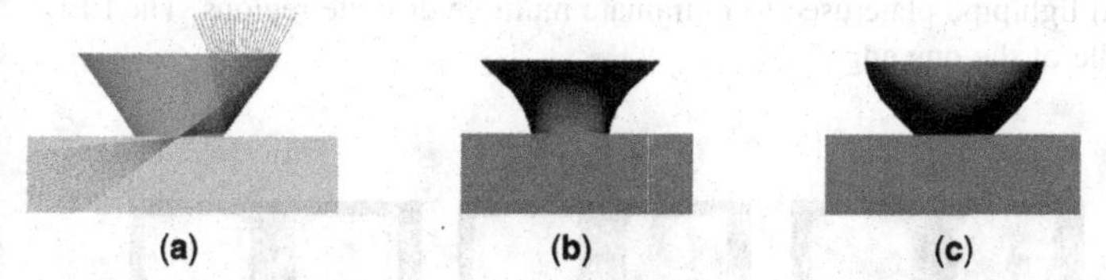

Figure 6.37 Inverted tapered cone with fan of rays (a) and two other extractor shapes (b and c).

to turn the light (e.g., Fig. 6.37). This inverted structure can be more difficult to fabricate than conventional structures, but it offers the potential for providing a backlight that does not use BEFs. Such a structure can be fabricated by bonding two separately produced pieces together. Some recent work has shown how to make the structure as a unitary element [59]. This type of extractor can also be optimized to provide a desired spatial luminance distribution.

To illustrate that this type of extractor concept can be optimized, we use the basic extractor geometry shown in Figure 6.37a where the top diameter is 45.5 μm, the height of the extractor is 36.5 um, and the angle of the side wall is 54.5°. These dimensions are a slightly scaled version of the sizes reported in Reference 59. All extractors in the backlight have the same size, but the placement of the extractors is varied to provide the required variation in extractor density. The sinusoidal shift placement approach shown in Figure 6.37c was used. As shown in Figure 6.38, a uniform output can be achieved.

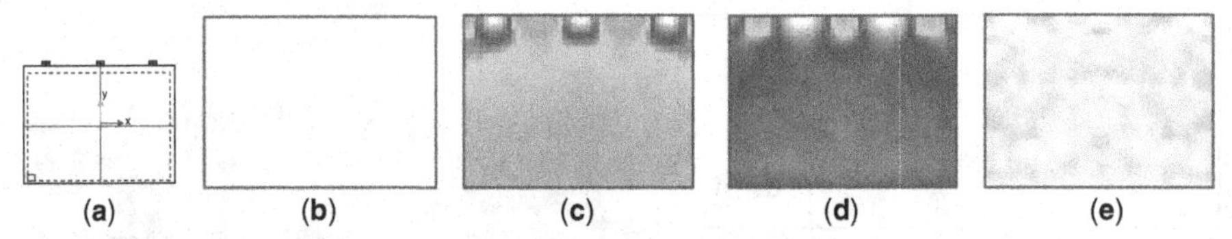

Figure 6.38 Three LED backlight geometry used with an inverted tapered cone extractor (a) with the starting uniform density (b) and the simulated output (c). After optimization, the extractor density (d), which provides the desired uniform output (e).

Figure 6.39 Keypad lightpipe plate used to illuminate multiple discrete regions. The LED is placed in the middle of the one edge.

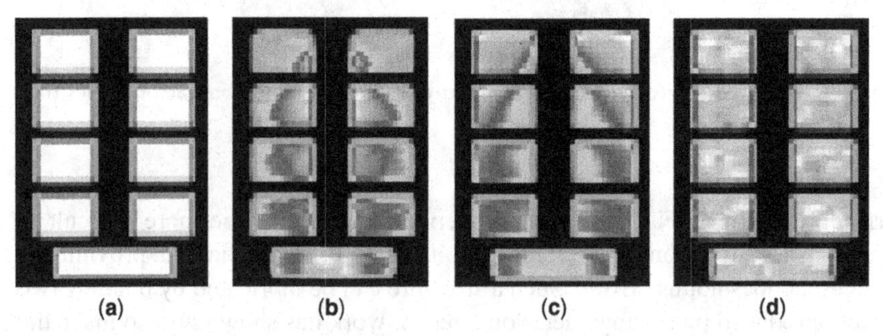

Figure 6.40 Starting extractor density (a) and the nonuniform output (b) for the keypad lightpipe. Optimized extractor density (c) and the uniform output (d).

6.7.8.6 Key Pads Different geometries can be used to illuminate key pads. One approach uses extractors placed below the key pads. This type of geometry can also be optimized. Shown in Figure 6.39 is a lightpipe with nine different rectangular regions, each of which needs to provide uniform output using a single LED.

The starting density pattern, which is uniform in the regions that need to be lit and is 0 in the regions that do not need to be lit is shown in Figure 6.40a. The output of this geometry is nonuniform, as shown in Figure 6.40b. After a couple of optimization iterations, the desired uniform output is achieved (see Fig. 6.40d) using the extractor density shown in Figure 6.40c. Binary logos are a related type of example and have been discussed previously (e.g., Reference 60).

6.8 NONUNIFORM LIGHTPIPE SHAPES

Some applications require the use of a lightpipe that has a shaped outline with interior holes so that mechanical structures, like knobs, can be placed in the interior of the lit area. Because the light has to travel around the holes, the extractor density pattern can be very nonuniform. As an illustration, the lightpipe shown in Figure 6.41 was created, and an extractor pattern applied to the back side. A single LED was placed along one edge of the lightpipe. With a uniform extractor pattern, the output is highly nonuniform. However, a couple of optimization iterations and the extractor pattern converged to a very uniform output, as shown in Figure 6.42.

When the receiver mesh used to collect Monte Carlo simulation rays is rectangular, the shaped lightpipe will have regions that only partially fill the bins used in the mesh (Fig. 6.43). This is different than the simple rectangular output case of a typical LCD backlight. When partially filled bins are used, it is important to make sure that the target distribution takes these partial bins into account. If not taken into account, then the optimized extractor pattern can have higher densities than needed near the perimeter of the lightpipe.

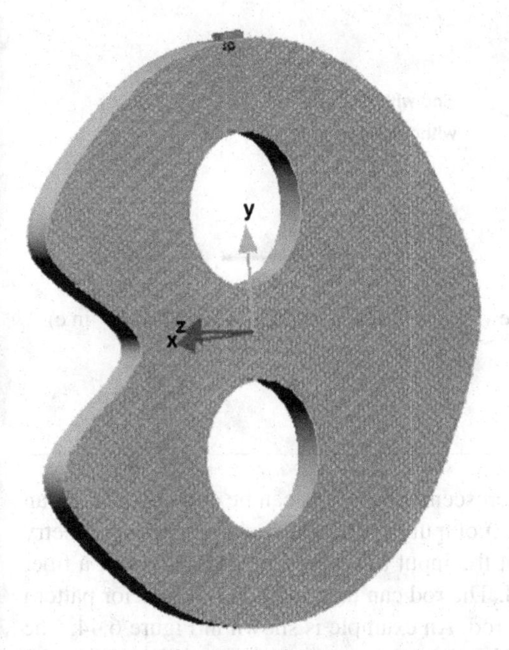

Figure 6.41 Shaped output lightpipe with interior holes. The LED is near the top edge.

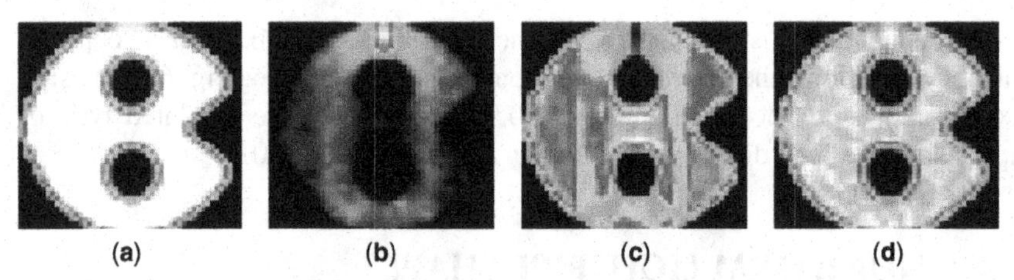

Figure 6.42 Starting extractor density (a) and the nonuniform output (b) for the shaped output lightpipe. Optimized extractor density (c) and the uniform output (d).

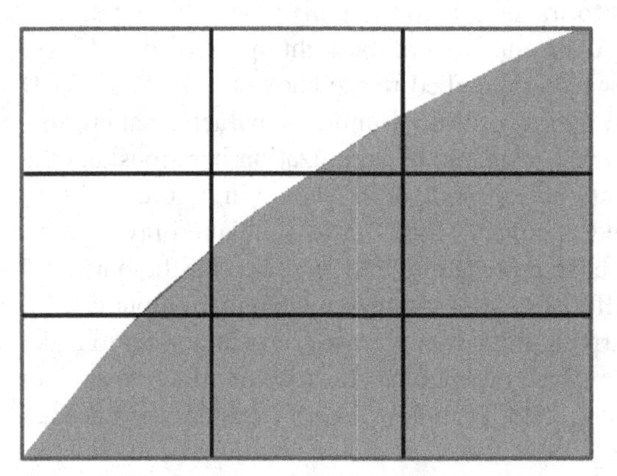

Figure 6.43 3 × 3 set of receiver bins placed over a small portion of a lightpipe. Partial bins need to be taken into account when defining the target distribution.

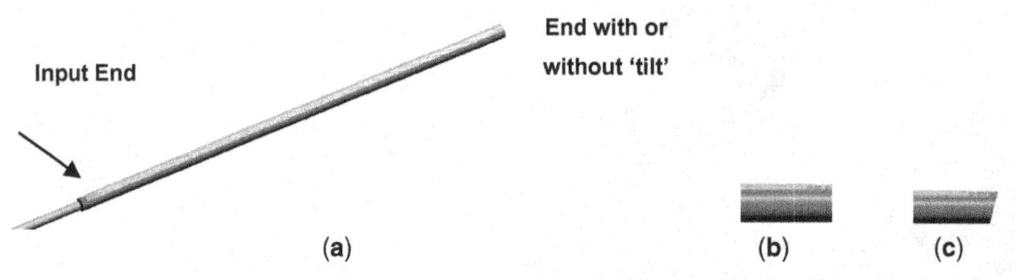

Figure 6.44 Rod luminaire (a) with a view of two different end tilts (0° in b and 15° in c).

6.9 ROD LUMINAIRE

LEDs can be used to replace long fluorescent tubes. This can be done by placing an array of LEDs along a line. If the LED output is sufficient, an alternative geometry places an LED or a group of LEDs at the input to a long rod. The output of a fiber can also be used to illuminate the rod. The rod can then include an extractor pattern to couple the light out the side of the rod. An example is shown in Figure 6.44. The extractor pattern can surround only a fraction of the rod perimeter, which allows the

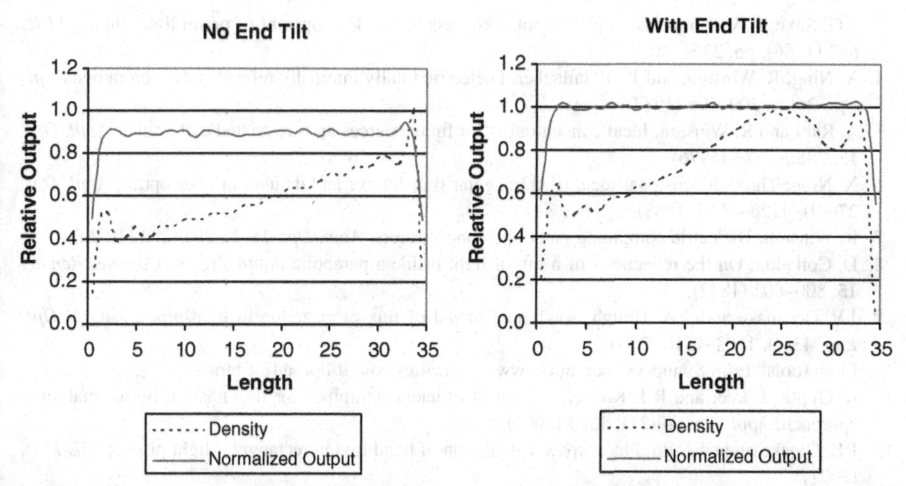

Figure 6.45 Output at the face of the rod luminaire and the extractor density for the case of no end tilt (left) and a 15° end tilt (right).

output to be more directional than a conventional fluorescent lamp since most of the light exits in the direction opposite to the extractor pattern.

It is generally desired that the output along the length for this type of luminaire be relatively constant. The extractor density can be adjusted to provide this desired uniformity. For the simple case of simply placing a mirror at the end of the rod opposite to the source, the required extractor pattern is roughly a monotonic pattern, with the maximum density at the end opposite to the source. This geometry allows some light to couple to the far end of the rod and then back to the source. If the source is not highly reflective, which is generally the case with a fiber illuminator, using a tilted end face can provide an increase in the extracted output. The optimized output for an end mirror place at 0° and tilted to 15° is shown in Figure 6.45. With the 15° end tilt, the extractor density pattern becomes more complicated at the far end. It can, however, be easily optimized. In the plots of Figure 6.45, the relative output is normalized to be 1 for the case of the 15° end tilt. As seen in the plots, there is a 10% improvement in the output by using the 15° end tilt.

ACKNOWLEDGMENTS

All results shown in this chapter were performed using LightTools from Synopsys Inc. (e.g., Reference 10).

REFERENCES

1. Apparatus for light dwellings or other structures, U.S. Patent 247,229 (1881).
2. L.A. Whitehead, R.A. Nodwell, and F.L. Curzon, New efficiency lightpipe for interior illumination, *Appl. Opt.* **21**(15), 2755–2757 (1982).

3. S.G. Saxe, L.A. Whitehead, and S. Cobb, Progress in the development of prism light guides, *SPIE 692* (1986), pp. 235–240.
4. X. Ning, R. Winston, and J. O'Gallagher, Dielectric totally internally reflecting concentrators, *Appl. Opt.* **26**(2), 300–305 (1987).
5. A. Rabl and R. Winston, Ideal concentrators for finite sources ands restricted exit angles, *Appl. Opt.* **15**, 2880–2883 (1976).
6. X. Ning, Three-dimensional ideal θ1/θ2 angular transformer and its uses in fiber optics, *Appl. Opt.* **27**(19), 4126–4130 (1988).
7. R. Winston, Dielectric compound parabolic concentrators, *Appl. Opt.* **15**(2), 291–292 (1976).
8. D. Colladon, On the reflections of a ray of light inside a parabolic liquid stream, *Comptes Rendus* **15**, 800–802 (1842).
9. J.V. Derlofske and T.A. Hough, Analytical model of flux propagation in light-pipe Systems, *Opt. Eng.* **43**(7), 1503–1510 (2004).
10. LightTools® from Synopsys Inc., http://www.opticalres.com/lt/ltprodds_f.html.
11. A. Gupta, J. Lee, and R.J. Koshel, Design of efficient lightpipes for illumination by an analytical approach, *Appl. Opt.* **40**(22), 3640 (2001).
12. P.L. Gleckman and J. Ito, Phase space calculation of bend loss in rectangular light pipes, *SPIE 1528* (1991).
13. H. Hatami-Hanza, P.L. Chu, and J. Nayyer, Low-loss optical waveguide-bend configuration with curved corner reflector, *Electron. Lett.* **25**(25), 2283–2285 (1992).
14. S.C. Chu and J.L. Chern, No-loss bent light pipe with an equiangular spiral, *Opt. Lett.* **30**(22), 3006–3009 (2005).
15. J. Chaves and M. Collares-Pereira, Ideal concentrators with gaps, *Appl. Opt.* **41**(7), 1267–1276 (2002).
16. T.L. Davenport, R.L. Hansler, T.E. Stenger, W.J. Cassarly, G.R. Allen, and R.F. Buelow, Changes in angular and spatial distribution introduced into fiber optic headlamp systems by the fiber optic cables, *SAE International Congress and Exposition*, Paper No. 981197 (1998).
17. M.M. Chen, J.B. Berkowitz-Mattuck, and P.E. Glaser, The use of a kaleidoscope to obtain uniform flux over a large area in a solar or arc imaging furnace, *Appl. Opt.* **2**, 265–271 (1963).
18. W.J. Cassarly, J.M. Davenport, and R.L. Hansler, Uniform lighting systems: uniform light delivery, *SAE 950904*, Vol. SP-1081 (1995), pp. 1–5.
19. M. Wagner, H.D. Geiler, and D. Wolff, High-performance laser beam shaping and homogenization system for semiconductor processing, *Meas. Sci. Technol. (U.K.)* **1**, 1193–1201 (1990).
20. P., Michaloski, Illuminators for microlithography, in F.M. Dickey, S.C. Holswade, and D.L. Shealy, eds., *Laser Beam Shaping Applications*, CRC Press (2005), pp. 1–51.
21. K. Iwasaki, Y. Ohyama, and Y. Nanaumi, Flattening laserbeam intensity distribution, *Lasers Appl.* **2**(4), 76 (1983).
22. K. Iwasaki, T. Hayashi, T. Goto, and S. Shimizu, Square and uniform laser output device: theory and applications, *Appl. Opt.* **29**, 1736–1744 (1990).
23. J.M. Geary, Channel integrator for laser beam uniformity on target, *Opt. Eng.* **27**, 972–977 (1988).
24. R.E. Grojean, D. Feldman, and J.F. Roach, Production of flat top beam profiles for high energy lasers, *Rev. Sci. Instrum.* **51**, 375–376 (1980).
25. M.R. Latta and K. Jain, Beam intensity uniformization by mirror folding, *Opt. Commum.* **49**, 27 (1984).
26. L.J. Coyne, Distributive fiber optic couplers using rectangular lightguides as mixing elements, *Proc—FOC 79* (1979), pp. 160–164.
27. C.M. Chang, K.W. Lin, K.V. Chen, S.M. Chen, and H.D. Shieh, An uniform rectangular illuminating optical system for liquid crystal light valve projectors, in *Proceedings of the Sixteenth International Display Research Conference*, Society for Information Display, Santa Ana, CA, pp. 257–260 (1996).
28. M.A. Handschy, J.R. McNeil, and P.E. Weissman, Ultrabright head mounted displays using LED-illuminated LCOS, *SPIE 62240S-6* (2006).
29. Y.-K. Cheng and J.L. Chern, Irradiance formations in hollow straight light pipes with square and circular shapes, *J. Opt. Soc. Am. A* **23**, 427–434 (2006).
30. C.M. Cheng and J.L. Chern, Optical transfer functions for specific-shaped apertures generated by illumination with a rectangular light pipe, *J. Opt. Soc. Am. A* **23**, 3123–3132 (2006).

31. Y.-K. Cheng, M.-H. Wang, and J.L. Chern, Irradiance formations of on-axis Lambertian point-like sources in polygonal total-internal-reflection straight light pipes, *J. Opt. Soc. Am. A* **24**, 2748–2757 (2007).
32. J.R. Jenness Jr., Computing uses of the optical tunnel, *Appl. Opt.* **29**(20), 2989–2991 (1990).
33. L.J. Krolak and D.J. Parker, The optical tunnel—A versatile electrooptic tool, *J. Soc. Motion Pict. Telev. Eng.* **72**, 177–180 (1963).
34. R.G. Waarts, D.W. Nam, D.F. Welch, D.R. Scrifes, J.C. Ehlert, W. Cassarly, J.M. Finlan, and K. Flood, Phased 2-D semiconductor laser array for high output power, *SPIE 1850* (1993) pp. 270–280.
35. D.S. Goodman, Producing uniform illumination, *OSA Annual Meeting*, Toronto (October 5, 1993).
36. D.E. Williamson, Cone channel condenser optics, *J. Opt. Soc. Am.* **42**, 712–714 (1952).
37. B. Fan, R.E. Tibbetts, J.S. Wilczynski, and D.F. Witman, Laser beam homogenizer, U.S. Patent 4,744,615 (1988).
38. K. Jain, Illumination system to produce self-luminous light beam of selected cross-section, uniform intensity and selected numerical aperture, U.S. Patent 5,059,013 (1991).
39. W.J. Cassarly and T.L. Davenport, Non-rotationally symmetric mixing rods, *SPIE* Vol. 6342 (2006).
40. W.J. Cassarly, Recent advances in mixing rods, *SPIE Vol 7103* (September 2008).
41. L. Garwin, The design of liquid scintillation cells, *Rev. Sci. Instrum.* **23**, 755–757 (1952).
42. R. Winston, J.C. Miñano, W.T. Welford, and P. Benítez, *Nonimaging Optics*, Academic Press, San Diego, CA (2004).
43. F. Fournier, W.J. Cassarly, and J. Rolland, Method to improve spatial uniformity with lightpipes, *Opt. Lett.* **33**, 1165–1167 (2008).
44. W.J. Cassarly, E. Fest, and D. Jenkins, Error estimation and smoothing of 2D illumination and chromaticity distributions, *SPIE* Vol. 4769 (2002).
45. C. Deller, G. Smith, and J. Franklin, Colour mixing LEDs with short microsphere doped acrylic rods, *Opt. Express* **12**, 3327–3333 (2004).
46. N. Davidson, L. Khaykovich, and E. Hasman, Anamorphic concentration of solar radiation beyond the one-dimensional thermodynamic limit, *Appl. Opt.* **39**, 3963–3967 (2000).
47. R. Leutz and H. Ries, Microstructured light guides overcoming the two-dimensional concentration limit, *Appl. Opt.* **44**, 6885–6889 (2005).
48. F. Zhang, C. Wang, R. Geng, Z. Tong, T. Ning, and S. Jian, Anamorphic beam concentrator for linear laser-diode bar, *Opt. Express* **15**, 17038–17043 (2007).
49. W.J. Cassarly, Backlight pattern optimization, *SPIE* Vol. 6834, Paper 191 (November 2007).
50. W.J. Cassarly, D. Jenkins, A. Gupta, and J. Koshel, Lightpipes: hidden devices that light our world, *Opt. Photonics News* **12**(8), 34–39 (2001).
51. R.R. Shannon, *The Art and Science of Optical Design*, Cambridge University Press, New York (1997).
52. T. Ide, H. Mizuta, H. Numata, Y. Taira, M. Suzuki, M. Noguchi, and Y. Katsu, Dot pattern generation technique using molecular dynamics, *J. Opt. Soc. Am. A* **20**, 248–255 (2003).
53. V. Ostromoukhov, Sampling with polyominoes, *ACM Trans. Graph.* **26**(3), article 78 (2007).
54. W.J. Cassarly and B. Irving, Noise tolerant illumination optimization applied to display devices, *SPIE 5638* (Feb. 2004) pp.67–80.
55. J.G. Chang and Y.B. Fang, Dot-pattern design of a light guide in an edge-lit backlight using a regional partition approach, *Opt. Eng.* **46**(4), 043002_1–043002-9 (2007).
56. J.G. Chang, M.H. Su, C.T. Lee, and C.C. Hwang, Generating random and nonoverlapping dot pattern for liquid-crystal display backlight light guides using molecular-dynamics method, *J. Appl. Physiol.* **98**, 114910 (2005).
57. J.G. Chang and C.T. Lee, Random-dot pattern design of a light guide in an edge-lit backlight: integration of optical design and dot generation scheme by the molecular-dynamics method, *J. Opt. Soc. Am. A* **24**(3), 839–849 (2007).
58. Electronic projection—fixed resolution projectors, ANSI/NAPM IT7, 228 (1997).
59. J.H. Lee, H.S. Lee, B.K. Lee, W.S. Choi, H.Y. Choi, and J.B. Yoon, Simple liquid crystal display backlight unit comprising only a single-sheet micropatterned polydimethylsiloxane (PDMS) light-guide plate, *Opt. Lett.* **32**(18), 2665–2667 (2007).

60. T.L. Davenport and W.J. Cassarly, Optimizing density patterns to achieve desired light extraction for displays, *SPIE 63420T* (July 2006).
61. J. Chaves and M. Collares-Pereira, Ultra flat ideal concentrators of high concentration, *Sol. Energy* **69**, 269–281 (2000).
62. M. Collares-Pereira, J.F. Mendes, A. Rabl, and H. Ries, Redirecting concentrated radiation, *SPIE 2538* (1995), pp. 131–135.
63. M.E. Knotts, Optics fun with gelatin, *Opt. Photonics News* **7**(4), 50–51 (1996).
64. W.J. Cassarly, Nonimaging optics: concentration and illumination, in *OSA Handbook of Optics*, 3rd edition, Vol. 2, McGraw-Hill, New York (2010), 39.1–39.51.
65. M.E. Barnett, Optical flow in an ideal light collector: the θ_1/θ_2 concentrator, *Optik* **57**(3), 391–400 (1980).
66. R.J. Koshel, Optimization of parameterized lightpipes, *SPIE 6342* (2006).
67. R.J. Koshel and A. Gupta, Characterization of lightpipes for efficient transfer of light, *SPIE 5942* (2005).
68. Y. Kudo and K. Matsumoto, Illuminating optical device, U.S. Patent 4,918,583 (1990).
69. K. Kreske, Optical design of a solar flux homogenizer for concentrator photovoltaics, *Appl. Opt.* **41**, 2053–2058 (2002).

CHAPTER 7

SAMPLING, OPTIMIZATION, AND TOLERANCING

John Koshel

This chapter introduces the reader to advanced topics of the design of illumination systems: sampling, optimization, and tolerancing. These aspects of illumination system design are still in their infancy, thus, it is expected that there will be a wealth of new material that will appear in the literature. Future editions of this book will expand upon these topics, including new material, while removing material that is dated. First, I provide an introduction to the three primary topics of the chapter. Second, I provide some design "tricks" that can assist the illumination designer: Monte Carlo modeling, reverse ray tracing, importance sampling, and a presentation on what is meant by the term "far field." These design "tricks" can be used to great effect to reduce modeling demands. Third, I address how one correctly samples an optical design in software. Sampling is done both by the number of rays traced and the binning (i.e., pixelization for surfaces or voxelization for volumes) at targets. Following sampling is a lengthy treatment on optimization, including the assignation and parameterization of variables, the figure of merit, optimization methods, and examples. The chapter ends with a preliminary discussion on tolerancing of illumination systems.

7.1 INTRODUCTION

Analogous to first-order (i.e., paraxial) design of lens systems, one sets up the basic characteristics of an illumination system, such as the spatial and angular aspects of the source(s) and the desired spatial and angular distributions at the target(s). The conversion of the source emission to the target distribution is driven by étendue, its conservation, and associated terms, like skewness. Starting in Chapter 1, the concept or definition of étendue is presented, and this term is touched upon in each succeeding chapter. Étendue and the related terms of skewness and concentration are the typical factors driving good design of nonimaging systems. A few examples are provided in Chapter 2, showing how to determine the "étendue budget," which delineates the maximum transfer efficiency assuming no additional losses in the optical system—that is, the only losses are attributed to étendue and/or skewness mismatch between the source and the target. This étendue mismatch is often caused

Illumination Engineering: Design with Nonimaging Optics, First Edition. R. John Koshel.
© 2013 the Institute of Electrical and Electronics Engineers. Published 2013 by John Wiley & Sons, Inc.

by a maximum extent of the transfer optics, and thus a maximum spatial or angular view of the source and/or target. This first-order design typically uses a simple source model, such as a point emitter or a small, finite source emitting into a Lambertian distribution. However, as in lens design, limiting a design to first order usually provides poor optical system performance, thus one must include the next order in the design process. In lens design, this is called third-order design, which addresses the third-order aberrations, including spherical, comatic, astigmatic, and so forth. A Cooke triplet has enough parameters (i.e., radii of curvatures and thicknesses) to correct third-order aberrations and chromatic aberration while providing the prescribed first-order properties. In nonimaging optics, third-order design means more aspects of the source are included, in particular a nonstandard angular emission distribution from a point or small-sized source. The optical surfaces are then tailored to efficiently transfer the source distribution into the desired target distribution. Continuing with the lens design analogy, one can increasingly add higher orders (e.g., fifth-order aberrations in imaging systems) in the design process by adding more aspects of the system. For example, early tailoring methods assumed planar sources that are spatially uniform and Lambertian. Analytic solutions can often be realized with these approximations, but real sources are nonplanar, non-Lambertian, and spatially nonuniform. The examples in Chapter 2 illustrate that while providing insight into the importance of étendue and skewness, the complexity of analytically calculating quantities is often difficult, if not impossible. Therefore, increasingly complex tailoring methods have been developed, such that the final optical design better meets its goals. Chapters 3 through 5 focused on such tailoring methods that innately strive to conserve étendue, which means that the goal of the final optical design is to provide the desired illumination distribution with high transfer efficiency from the source(s) to the target(s). Since around 1990, one can follow the increasing power and complexity of these tailoring methods, including:

- Point source with a nonstandard angular emission distribution [1]
- Finite source with an analytic emission distribution [2], and
- Realistic source models with a prescribed emission distribution [3–5].

The increasing power and complexity of these methods leads to nonanalytic solutions, such that the final optics are numerical structures or meshes. For example, most state-of-the-art nonimaging optics use nonuniform, rational b-splines (NURBS) to describe the surfaces. These tailoring methods cannot provide solutions for all applications nor can they be assured to be the best solutions; therefore, as in lens design, optimization methods are employed to potentially find a solution, refine the solution, and finally maximize performance. Simply, tailoring and optimization work hand-in-hand to provide the best system to meet the design requirements. Optical analysis software has recently included optimization routines to better design illumination systems.

Next, upon completion of the nominal optical design, including optimization, time must be invested to tolerance the system. Until this time, one typically assumes perfectly made optical surfaces, no source variation, and alignment is perfect between the various components. Of course, these are not realistic, and while illumination optics are not drastically affected by the magnitude of errors seen in imaging systems, the ranges of parameter variation (i.e., errors) are much larger. Illumination optics are being made in comparatively short times—seconds compared with hours, days, or longer for imaging optics. For example, plastic injection molding is a standard process used for nonimaging optics, where process times as short as 5–30 seconds are typical. So rather than the $\lambda/4$ errors in imaging circles, micron-plus errors are realized (e.g., the range of placement of the die in an LED has decreased from 100 μm to around 40 μm over the past decade). Unfortunately, the tolerancing standards used by the imaging community are not applicable to the nonimaging community. The standards are written for distinct parameters, such as radius of curvature, thickness, aspheric coefficients, and so forth. With the numerical constructs for nonimaging surfaces, one must first determine how to parameterize the surfaces, denote the optical axes for these surfaces, and finally determine the sensitivity of these surfaces, as their various parameters are varied. Tolerancing of illumination systems is a time-consuming process since the designer must do a lot of the setup and analysis without automated software tools.

At this time, optimization and tolerancing rely on ray tracing techniques. Thus, the next two sections discuss aspects of ray tracing, including Monte Carlo sources and ray propagation, "tricks" to improve the accuracy of ray tracing while reducing the time to convergence, and how many rays must be traced to reach this convergence. Convergence implies that an accurate determination of both the transfer efficiency and the illumination distribution at the target has been realized. Section 7.4 then presents aspects and nuances of optimization of stochastic systems, for example, the ray tracing process in nonimaging systems. Optimization of nonimaging systems has been readily explored over the last 20 years, and it remains a hot topic in the illumination field. At this time, robust algorithms are being included in commercially available optical design and analysis software codes. Finally, this chapter ends with an introduction to tolerancing methods of nonimaging systems. Research into the tolerancing of nonimaging systems has been increasingly studied over the past 10 years. Ultimately, tailoring methods will integrate optimization and tolerancing into their algorithms, such that the final system design includes the nominal source model and the expected range of source variation, optical surface error, and fabrication variation.

7.2 DESIGN TRICKS

Because analysis of nonimaging optical systems is often based on ray tracing, which is a process that requires a sizable amount of time and is limited by the speed of the computer processors being used, it is important for the designer to understand methods to improve modeling efficiency. It is imperative that the designer understands how the software traces rays through the optical system and ascertains the illumination distribution at the targets. First and foremost is the concept of nonsequential ray tracing, which along with source setup is the focus of the next section. Next is the method of reverse ray tracing, in which the rays are traced from the target(s) back to the source(s). In systems with scatter, such as diffuse surfaces, importance targets or sampling can be used to more efficiently trace rays from scattering surfaces to the target (or source in reverse ray tracing). Finally, this section ends with a discussion of what the term "far field" implies. Note, that like the discussion in Chapter 1, the correct terminology is far-field irradiance, not intensity, distribution. This facet is discussed below.

While there are four subsections herein, there are numerous other techniques that can be employed to better the performance of computer ray tracing. This section is an introduction to concepts and methods that assist the designer in better modeling of illumination systems; however, a number of these other methods are application dependent and thus beyond the scope of this introductory material. Additionally, the reader is encouraged to research new design tools for their given systems and then report the results at conferences or in the literature. Tool development is an active area in the illumination field.

7.2.1 Monte Carlo Processes

Monte Carlo means random or stochastic. Using stochastic processes is the standard for modeling illumination systems with ray sets. Random ray sets are developed for sources, and then Monte Carlo ray tracing is used to determine the illumination distribution at the target(s). These two Monte Carlo processes are fundamentally different, and commercial optical analysis software have slightly different implementations. Therefore, it is important for an illumination system designer to understand these two Monte Carlo processes (among others in the software packages). In the next two subsections, these two are described along with their respective non-Monte Carlo methods.

7.2.1.1 Monte Carlo Sources Ray tracing methods originally implemented grids in Cartesian, polar, staggered rectangular, hexagonal, and other periodic patterns in order to approximate the emission characteristics of a source. Using these sources to model the performance of an optical system, especially nonimaging ones, often resulted in target distributions that were adversely affected by the period of the source ray pattern. Thus, rather than a periodic pattern, a random set of rays is placed on the surface source (e.g., the coil of an incandescent filament) or within the source volume (e.g., the plasma of an arc source). Additionally, the angle through direction cosines for each ray is randomly set based on the applied distribution (e.g., Lambertian). The process of modifying the spatial and angular emission characteristics of a source is often called apodization.

There are essentially two methods of setting up Monte Carlo sources:

- *Random.* Based solely on the spatial and angular emission aspects of the source (either the functional distributions or their limits), a random set of rays is appended to the emitting surfaces or volumes of the source. A small ray set size has the potential to locate too many rays at a prescribed spatial position and/or too many rays are setup to be within a certain angular range.
- *Sobol* [6]. Essentially, this method subsamples the spatial and angular emission characteristics of the source, such that a more uniform representation of

the source emission in both position and angle is realized. This method combines aspects of a grid source and a random source.

Note that the power of each ray must then be set. If the angular and spatial emission distributions are used to select the ray direction cosines and positions, respectively, then each ray is given the same power. In this case, if M rays are created for a source of flux Φ, then each ray is given the power of Φ/M. Alternatively, the limits of the spatial and angular emission distributions can be used, and then the power of each ray is set according to the functional distribution. In this case, the initial flux of each ray is dependent on the value of the apodization function. For example, M rays are created for a spatially uniform planar source that emits into a Lambertian angular distribution. Each point in the source emits into a hemisphere with a Lambertian angular distribution that decreases as the cosine of the angle from the normal to the plane, θ. Each generated ray is randomly positioned on the plane and its emission angle within the limits of the hemisphere. The flux of each ray is set with

$$\Phi_i = \Phi \frac{I(\theta_i)}{\sum_{i=1}^{M} I(\theta_i)} = \frac{\Phi \cos\theta_i}{\sum_{i=1}^{M} \cos\theta_i}, \tag{7.1}$$

where i is the ray number within [1, M], Φ is the total source flux, and $I(\theta)$ is the intensity distribution of the source. The denominator in this equation ensures normalization of the source such that the sum of the ray fluxes is equal to Φ. The right-hand part of Equation (7.1) is for the case of Lambertian emission over the entire source. For different angular emission patterns, substitute the functional form of the intensity into the middle part of Equation (7.1). If the source has spatial variation while the angular distribution is isotropic, then substitute the irradiance $E(\theta_i)$ for $I(\theta_i)$. For sources that are apodized as a function of both position and angle, then substitute the radiance $L(\theta_i)$ for $I(\theta_i)$. For the latter sources, the calculation required with the radiance function can be complicated, so it is often best to employ optical design and analysis software.

It is typical to use the actual spatial and angular distributions to determine the source ray set, while giving each ray the same flux of Φ/M. However, it is important to determine which of these two methods (or a hybrid) is being used in your software.

7.2.1.2 Monte Carlo Ray Tracing To explain the nuances of ray propagation, an example of a ray normally incident on an uncoated surface with air on one side and a material of index 1.5 on the other is used in this section. The standard for ray tracing in most optical analysis software is to allow ray splitting. Ray splitting means that one incident ray can be turned into multiple rays after interaction with a surface (e.g., transmission and reflection with the change of index of refraction) or volume (e.g., scattering). Thus, in the example, two rays are propagated after the bare surface, where the reflected one has 4% of the original ray power, while the transmitted has 96%. At successive ray-surface interactions, the incident rays are split again, which leads to a rapidly increasing number of rays, including ones with very

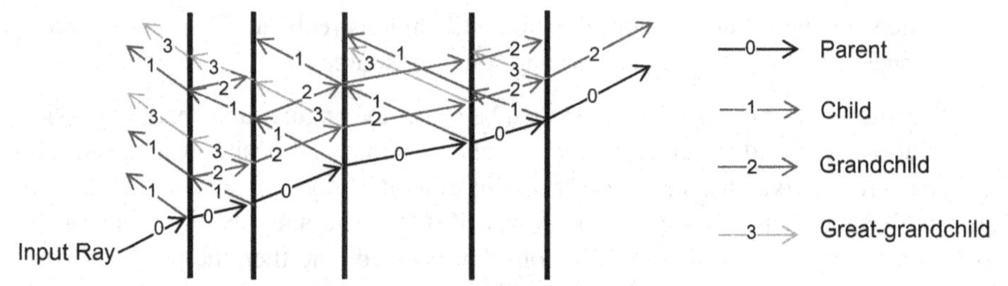

Figure 7.1 Layout of a basic optical system showing the generation of successive generations of rays. It assumes that the transmitted ray is the earlier generation. The colors and numbers indicate the generation order past the parent, which is shown in black and is generation 0. Note that only through the great-grandchild generation (3) is shown, and that all potential later generation rays are not shown.

little power, being traced. There are methods to limit the number of rays generated when ray splitting is invoked by controlling the ancestry generation. The initial source ray is called the parent ray, and after the surface interaction, one of the rays retains this parent status, while the other one is called the child ray. Typically, the ray with the most power is called the parent ray, but such depends on the software and how the user sets up the propagation characteristics. Further propagation and splitting of these rays, as shown in Figure 7.1, at successive surfaces provides:

- Parent ray leads to parent and child rays
- Child ray leads to child and grandchild rays
- Grandchild ray leads to grandchild and great-grandchild rays, and so forth.

As can be seen by Figure 7.1, the proliferation of rays can be dramatic through an optical system modeled by ray tracing. One input ray through a multi-surface optical surface can lead to many rays; therefore, this is the primary reason to include limits on ancestry ray generation. By limiting ancestry, the designer can stop ray-splitting generations beyond a desired level (e.g., grandchild). There are other limits to control ray propagation and generation, especially when splitting is implemented, including stopping rays that have below prescribed absolute or relative flux, one that have interacted with a given object greater than a provided number of times in a row, or limiting the total number of interactions that a ray may undergo during modeling.

Ray splitting is a powerful method to sample all physical processes in an optical system model; however, it is susceptible to the generation of a multitude of rays, many of which do not have much power and at the added expense of increased modeling time. Systems with a multitude of ray-surface interactions are especially susceptible to this problem. Such systems include complex imaging systems, light-pipes, and, in particular, backlit displays that employ a side-lit lightguide. A different method of ray tracing is required to stop the generation of rays and reduce the modeling time while retaining accuracy. Monte Carlo ray tracing is such a method. At each interface in the optical system, volume scattering event, and so forth, a single ray is continued along its propagation path. Random number generation determines the path that each ray follows in the optical system. Using the example at an uncoated optical surface, a single ray can follow either the refractive path or reflective path at the point of incidence. The probability for each path is determined, and the randomly generated number determines which path the ray takes. On average, the reflection path is followed 4% of the time, and the transmission path is followed the remaining 96% of the time. For this reason, Monte Carlo ray propagation is often called "one ray in, one ray out," which means that if one traces *M* rays from the source, a total of *M* rays are traced in the analysis. This process is easy to implement with commercial software, but there are a number of other aspects that must be considered to fully explain Monte Carlo ray propagation:

- *Continuation of Process.* At each ray-surface interface in the optical system, this process is continued until the ray is absorbed, does not strike another entity, or other criteria, such as maximum number of surface interactions is achieved.
- *Potential Ray Paths.* Not only reflection and refraction are included in the determination of the final ray path, but also scatter, diffraction, and more complicated optical phenomena, such as birefringence, can lead to a multitude of ray paths at each ray-surface interaction.
- *Inclusion of Absorption.* the original power in the incident ray less the surface absorption is continued in the single ray leaving the optical surface. One could consider adding the absorbed "ray path" to the Monte Carlo process, but to date, this "path" has not been added to the process.
- *Inclusion of Ray Splitting.* There are times when ray splitting is beneficial in proper modeling of an optical surface or object, such as for integrating spheres and cavities. This point is discussed in Section 7.2.3.

7.2.2 Reverse Ray Tracing

In most ray-trace modeling situations, it is logical to trace from the source(s) to the target(s). First, it makes intuitive sense to the designer. Second, there are accurate source models and geometries based on geometrical models and/or measurements of the sources. Third, software assists with all facets of the modeling process, including locating the source as desired and well-developed tools to perform the analysis in the forward direction. However, there are times it makes sense to trace from the target(s) to the source(s), or a hybrid where the rays are traced in both the forward and reverse directions to a prescribed location in the model.

The general rule is that ray tracing should occur in the direction with the largest étendue. Consider Figure 7.2, which uses a "black box" optical system with the source on the left and the target on the right. The areas of the source and the target are the same, but the solid angle subtense on the source side is greater than that on the target side. Thus, the source étendue is greater than the target étendue, meaning that this system does not conserve étendue leading to transfer inefficiency. Thus, upon ray tracing in the forward direction, a number of the rays will not strike the target, so it is better to trace in the backward direction from the target to the source.

An example is done to illustrate the utility of reverse ray tracing. Consider Figure 7.3b, in which the generic optical system of Figure 7.2 is replaced by a

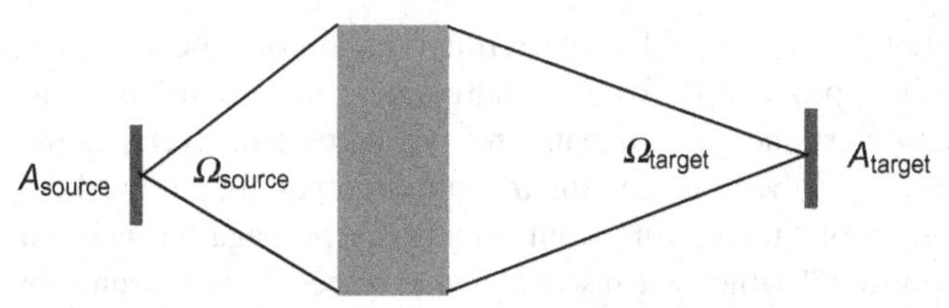

Figure 7.2 Étendue is the driving factor in determining in which direction is the most efficient to trace rays. In this figure, the areas of the source (A_{source}) and target (A_{target}) are equal, so the subtended solid angle is the determining factor. The source is closer to the optical system, so $\Omega_{source} < \Omega_{target}$ is more efficient to trace in the reverse direction.

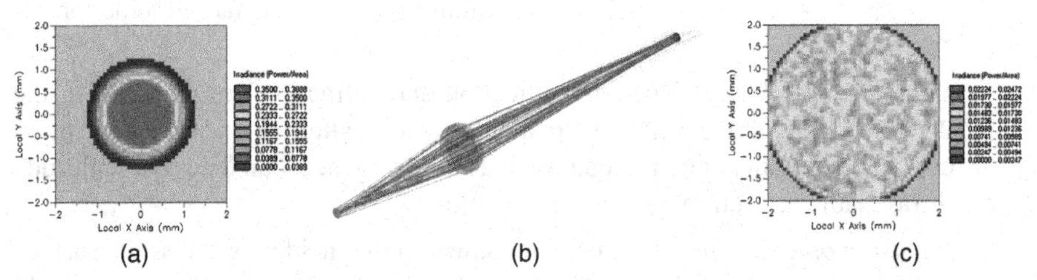

Figure 7.3 Modeled example where the generic optical system of Figure 7.2 is a singlet and the source is less than half the distance to the lens compared with the target: (a) the irradiance distribution from a reverse ray trace using 1M rays, (b) the layout of the singlet, 4-mm diameter source (left, in green, with forward rays shown in green), and 4-mm diameter target (right, in red, with reverse rays shown in red) for the case presented here, and (c) the irradiance distribution from a forward ray trace using 1M rays.

singlet ($R_1 = -50$ mm, $R_2 = 50$ mm, $t = 5$ mm, $CA = 20$ mm, and $n = 1.5$). The supposed unitary magnification of keeping the source and target the same size means the difference in solid angle is the result of a user-entered positive defocus of the target (i.e., the target is further away the lens compared with the source). In this example, the source is located 68 mm from the front surface of the lens and the target is located 145 mm from the back surface of the lens. For the purposes of this discussion, assume that the designer does not know the numerical aperture of the optical system (i.e., lens), and the emitter emits into a hemispherical Lambertian distribution with a 4-mm diameter perfect absorber on the other side of the optic. There are several ways that the designer could model this system with ray tracing:

1. Tracing the into a hemisphere in the forward or reverse directions:
   (a) Forward direction: the transfer efficiency is found to be 0.46%, while
   (b) Reverse direction: the transfer efficiency is found to be 0.47%;
2. Tracing into the clear aperture only in the forward or reverse directions:
   (a) Forward direction: the transfer efficiency is found to be 23.6%, while
   (b) Reverse direction: the transfer efficiency is found to be 100%; and

3. Tracing into the respective pupil in the forward or reverse directions:
   (a) Forward direction: the transfer efficiency is 100%, while
   (b) Reverse direction: the transfer efficiency is 100%.

The first method is brute force, thus, the transfer is expected to be low in comparison with methods that use characteristics of the optical system. For example, in the second method, the designer uses a quasi-Lambertian ray set that fills the clear aperture of the singlet. Due to the defocus in this example, the forward ray trace transfer efficiency suffers. The loss in the forward direction is due to rays missing the target since the actual pupil in the forward direction is smaller than the lens clear aperture. By using the entrance pupil position and diameter in the forward direction and the exit pupil position and diameter in the reverse direction, one can further improve a given ray trace.* Note that this simple example is intended to be illustrative, since most designers would set up their analysis to properly sample the respective pupil of an optical system. However, it is often difficult if not impossible to determine the "pupils" in optical systems, especially those with large acceptance angles. For example, in wide field-of-view lenses, the stop and thus pupil positions and diameters vary as a function of field angle [7]. In nonimaging systems, this result is further complicated, in particular, systems comprised of numerous paths from the source(s) to the target(s). Returning to the example, Figure 7.3a (reverse tracing) and 7.3c (forward tracing) show that the irradiance distributions are similar, but that Figure 7.3c is less sampled due to fewer rays striking the target. Figure 7.3a shows the effective area of the source that is transmitted through the singlet, which means it is beneficial to determine the étendue of the lens (i.e., as per Chapter 2) before analysis is undertaken. While this example is simple, it does illustrate the concept of how to select the preferred ray tracing direction in optical systems when one knows only the basic characteristics of the optical system (e.g., case 2 above). If you know more aspects of the optical system (e.g., case 3 above), then forward, reverse, or hybrid ray tracing provide the same results.

In reverse ray tracing, one must accurately account for the radiance function of the source(s), since the rays are traced backwards. Reverse ray tracing is done by turning the target(s) into emitters and essentially turning the source(s) into detectors; however, the original source(s) maintains its radiance emission characteristics. When a reverse ray strikes the source, the spatial and angular aspects of the point of incidence sample the radiance emission distribution. The flux that would have been assigned to a forward ray is then assigned to this reverse ray, such that the actual source flux "flows" backward on the reverse rays that are traced. One must account for the losses along the ray path, including absorption, ray splitting, scatter, and so forth, so, normalization is required at the end of the whole reverse ray trace.

* Note that the étendue is proportional to $\sin^2\theta$. Using the on-axis half angles of $\theta_{forward} = 8.07°$ and $\theta_{reverse} = 3.88°$, one can calculate the ratio of solid angles in the two directions: $\Omega_{source}/\Omega_{target} = 4.31$. This value is approximately the multiplicative factor in the modeled transfer efficiencies of case 2 (23.6–100%). The discrepancy is due to the variation of subtended solid angle across the source and target sides of the optical system.

As shown in the case of not being able to use even basic characteristics of the optical system, ray tracing in the forward or reverse directions provides the same results. The basic characteristics, such as angular subtense in the two ray trace directions, can greatly assist in the best choice of ray trace directions. A discussed, in real systems, it is not only too complex to determine accurately what the étendue is in the two potential ray trace directions, but the basic characteristics are hard to discern. This result is especially true in systems that benefit from hybrid ray tracing, that is, those in which the designer benefits from both forward and reverse ray tracing. For example, consider a side-backlit display that uses small sources emitting into a Lambertian distribution (e.g., a tubular cold-cathode fluorescent lamp or LEDs). Since the display is designed around ejecting the light toward viewers, one likely first considers tracing the rays in the forward direction; however, after the display surface, the optical system is the visual system(s) of the observer(s). In this case, the étendue is small, since the viewers sparsely inhabit the field of regard of the display. A compromise is to trace the rays from the source to the exit surface of the display and trace rays in reverse from the eyes of the observer(s) to the display exit surface. This hybrid approach essentially makes the whole backlit display the source and ensures efficient ray tracing to model the appearance of the display to viewers. Additional ray tracing shortcuts assist in this design process, especially for systems that have diffuse or scattering surfaces. Diffuse surfaces can use a technique called importance sampling to more effectively sample the system with rays. Importance sampling is the focus of the next section.

7.2.3 Importance Sampling

If there is a diffuse surface or volume near the end of a given propagation direction, forward or reverse, then one can use the scatter properties of this entity to improve upon the ray transfer efficiency of the model. In optical design and analysis programs, the scatter profile can be used to select which rays are further propagated in the model. As per Figure 7.4, in which rays from a source strike a diffuser, there is 100% ray-transfer efficiency to the target by using the scatter distribution of the diffuser. This process uses what is called an importance target or edge within the

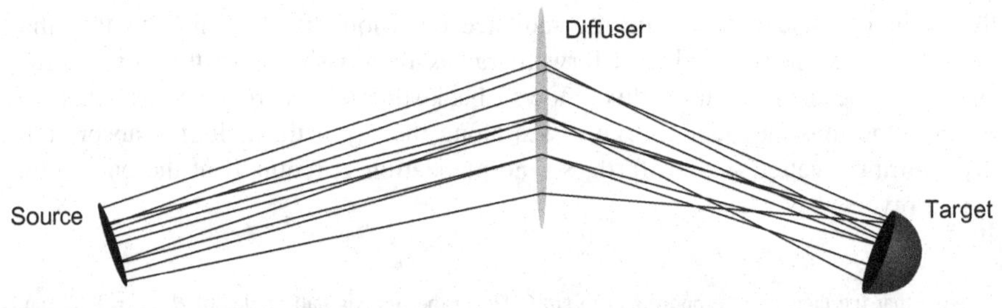

Figure 7.4 Illustration of importance sampling with rays from the source (red, left) incident on a diffuser (grey, center). The scatter properties of the diffuser are used to redirect the rays to an importance edge placed at the target (blue, right). Each scattered ray is directed to a randomly determined position in the importance edge at the target.

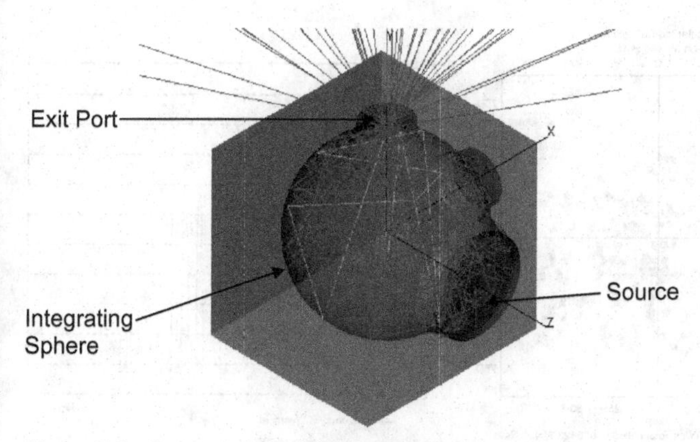

Figure 7.5 Layout of the 2-in diameter integrating sphere representative of model 3P-GPS-020-SL from LabSphere [8]. Also shown is a single source ray that generates many child rays that are directed to the exit port using the importance sampling technique.

software model. For the rays that are incident on the diffuser, 100% of them are redirected to the importance target in a Monte Carlo approach. However, the flux of each ray is adjusted to properly sample the scatter properties (e.g., BSDF) into the solid angle subtended by the target. Therefore, note that the term "ray transfer efficiency" rather than "transfer efficiency" is used here. In the previous section, the transfer efficiency implied the "flux-transfer efficiency," such that the percentages denoted the amount of source flux that was transferred from the source to target (forward) or target to source (reverse).* In this section, the ray transfer efficiency indicates the number of rays that make it from the source to target (forward) or target to source (reverse). The process of using the scatter properties of an entity for importance sampling means that the scatter losses have to be taken into account.

Importance sampling is important for investigating scattered light aspects of systems, such as telescopes. It is also important for illumination components that rely on scattering for their performance, such as diffusers and integrating spheres. In fact, the "one ray in, one ray out" ray tracing technique described in Section 7.2.1.2 should be removed for such illumination components. An integrating sphere is used to illustrate the importance of using importance sampling and the removal of "one ray in, one ray out." As shown in Figure 7.5, consider a 2-in diameter integrating sphere with an assumed 98% reflective Lambertian scattering property on the interior of the sphere† [8]. The source emits into the entrance port of the sphere, while an exit port is located at 90° to the entrance port. Also shown in Figure 7.5 is the result for a single ray traced from a source and the child rays that arise due to importance sampling and the removal of "one ray in, one ray out."

* Hereafter the term transfer efficiency means flux-transfer efficiency while the term ray transfer efficiency is explicitly stated.

† This model is representative of LabSphere integrating sphere model 3P-GPS-020-SL coated with Spectralon and the internal beam block of the exit aperture is not included[8].

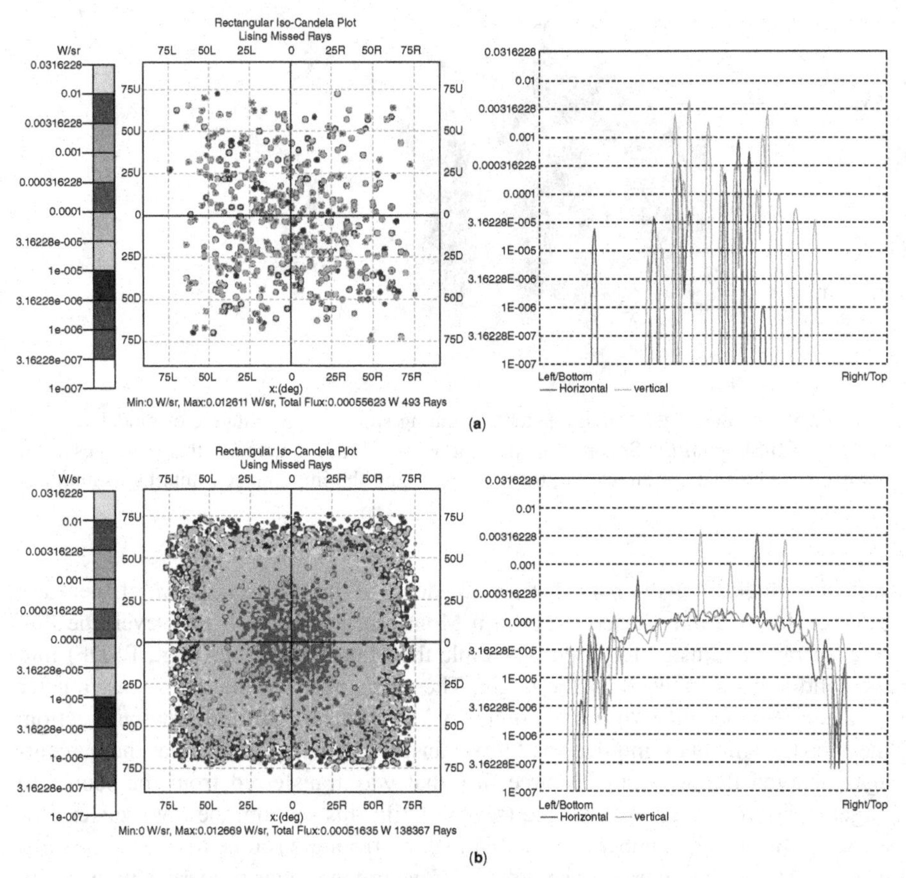

Figure 7.6 Intensity results for the integrating sphere of Figure 7.5 for (a) "one ray in, one ray out" Monte Carlo ray tracing and (b) importance sampling with Monte Carlo ray tracing. On the left for each of the subfigures is shown the intensity distribution in a Cartesian angular grid, while the right hand is shown the cross sections along the crosshairs (blue = horizontal and green = vertical).

This method is compared with the standard Monte Carlo ray tracing method of Section 7.2.1.2:

- *"One Ray In, One Ray Out"* (Section 7.2.1.2). Each time a ray is incident on the interior of the integrating sphere, it is randomly redirected according to the Lambertian properties of the scattering surface. By iteratively bouncing around in the integrating sphere, there is a probability that the ray goes through the exit port. Upon tracing 4000 source rays, 483 rays exit the sphere with the resulting intensity distribution of Figure 7.6a, shown with a vertical log scale.
- *Importance Sampling* (Section 7.2.3). Each time the parent ray is incident on the interior surface of the integrating sphere, a single child ray is directed to

the exit port while the parent ray is allowed to scatter into an angle except toward the exit port. Upon tracing 4000 source rays, 138,367 rays exit the sphere with the resulting intensity distribution of Figure 7.6b, shown with a vertical log scale.

Note that importance sampling still takes advantage of Monte Carlo ray tracing, but each time a parent ray is incident on the integrating sphere interior, two rays are propagated. The parent ray is allowed to continue to further sample successive reflections from the sphere, while a child ray is directed toward the exit aperture. Thus, though there is ray generation with this method, it better samples the emission from the integrating sphere. Note that this example also illustrates why there is an obstruction to block direct illumination of the exit aperture from the source. The peaks in Figure 7.6b illustrate this direct illumination, which are removed with the beam block to ensure that all source radiation interacts with the coating of the integrating sphere at least once.

7.2.4 Far-Field Irradiance

In the far-field limit, the source is approximated by a point source, so the rays being used to analyze an optical design of no (or little) crossing. In this limit, the spatial distribution of radiation has evolved into a static pattern in a given plane. Note that the angular distribution is always static after the last object of an optical system. Thus one reaches the far-field limit, or far-field irradiance, when the shape of the irradiance distribution at a plane matches that of the intensity distribution.* The term "matching" is vague since subjective criteria often define the allowed variation between the two distributions. A good rule of thumb is less than 10% variation between the two distributions. Figure 7.7 shows the case for a system analogous to that in Figure 7.3b, for which the target is moved to positions prescribed distances away from the nominal image plane. Figure 7.7a is the static, normalized intensity distribution, while the panels b–f figures on the left are the normalized irradiance distributions at the stated distance beyond the image plane, and the panels b–f figures on the right are the relative percentage comparisons of the two distributions

$$\Delta(i,j) = 100\frac{\tilde{E}(i,j) - \tilde{I}(i,j)}{\tilde{I}(i,j)}, \tag{7.2}$$

where Δ is the relative difference, $\tilde{E}$ is the normalized irradiance distribution, $\tilde{I}$ is the normalized intensity distribution, and i and j are the sampling counters of the distributions in the horizontal and vertical directions, respectively. In each plot of Figure 7.7, $i = [1, 51]$ and $j = [1, 51]$. The normalization process is to make the peak irradiance and intensity in a given distribution equal to 1. The units of Equation (7.2)

* The term far-field intensity is often used; however, this term is expressly incorrect. The intensity pattern is always a far-field distribution since it does not evolve with distance. Rather, one is actually referring to the far-field irradiance distribution. The incorrect usage of intensity for irradiance is discussed in Section 1.6.1, the footnotes, and the references therein.

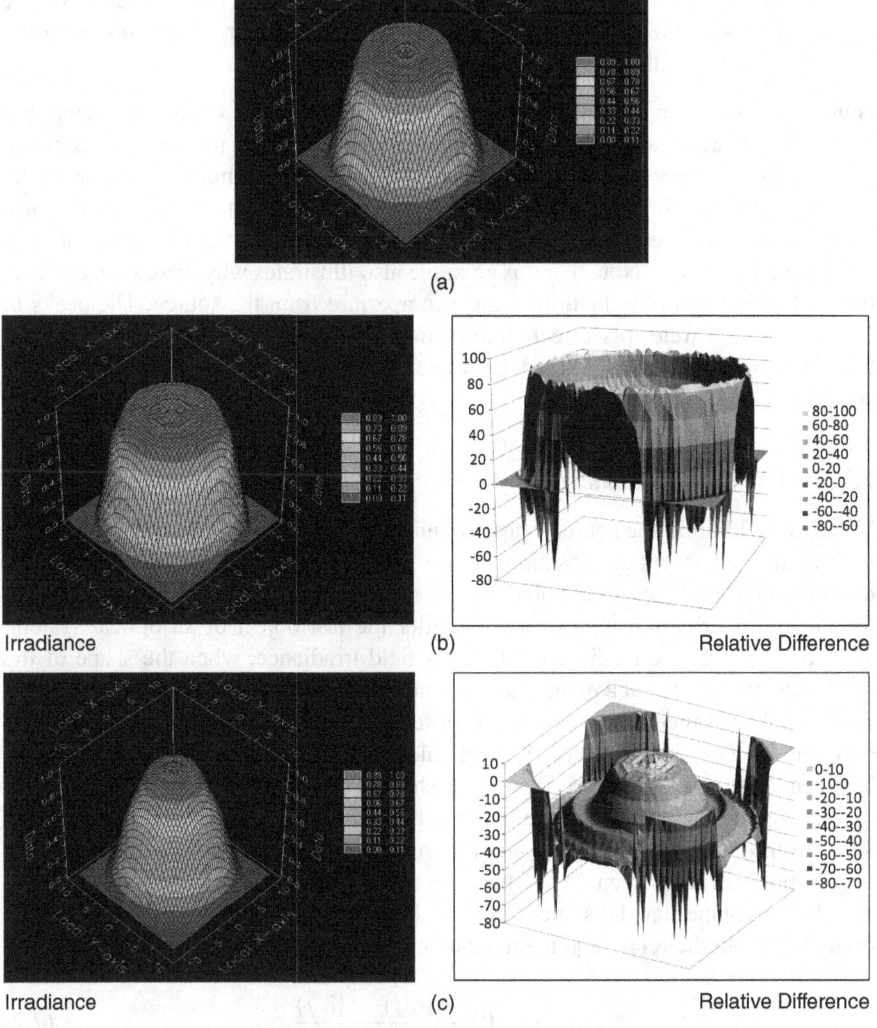

Figure 7.7 Comparison of the (a) intensity distribution for a forward ray trace of the system of Figure 7.3b to the irradiance distribution at several positions beyond the best focus plane (b) at nominal focus, (c) 100, (d) 500, (e) 1000, and (f) 5000 mm. The right-hand figures are the irradiance distributions while the left-hand figures show the Equation (7.2) percentage relative difference between the 51 × 51 bins that comprise each plot.

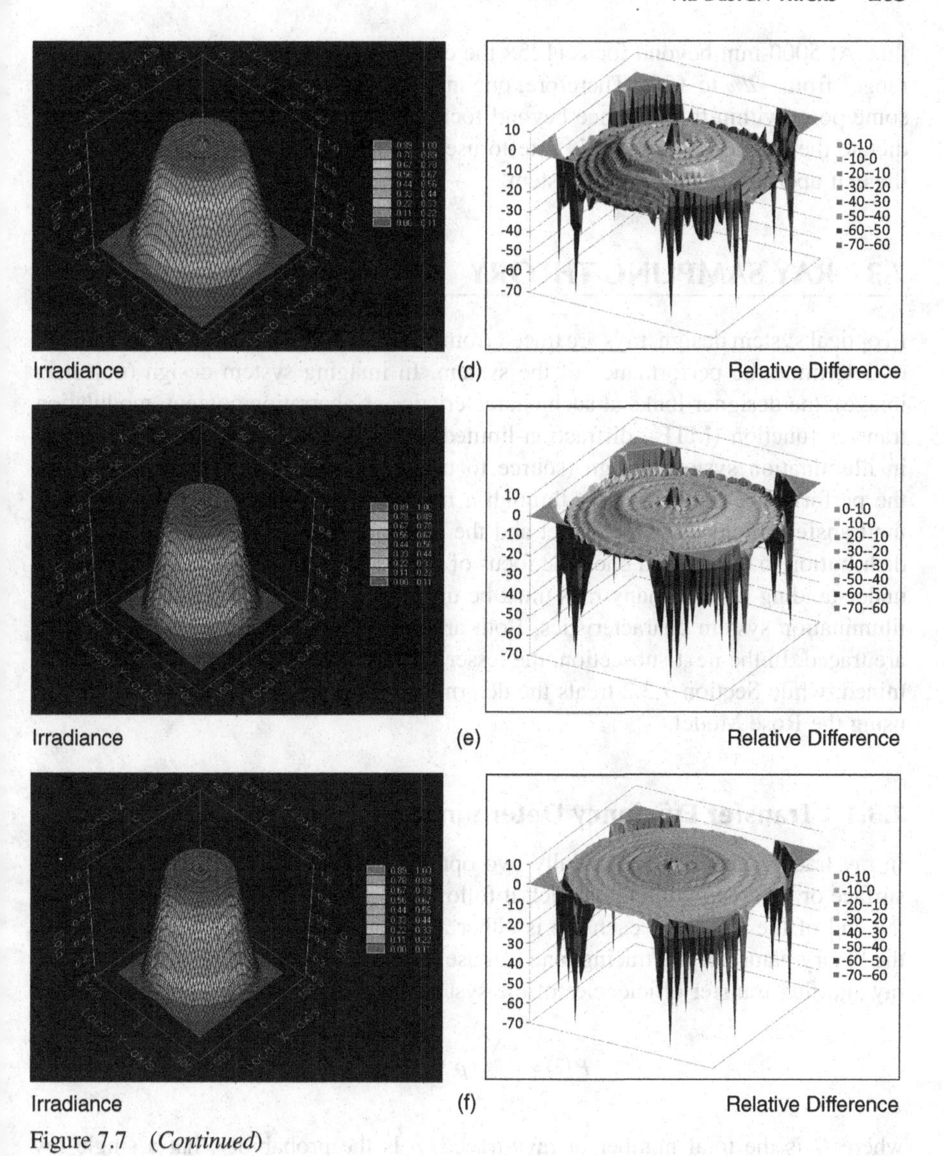

Figure 7.7 (*Continued*)

do not agree with one another, but one is simply comparing the shapes of the two distributions, not the values. Additionally, there is expected to be significant variation at the edges of the illumination distribution due to the values being small. Thus, one should place their attention on the central regions of each plot.

At nominal focus (Fig. 7.7b), the relative difference ranges up to +100%. This relative difference between the intensity and irradiance plots decreases as the distance from best focus is increased. At 1000-mm beyond focus (25× the clear aperture of the lens), the variation ranges from −10% to +3% over the region with substantial flux. At 5000-mm beyond focus (125× the clear aperture of the lens), this variation ranges from −2% to +2%. Therefore, one may say the far field has been attained at some point within this distance beyond focus. System and designer considerations dictate the final choice, but it is safe to use a value of 100× the greatest extent of the exit aperture of the optical system.

7.3 RAY SAMPLING THEORY

In optical system design, rays are traced from the source (object) to the target (image) to determine the performance of the system. In imaging system design (object to image), the designer looks at such characteristics as aberration content, modulation transfer function (MTF), diffraction-limited performance, and so forth. Likewise, in illumination system design (source to target), the optical designer maximizes the performance of the system through a number of constructs, but in particular, the transfer efficiency to the target and the agreement of the obtained illumination distribution to the desired one. The focus of this section is to provide the theoretical understanding of how many rays must be traced to accurately determine these two illumination system characteristics. Both are governed by the number of rays that are traced. In the next subsection, the lesser criterion of transfer efficiency is determined, while Section 7.3.2 treats the determination of the illumination distribution using the Rose Model.

7.3.1 Transfer Efficiency Determination

In ray tracing, there are essentially two options for each ray as it interacts with a surface or volume within the model: it follows a path to the target or it does not. At the end of the ray trace, each ray is either at the target or it does not, thus, due to the binary nature of ray tracing, one can use the binomial distribution to explain the ray and flux transfer efficiencies of the system

$$P(i) = \binom{N}{i} p^i (1-p)^{N-i}, \tag{7.3}$$

where N is the total number of rays traced, p is the probability that a single ray makes it from the source to the target, i is the number of rays that make it from the source to the target, and $P(i)$ is the probability that i rays make it to the target. The first term in Equation (7.3) is called "N choose i," where this term is defined as

$$\binom{N}{i} = \frac{N!}{i!(N-i)!}, \tag{7.4}$$

Assuming a discrete system, the expected ray-transfer efficiency is

$$\bar{\eta}_{\text{ray}} = \frac{\text{round}(Np)}{N}, \tag{7.5}$$

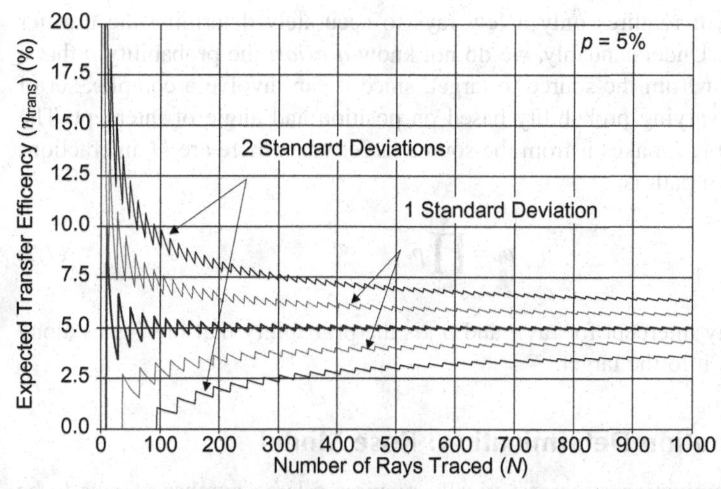

Figure 7.8 The expected ray transfer efficiency using the binomial distribution when N rays are traced. The blue curve shows the average value, while the red curves show the ray on ±1 standard deviation (i.e., 68% of the traces with N rays)and the black curves show ±2 standard deviations (i.e., 95% of the traces with N rays).

and the standard deviation around the expectation is

$$\bar{\eta}_{\text{ray}} \pm k\sigma = \frac{\text{round}\left(Np - k\sqrt{Np(1-p)}\right)}{N}, \tag{7.6}$$

where k is the confidence factor and the round terms in the numerator of each equation denotes the discrete nature of the rounding process—that is, a partial ray cannot make it from the source to target. For a source in which all the rays are given the same arbitrary power, then Equation (7.5) also represents the flux-transfer efficiency. If each ray starts with a different power, then the flux transfer efficiency is found by including the incident ray power as the respective weight factor.

As an example, consider a system that has an unknown $p = 5\%$ chance that each ray makes it from the source to the target. Figure 7.8 shows the resulting determination of the ray transfer efficiency. There are three colored curves in this plot:

- *Blue Curve.* The determined ray transfer efficiency after tracing N = horizontal axis value number of rays.
- *Red Curves.* The envelope for confidence value of $k = 1$, which means that around 68% of the ray traces with N = horizontal axis value rays lies between these two curves.
- *Black Curves.* The envelope for confidence value of $k = 2$, which means that around 95% of the ray traces with N = horizontal axis value rays lies between these two curves.

As can be seen, it requires only a few rays to accurately determine the transfer efficiency terms. Understandably, we do not know *a priori* the probability p that a given ray takes it from the source to target, since it can involve a complex set of paths that have varying probability based on position and angle of intercept. The probability that ray i makes it from the source target when there are M_i interactions along the selected path is

$$p_i = \prod_{j=1}^{M_i} p_{i,j}, \tag{7.7}$$

where j is the ray intercept for ray i and $p_{i,j}$ is the probability that ray i goes along a path that takes it to the target.

7.3.2 Distribution Determination: Rose Model

Modeling of illumination systems typically requires a large number of rays to be traced in order to accurately determine the distribution at the target; however, until recently, the guidance was to trace as many rays as possible [9]. A better solution is to base the number of rays on the signal-to-noise ratio (SNR), where a higher SNR is better. A model based on photon statistics analogous to the algorithm that Albert Rose developed for low light-level television camera tubes is developed in this section [10]. Hereafter, this model is called the Rose model. For Rose, in low-light level environments, the stochastic photon distribution detrimentally affects image quality and the ability to pass information. From his modeling, he was able to determine the best light levels for imaging of scenes with prescribed equipment. In ray tracing, each ray is analogous to a photon, and the determination of illumination features (signal) within a background (noise) provides us the SNR. There are three important parameters for the determination of the required number of rays [9]:

1. *Image Resolution.* The smallest required feature size of area A in a background of area A_0. This feature size is often dictated not only by engineering requirements, but also aesthetics. Additionally, this feature size determines the size of the pixels or sampling at the target, and thus the total number of pixels at the target.
2. *Contrast.* The ratio of the background illumination distribution E to that of a bright or dark feature with illumination distribution $E + \Delta E$:

$$C = \frac{\Delta E}{E}. \tag{7.8}$$

   Note that E is denoted as the irradiance through the remainder of this section, but the terms E and ΔE can denote any radiometric or photometric term (e.g., intensity, radiance, or luminance).
3. *SNR.* The allowable "fluctuations" in the illumination distribution. These fluctuations in a random ray trace distribution can lead to false features and/or obscured features, especially as the number of rays traced is reduced. In order

to discern the feature, its illumination level compared with that of the background must be greater than the statistical fluctuation. For this development, we use a normal distribution to model the variation across the illumination distribution,

$$f(x) = \frac{1}{\sqrt{2\pi}\sigma} e^{-(x-\mu)^2/2\sigma^2}, \quad -\infty < x < \infty, \tag{7.9}$$

where σ is the standard deviation of the background and μ is the average background level.

The variable x is the variation of the irradiance, E, due to the random process of its determination. With random noise, the SNR is given by

$$SNR = \frac{\Delta E}{\sigma}. \tag{7.10}$$

In order to discern the features on a constant background, the feature irradiance must be greater than the standard deviation, $\Delta E \geq \sigma$; however, this means that at the lower limit the SNR is equal to 1. Thus, it is better to be several multiples, k, above the standard deviation,

$$\Delta E \geq k\sigma, \tag{7.11}$$

where k is the confidence factor and denotes the number of standard deviations the signal is above the background. The equal sign in Equation (7.11) denotes that we are just able to discern the feature on the background, and, thus provides the lower bound on the number of rays that must be traced. Next, no matter how many rays are traced, there is the potential that a false ray rays is obtained. The probability, P_{false}, that such happens is the ratio of the background size to that of the desired feature size multiplied by the probability distribution

$$P_{\text{false}} = 2\frac{A_0}{A} P_N(k\sigma), \tag{7.12}$$

where the factor of 2 denotes bright and darker features exist, and $P_N(k\sigma)$ is given by

$$P_N(x) = 1 - \int_{-\infty}^{x} f(u)\,du, \tag{7.13}$$

which upon using Equation (7.4) within Equation (7.8) gives

$$P_N(x) = 1 - \frac{1}{\sqrt{2\pi}\sigma} \int_{-\infty}^{x} e^{-(u-\mu)^2/2\sigma^2}\,du. \tag{7.14}$$

This equation indicates the probability that the illumination variation is greater than $k\sigma$ in a feature of area A over a background of A_0. P_{false} is set by the designer to the desired level of accuracy, such as 1%, which gives $P_N(k\sigma)$. Solving for k in $P_N(k\sigma)$ is numerical in nature, but it is given by the error function (erf) or complementary error function (erfc):

$$\text{erf}(x) = \frac{2}{\sqrt{\pi}} \int_0^x e^{-v^2}\, dv \tag{7.15a}$$

and

$$\text{erfc}(x) = 1 - \text{erf}(x) = \frac{2}{\sqrt{\pi}} \int_x^\infty e^{-v^2}\, dv. \tag{7.15b}$$

Using the two forms of Equation (7.15), one finds for the probability distribution

$$P_N(x) = \frac{1}{2}\left[1 - \text{erf}\left(\frac{x-\mu}{\sqrt{2}\sigma}\right)\right] = \frac{1}{2}\text{erfc}\left(\frac{x-\mu}{\sqrt{2}\sigma}\right). \tag{7.16}$$

This result is substituted into Equation (7.7)

$$P_{\text{false}} = \frac{A_0}{A}\text{erfc}\left(\frac{x-\mu}{\sqrt{2}\sigma}\right). \tag{7.17}$$

Therefore, when $x - \mu = k\sigma$, we are at the lower limit as given by Equation (7.11)

$$P_{\text{false}} = \frac{A_0}{A}\text{erfc}\left(\frac{k}{\sqrt{2}}\right). \tag{7.18}$$

The solution for k can be found from lookup tables, root-finding techniques, or software with embedded error function calls. This value of k is the required SNR to discern a feature of size A within a background of size A_0. Rose postulated that the minimum number of photons, M_{min}, to detect a feature of size A and contrast of C on a background area of A_0 is [10]

$$M_{\text{min}} = \frac{A_0}{A}\frac{k^2}{C^2}. \tag{7.19}$$

Equation (7.19) assumes that all rays from the source are incident on the target, but this expectation is unrealistic. One includes the ray transfer efficiency to correctly account for losses with propagation in the optical system

$$M_{\text{min}} = \frac{1}{\eta_{\text{ray}}}\frac{A_0}{A}\frac{k^2}{C^2}, \tag{7.20}$$

where the bar over the η_{ray} term, as per Equations (7.5) and (7.6), is not included because this ray transfer efficiency is typically determined via modeling. Note that Equation (7.5) has the following trends:

- Decreasing the feature size (A): inverse linear increase in the number of rays
- Increasing the background size (A_0): linearly increases the number of rays
- Increasing the SNR (k): quadratic increase in the number of rays
- Decreasing contrast (C): inverse quadratic increases in the number of ray, and

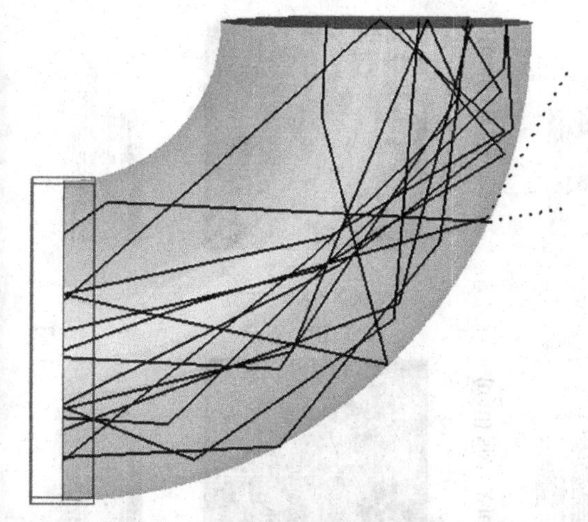

Figure 7.9 Layout of the common center, single-bend lightpipe with an inner radius of 10 mm, outer radius of 20 mm, thickness of 10 mm, and an index of 1.5. Representative rays are traced from the source on the left to the exit aperture at the top.

- Decreasing ray transfer efficiency (η_{ray}): inverse linear increase in the number of rays.

Note that often, the specifications of the feature and background sizes are given in pixels. This choice is due to the software modeling programs, which bin the illumination distribution data into a Cartesian grid. Thus, by increasing the resolution of the background grid while keeping the number of feature pixels consistent, then there is a quadratic increase in the number of rays required to properly sample the system.

As an example of the utility of this method, consider the lightpipe as shown in Figure 7.9. The lightpipe has a single common-center bend of 90°, a circular cross section, an inner bend radius of 10 mm, and outer bend radius of 20 mm, an index of refraction of 1.5, and it is immersed in air. The input (lower left) and output (upper right) are assumed to be perfectly transmissive, while the side walls are assumed to be uncoated. At the exit port, a Cartesian grid of 51 × 51 (A_0) circumscribes the circular cross section of the lightpipe. We are interested in irradiance feature sizes of 5 pixels (A) with a contrast of 10% (C) compared with the background. The probability that we obtain a false result is 1% (P_{false}). Using Equation (7.18), we determine $k \approx 4.27$. Additionally, through preliminary small ray traces, we determine that $\eta_{\text{ray}} \approx 50.3\%$. Substituting these values into Equation (7.20), we determine that we need to trace nearly 1.9 million rays to correctly sample this system. Figure 7.10 shows a series of plots of the irradiance distribution at the exit port when 20 to 2M rays are traced. The scale in Figure 7.10f is such that each level is about 5% of the peak irradiance, so the distribution is expected to be accurate over two levels.

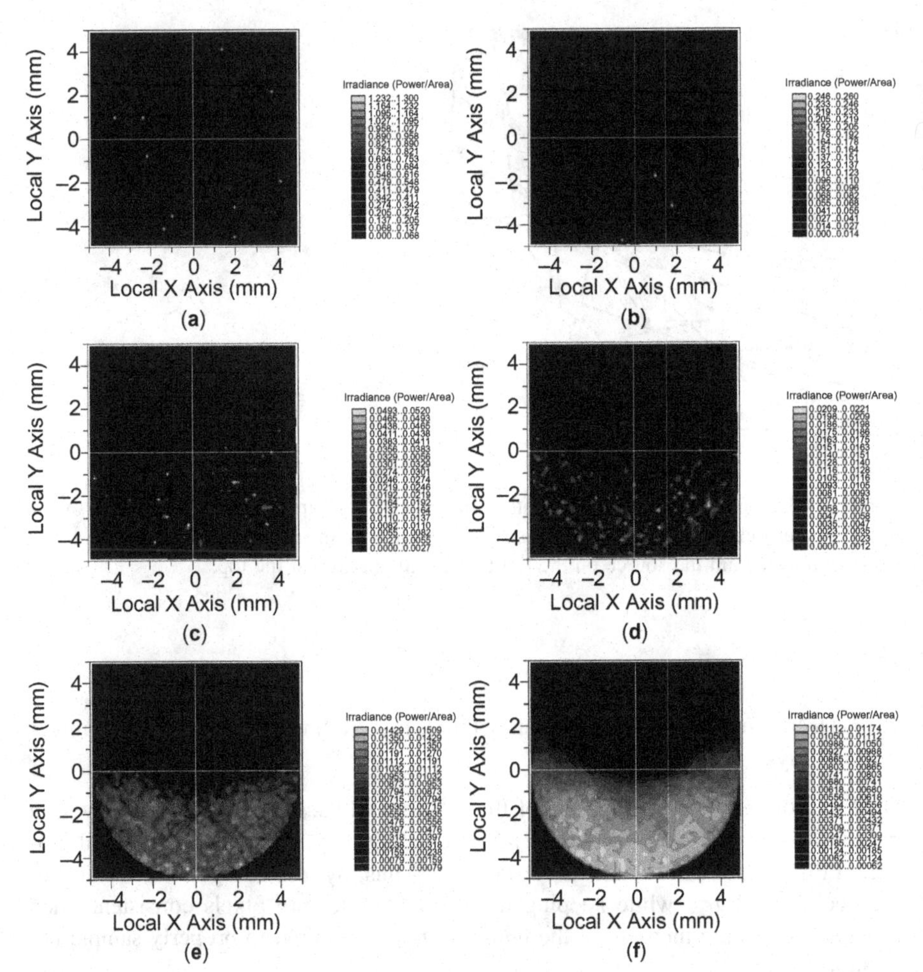

Figure 7.10 Ray trace results of tracing a set number of rays in the common-center bend lightpipe of Figure 7.9 with (a) 20, (b) 200, (c) 2000, (d) 20,000, (e) 200,000, and (f) 2,000,000 rays.

7.4 OPTIMIZATION

Over the past couple of decades, the need for illumination system optimization has grown within the optical design community; however, optimization of such systems is still in its infancy. There are a number of factors that give rise to this predicament that time and additional research will address. These factors include [9]

- *Sampling Requirements.* As described in Section 7.3, a large number of rays must be traced to properly sample the illumination distribution at the target.
- *System Parameterization.* This aspect is two pronged in that there is a dearth of parameterization standards and protocols for nonimaging surfaces, and the results are often described with numerical constructs that arose out of the CAD industry.

- *Object and Surface Interference.* Algorithms for limiting overlap or interference of objects and surfaces with other objects or themselves need to be developed for automated optimization techniques.
- *Optimization of stochastic Systems.* Due to sampling with discrete ray sets, there is the opportunity to have the optimization algorithm controlled to some extent by noise.
- *Merit Function Designation.* Unlike imaging optic optimization, the illumination system requirements are more nebulous and typically require the designer to individually provide the merit function.
- *Fabrication and Tolerances.* The success of designs is often driven by the ability, or inability, to make the optics within the desired tolerance range in a suitably short time.

Furthermore, and at the crux of the complexity, there are a limited, but growing, number of illumination design methods, but as indicated in Chapters 3–6, the methods are increasingly robust and powerful. People entering the design ranks of the illumination field must first learn of the various types of illumination optics and their utility for their applications. The designer must then choose *a priori* how to tackle a design problem with, for example, a lightpipe, lens, lens array, reflector, Fresnel optic, hybrid optic, or some combination of these optics. Ultimately, the selection of the optical component is often based not on obtaining optimal performance, but rather on manufacturability, cost, familiarity, or the simple ability to just meet system requirements. The limitations of the selected design method, the specified fabrication method, and the inherent tolerances drive the performance of the overall system, not the skills of the designer or the power of the software.

In conclusion, illumination system optimization is more complex and requires more input and insight from the optical designer in comparison with imaging system optimization. This state will evolve over the next few decades with the illumination designer obtaining more robust, powerful, and accurate tools. Until this time, it is prudent for someone requiring nonimaging optic optimization to learn as much as possible about the various aspects of the optimization process. The various sections of this chapter address the bulleted items above:

Sampling requirements: see Section 7.3, in particular Section 7.3.2

System parameterization: see Section 7.4.1

Object and surface interference: see Sections 7.4.1 and 7.4.4

Optimization of stochastic systems: see Sections 7.3.2, 7.4.2, and 7.4.3

Merit function designation: see Section 7.4.2, and

Fabrication and tolerances: see Section 7.5.

7.4.1 Geometrical Complexity

First, consider the differences between an imaging system (a) and a nonimaging system (b) as illustrated in Figure 7.11. Figure 7.11a shows three lens elements that work in series meaning that there is a distinct sequence of ray propagation through the lenses. Lens design codes were first developed around this sequential optics

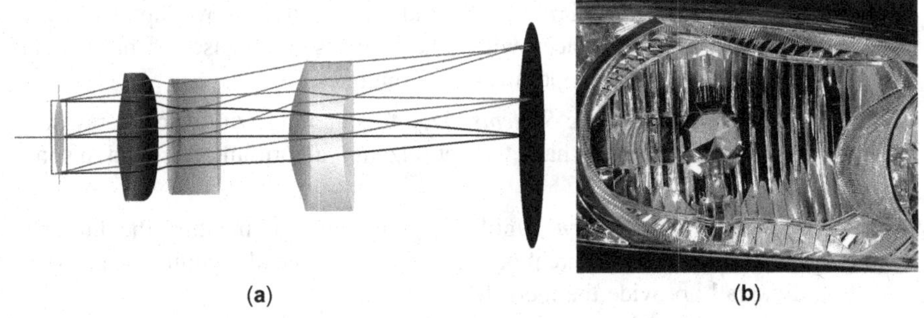

Figure 7.11 Examples of (a) sequential optical design (imaging lens) and (b) nonsequential optical design (low-beam headlamp).

or ray tracing process. Of course, lens design codes have the ability to not follow this sequential ray propagation method, but sequential optical design is the basis of imaging design. Counter to imaging design, consider an automobile forward lighting system assembly that includes many subsystems that individually do not work in series [11]: high-beam and low-beam headlamps, turn signal, and sidemarker. For the low-beam headlamp shown in Figure 7.11b, each facet of the reflector is working in combination with the other facets. Another term for this is a parallel optical design. Rays can take a multitude of paths through the system to the target, thus nonsequential ray tracing is employed to characterize the performance. Nonsequential ray tracing is the basis of optical design and analysis software.

Designing these parallel optical designs with nonsequential analysis can be difficult because of the interdependent ray paths. Additionally, complex shapes for the optics are often demanded. Standard shapes, such as conics, aspheres, and so forth, can be used, but for better performance, nonanalytic surfaces, such as found in CAD software, are required. These CAD constructs are numerical in nature, so methods to parameterize them must be established. Finally, during optimization, the individual parts of the parallel optics have to be restricted from interfering with one another or eve with themselves. These three aspects are expanded upon in the next three subsections.

7.4.1.1 CAD Geometry Computer-aided design or CAD is a necessary tool in the design of current illumination optics. First, CAD geometries often provide the starting point for the illumination system design. Second, the design methods explained in earlier chapters are better fit with the standards in CAD software. NURBS are the best example of CAD geometry that is both a standard and the basis for explaining the shape of novel illumination optics [12]. The native CAD, IGES, and STEP geometry is imported via NURBS into the optical design and analysis software. Optical properties are assigned to the imported CAD geometry, sources are positioned, target surfaces are designated, and ray tracing is performed. In conclusion, an illumination designer must have good understanding of CAD and at minimum a basic understanding of NURBS. For further information on NURBS, consult Reference [12], and on CAD, consult Reference [11] and the references therein.

7.4.1.2 Variables and Parameterization After selecting the design method and completing the initial design of the illumination system, one typically turns to optimization to refine the system performance, include realistic aspects of the system (e.g., nonuniform, non-Lambertian emission from the source), and maximize performance. First, the variables, including source position and orientation, surface positions and slopes, and optical characteristics, such as reflectivity, must be determined. There is low difficulty with determining the important parameters, but often the set is large. The importance of each variable should be determined in order to reduce the convergence time of the optimization. A series of studies is undertaken to determine the effectiveness of each parameter in the reduction of the merit function, often called the figure of merit (FOM). Unfortunately, this process does not ensure that all the correct variables or the method to best parameterize them are found. It is typical to define more variables as the optimization process progresses.

Next, while the determination of the important variables is not an overly cumbersome process, the correct method to parameterize them is typically tedious and time demanding. This difficulty arises for two reasons: there is limited development of parameterization techniques for illumination optics, and a number of illumination optics are developed using numerical techniques, such as NURBS.* For the former, consider lightpipe design, which has a limited number of design methods, and until recently did not have the semblance of any established parameterization [13, 14].

For the limitations from CAD representation, consider the facets of the reflector in Figure 7.11b. Each of these facets is described by a complex formulation, typically NURBS, that is tailored in the initial design process to spread the light from the source, located under the central bulb shield in the figure, over the designated target region with a prescribed distribution. At this point, the shape of the facets is not optimal, but based on a set of equations, an approximation of the source emission, and typically nonscattering surfaces. The optimization procedure is then used to alter the shape of each facet in order to improve upon the illumination distribution. In this example, the reflector facets are CAD constructs, which are difficult to parameterize since they are implicit surfaces. Explicit surfaces, such as polynomials, are parameterized through polynomial coefficients. Implicit surfaces have numerical control points and knots that parameterize the shape of the object. For the former, one alters the coefficients that explain the surface shape, while for implicit surfaces, one changes the location of the control points, the weighting factors of the control points, the degree or order of the construct, or the value of the knots. As can be seen, there are a number of potential variables with an implicit surface, such as one defined by NURBS. Even in the cross-sectional case where a NURBS surface is described by a swept or revolved curve, there are a number of control points with three potential variables per point: two positions and one weight. The positional terms control the location of the control point in space, while the weight term

* In this section, an illustrative introduction to NURBS is provided. By no means is it inclusive, but rather to give the reader a basic understanding of NURBS and their use in optical design. Please consult the literature for detailed information about the mathematics of NURBS [12].

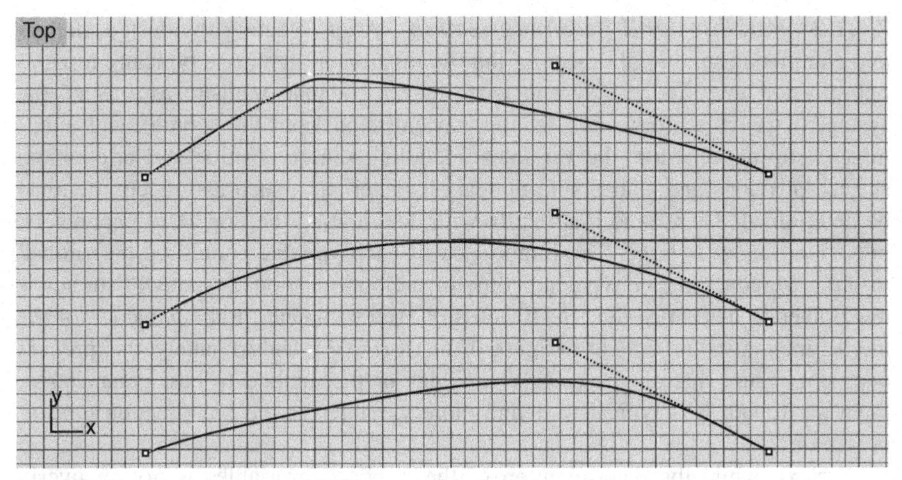

Figure 7.12 Three example NURBS curves each made with four control points that were initially the same with control point weight values of 1.0. The second point (shown in yellow) for the top curve has a weight factor of 10.0, while the same point for the bottom curve has a weight factor of 0.1.

describes how the control points affect the surface shape. An always positive, low weight value (i.e., $w \rightarrow 0$) means a given control point has reduced impact on the shape, while a large weight value (i.e., $1/w \rightarrow 0$) means the control point "pulls" greatly on the shape. Figure 7.12 shows three representative NURBS curves and their control points. Initially, the curves were exactly the same, except offset in the vertical direction. Each control point had a weight value of 1.0, but the second point in the top and bottom curves have their weights changed to 10.0 and 0.1, respectively. One way to think of the utility of the control point weight value is to treat it as if the weight value is the mass of an object that is affecting the shape of an orbit. For the second point of the top curve in Figure 7.12, the weight ("mass") is 10× and 100× than for the same point in the other two curves, respectively. Thus, it pulls on the curve ("orbit") more strongly than the other points.* Likewise, the second point of the bottom curve has a low weight ("mass") so it has weak pull on the curve ("orbit") compared with the other curves.

For the four-control point, NURBS curves, which is not many points by any means, shown in Figure 7.12, one has up to 12 variables. By reducing the number of control points of the NURBS entity, one loses the ability to direct the rays after interaction with the surface. Other NURBS optimization techniques have investigated control of the local slope, which is a combination of the positional parameters and the weight value [15].

In conclusion, this area of illumination system optimization is in need of further development. It is found that each illumination optimization problem should

* Note that the two points on the ends of each curve have enforced that the curve goes through each of the end points. This result is accomplished with the knot vectors, which are not explained herein. Please consult Reference [12] for further insight.

be considered separately since the requirements and variables are expected to be widely differing [16, 17]. A number of issues exist, such as increased parameterization techniques of explicit surfaces, refined method to parameterize NURBS or other numerical constructs, and automation in the optical design and analysis software to include these variables in the optimization. Fractional optimization addresses the object interference issue described in the next section while providing a novel controlled parameterization method [15]. Fractional optimization is fully presented in Section 7.4.4.

7.4.1.3 Object Overlap, Interference, Linking, and Mapping During optimization of nonsequential systems, an added level of complexity can occur that inhibits convergence. Spatial parameters, such as position, size, and shape can be changed during the process. Without careful control of overlap or interference, nonphysical systems can be generated for which the FOM is still computed. Overlap is when an object interferes with itself, while interference is when one object interferes with another.

For an example of overlap, consider the hybrid LED collimator shown in Figure 7.13.* The initial 3D, revolved optic is shown at the top left (a) while its 2D, NURBS cross-sectional curve is shown at the top right (b). After optimization, as designated by the red arrow pointing downward, we obtain the 3D shape at the bottom left (c) and the 2D cross sectional curve in the bottom right (d). Note that the red circles denote the overlap of the curve and the resulting revolved surface. The optimizer found this solution as explicitly valid, but realistically, this solution is completely infeasible. This result is analogous to negative edge or center overlap in lens design optimization, in which the back surface projects across the front surface. In lens design code, this error is easily circumvented by checking that neighboring curves or surfaces do not overlap in the sequential train describing the system. This step is more difficult to solve in nonsequential systems, since the overlap could be with a multitude of curves or surfaces that comprise a given object. A technique called fractional optimization that alleviates the possibility of this occurring during optimization is presented in Section 7.4.4.

An example of object interference is shown in Figure 7.14. Once again, the hybrid optic of Figure 7.13 is used, but in this case the LED axial position is allowed to vary during optimization while the optic shape is held static. The optimization code finds a valid solution by locating the LED partially within the hybrid optic. If the optimization algorithm is not written to change the size of the recess, then a completely infeasible solution is found. Linking and mapping, which are described next, provide rudimentary techniques to alleviate object interference. Fractional optimization can also be used to alleviate the potential of object interference during optimization.

Linking and mapping are two techniques to alleviate the potential of object interference. Linking is the connection of an optimization variable of one object to nonoptimization parameters of another object, while mapping is a nonlinear

* This optic and effective methods of optimizing it are greatly expanded upon in Section 7.4.4 on fractional optimization.

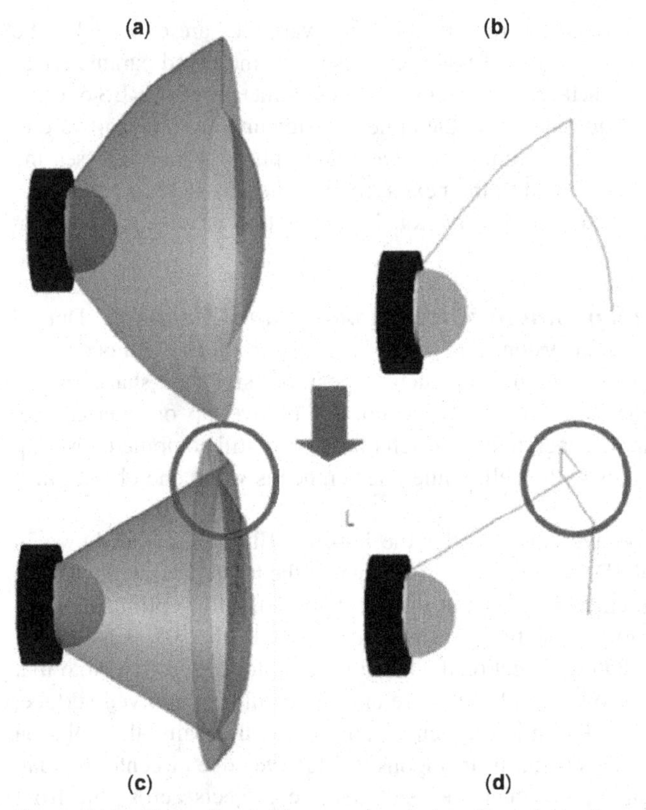

Figure 7.13 Example of object overlap during the optimization of a hybrid LED collimator, including (a) the initial 3D revolved optic shape, (b) the 2D cross-sectional curve used to create (a), (c) the optimized 3D revolved optic shape, and (d) the 2D cross-sectional curve used to create (c). The red circles in panels c and d indicate the problematic overlap region. Note that the LED is placed in a recess of the hybrid optic. The curve comprising this recess is not shown in any of the subfigures.

relationship of an optimization variable of one object to nonoptimization parameters of another object. For example, Figure 7.15 shows a simple example of linking that is used in the example of Figure 7.14. The LED is comprised of several objects, including the dome, body, die, and ray set. In the example shown in Figure 7.15, the ray set (shown in orange) is one object, while the physical structure, including the dome and body, is another. The axial position of the ray set, as designated by the orange arrow, is the actual variable, but to maintain reality, the LED geometry must move with the rays. This example may appear trivial at first, but it is imperative that in software, the designer ensures the link between these objects. Many illumination system optimizations fail because of this simple oversight. The same linking could be done with the optic in Figure 7.14, such that the optic moves in conjunction with the LED; however, this additional link would inhibit the utility of

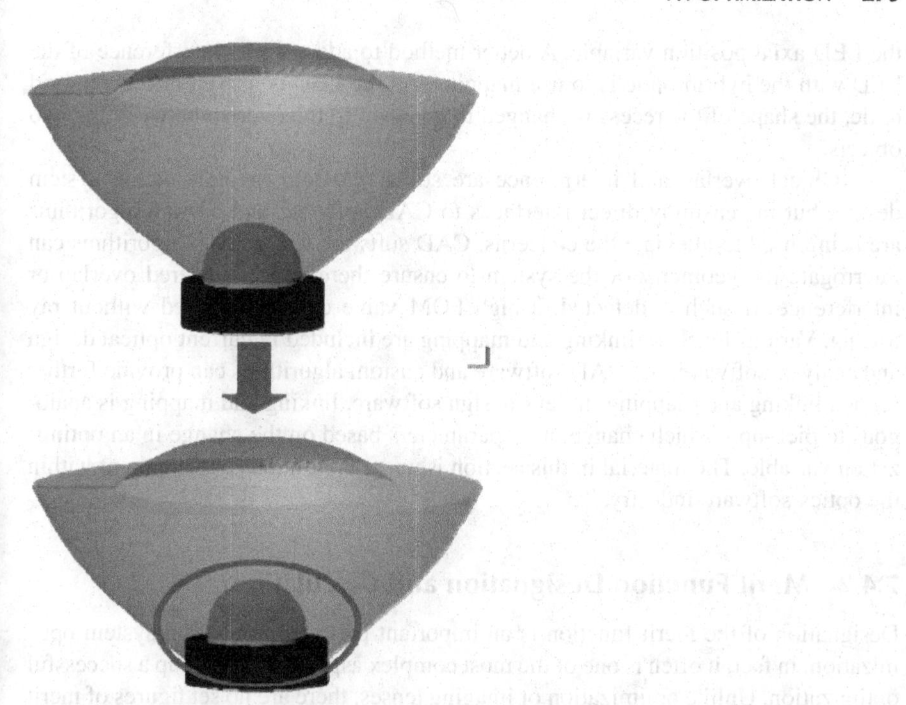

Figure 7.14 Example of object interference during the optimization of a hybrid LED collimator, including (a) the initial 3D revolved optic shape with an LED placed in an air-filled recess and (b) the optimized 3D revolved optic shape with the LED partially residing within the optic as indicated by the red ellipse.

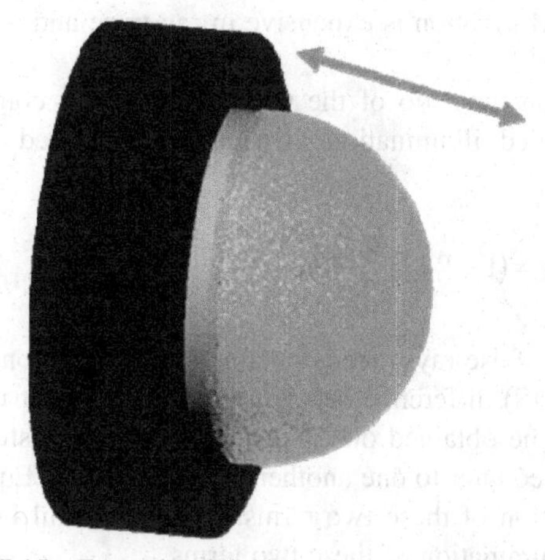

Figure 7.15 Example of linking when the variable is the axial position of the rays (orange dots with the variable designated by the orange arrow), such that the LED geometry, including the dome and body, are moved in conjunction with the rays.

the LED axial position variable. A better method to address this interference of the LED with the hybrid optic is to use mapping. As the LED is moved into the hybrid optic, the shape of the recess is changed to conform to the overlap between the two objects.

Object overlap and interference are still a problem in illumination system design, but increasingly direct interfaces to CAD software and custom algorithms are being used to alleviate the concerns. CAD software and custom algorithms can interrogate the geometry of the system to ensure there is no undesired overlap or interference. If such is detected, a high FOM value can be assigned without ray tracing. Various levels of linking and mapping are included in current optical design and analysis software, but CAD software and custom algorithms can provide further refined linking and mapping. In lens design software, linking and mapping is analogous to pick-ups, which change other parameters based on the change in an optimization variable. The material in this section is under continuous development within the optics software industry.

7.4.2 Merit Function Designation and Calculation

Designation of the merit function is an important part of illumination system optimization, in fact, it often is one of the most complex aspects of setting up a successful optimization. Unlike optimization of imaging lenses, there are no set figures of merit that can be used for nonimaging systems. One has to first ascertain the important aspect of the problems, such as the transfer efficiency and agreement to the desired illumination distribution at the target. Other factors, such as cost, volume, color distribution, and tolerance to fabrication errors are included in the merit function. The interim results during an optimization run indicate if the chosen FOM is trending in the correct direction. If they are not, it behooves the designer to halt the optimization and adjust its parameters or add to them. This step should be done since the overhead for calculating each FOM iteration is expensive in ray trace and analysis time.

A common FOM is based on the ratio of the transfer efficiency compared with the agreement of the modeled illumination distribution compared to the desired one

$$\mathrm{FOM} = (1 + P_{\mathrm{false}})\frac{\Delta f_{\mathrm{RMS}}}{\eta}, \tag{7.21}$$

where P_{false} is the probability that a false ray trace is obtained as per Section 7.3.2, Δf_{RMS} is the root mean square (RMS) difference between the modeled distribution (e.g., irradiance or intensity) and the obtained one, and η is the flux transfer efficiency. The terms Δf_{RMS} and η run counter to one another, and the form of Equation (7.21) does not require normalization of these two terms. Another useful FOM is the weighted (root) mean square summation of these two terms

$$\mathrm{FOM} = (1 + P_{\mathrm{false}})\left(w_f^2 \frac{\Delta f_{\mathrm{RMS}}^2}{f_{\mathrm{peak}}^2} + w_\eta^2 (1 - \eta)^2 \right), \tag{7.22}$$

where w_f and w_η are the weight terms for the distribution agreement and transfer efficiency terms, respectively, and f_{peak} is the peak value of the required illumination distribution. The ratio of the agreement of the distribution to the peak of the distribution provides suitable normalization such that the two weight factors can be better set to select the more important optimization criterion. Note that in both Equations (7.21) and (7.22), the term Δf_{RMS} is given by

$$\Delta f_{RMS} = \sqrt{\sum_{j=1}^{N_j} \sum_{i=1}^{N_i} \left[f_{ray}(i, j) - f_{goal}(i, j) \right]^2}, \tag{7.23}$$

where f_{ray} is the distribution at bin/pixel i, j determined though ray tracing, and f_{goal} is the desired illumination distribution, N_i is the maximum number of horizontal pixels, and N_j is the maximum number of vertical pixels. I have had successful optimizations with these two FOMs, but there are innumerable others, for example, consider those proposed by Cassarly [18]. The inclusion of the P_{false} term in both FOMs provides for increasing the number of rays to increase during the ray trace so that P_{false} goes to a smaller value.

The standard for stopping an optimization is when the FOM values have converged to a value; however, using such is not suggested for systems with stochastic noise. In optimizations on such systems, the FOM for a given system can vary from one ray trace to another due to the variation of the randomly determined ray set and randomly determined ray path followed due to Monte Carlo ray tracing. Rather, it is beneficial to use convergence of the variables. This choice is available in commercial optical design and analysis optimization tools.

7.4.3 Optimization Methods

There are numerous optimization methods that can be used in illumination system optimization, including the simplex method [19], simulated simplex [20], optimal simplex [21], polynomial fitting with damped least squares [16], and global synthesis method [17]. These methods are robust, do not adversely converge, and provide the ability to constrain the optimization. Additionally, due to the stochastic nature of the optimization of illumination systems, the utility of derivatives is limited. Due to the noise inherent in ray tracing analysis, the derivatives of the merit space can vary at each point appreciably. Therefore, I have found the simplex method in its various forms to be a good choice for illumination system optimization. The simplex method is akin to a ball rolling down a hill in N-dimensional space.

Constrained optimization is important so the user can restrict the merit space to realizable systems. There are two types of constraints: user and geometry. In user constraints, the designer provides ranges of values for the variables. For geometry constraints, one has a system that fails due to geometrical concerns, such as overlap or interference. For the latter, it is useful to assign a failed system with a high FOM without performing the time-consuming ray traces and analysis. For user constraints that impact upon a user-set boundary, such as size of the optic, then it is useful to finish the ray trace. There are two options for the handling when the optimizer runs up to the boundary on at least one variable:

- Solve for FOM and assign a penalty with multipliers of the FOM or
- "Reflect" spurious point(s)/object(s) into allowed space. For example, reflect only halfway back into the FOM space.

$$x' = x_{\text{lim}} - \frac{x - x_{\text{lim}}}{2}, \tag{7.24}$$

where x_{lim} is the upper or lower limit for the variable with value x that lies outside the allowed limits, and x' is the new value for the variable that lies within the allowed range. Equation (7.24) ensures that any system modeled lies with the allowable ranges of the variables, and the boundaries are slowly approached with a method akin to Zeno's paradox.

7.4.4 Fractional Optimization with Example: LED Collimator

As presented in Section 7.4.1.3 and specifically in Figure 7.13, optimization of nonimaging systems presents challenges to maintain realistic systems that can be manufactured.* Standard user constraints cannot fully account for the potential overlap of nonsequential optics; therefore, a better method to control the merit space is required to design these demanding optics. The hybrid LED collimator of Section 7.4.1.3 is used for the remainder of this section to highlight the utility of dynamic variables that are dependent upon one another to control overlap and interference.

The hybrid LED collimator is shown in Figure 7.16. There are four sections comprising this optic:

- *LED with Recess.* A Lumileds LXHL-PL01 LED (Philips Lumileds Lighting, San Jose, CA) is placed in a conforming recess. The recess has a small air gap of 0.01 mm between the LED dome and the hybrid optic. The shape of this recess is not changed during this example. The intensity pattern for this LED is shown in Figure 7.17. A ray set available from the manufacturer is used for this study.
- *Side Wall.* The side wall is initially parabolic with its focus at the center of the LED die. The goal of this surface is to use TIR to collimate the incident rays. This shape is developed by revolving a NURBS curve, made with two end points and one control point, around the optical axis of the optic. During this study the shape and size of the side wall is a primary focus of the optimization.
- *Annular Ring.* For the rays that are incident on the parabolic side wall, the rays are not deviated from their collimated path out of the hybrid optic. During this study the shape of the annular ring is not changed (i.e., it will remain planar) but its size is allowed to vary. The annular ring surface is made by revolving a line that connects the side wall and central lens cross-sectional curves around the optical size of the system.
- *Center Lens.* The center lens is initially collimating the direct radiation from the LED die. This shape is developed by revolving a NURBS curve, made

* This section is greatly based on my proceedings paper on fractional optimization [15].

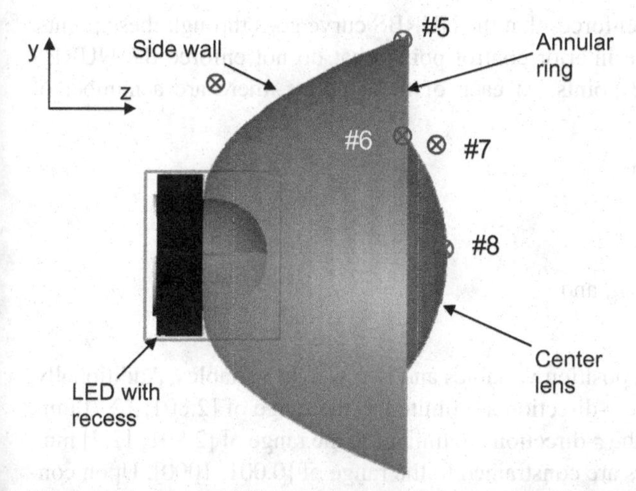

Figure 7.16 Hybrid LED collimator with four regions: LED with recess, side wall, annular ring, and center lens. The numbers indicate the point number where a red X-circle denotes an end control point and a blue X-circle denotes an interior control point.

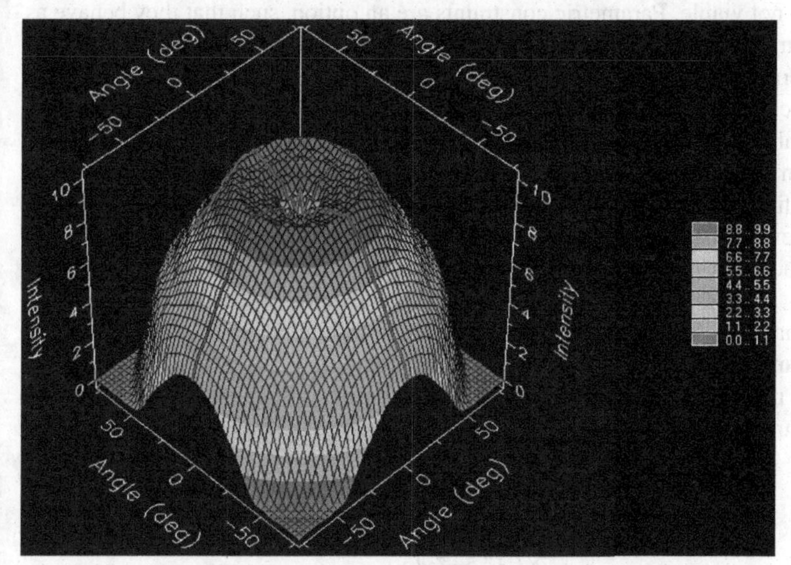

Figure 7.17 Luminous intensity distribution for the Lumileds LXHL-PL01 LED used in the hybrid optic design of Figure 7.16.

with two end points and one control point, around the optical axis of the optic. The shape and size of the center lens is a focus for the remainder of this section.

In Figure 7.16, a number of NURBS control points that govern the shape of the hybrid optic are shown. The red X-circles denote end control points for which the choice of the knot vector enforces that the NURBS curve goes through these points. The blue X-circles denote interior control points that do not enforce the NURBS curve goes through these points. At each of these points, there are a number of variables

- Point 4: y_4, z_4, and w_4
- Point 5: y_5 and z_5
- Point 6: y_6
- Point 7: y_7, z_7, and w_7, and
- Point 8: z_8.

This selection gives eight position variables and two weight variables. Additionally, the transverse terms in the y-direction are limited to the range of [2.801, 12.7] mm, the longitudinal terms in the z-direction are limited to the range of [2.801, 12.7] mm, and the weight parameters are constrained to the range of [0.001, 1000]. Upon conclusion of optimization, it is typical to obtain a system that looks like that of Figure 7.13c,d. The various NURBS curves that define this optic cross each other. Such an optic is unrealistic and cannot be manufactured.

Thus, simple constrained optimization of point-variable defined illumination systems is not viable. Parametric constraints are an option, such that they behave as pick-ups in lens design. However, these parametric constraints are typically limited to static terms rather than dynamic variables. Thus, more complex constraints, such as dynamic parameterizations are required. The dynamic variables are dependent on other variables, such that a global maximum and minimum are set for each system, which then cascades down to interior points, as shown in Figure 7.18. Considering the simplified picture with three points, point 1 (only the z value is used here), z_1, lies between the maximum and minimum values of the system. Point 2, z_2, lies within point 1, such that its maximum value is that of z_1, and its minimum is the system minimum. Finally, point 3, z_3, lies within points 1 and 2, such that its maximum value is that of z_2 and its minimum is the system minimum. Thus, the constrained variable range of z_1 is $[z_{min}, z_{max}]$, while z_2 is $[z_{min}, z_1]$, and z_3 is $[z_{min}, z_2]$. Normalization of the ranges can be done by dividing through by z_{max}, such that the optimization variable ranges are:

$$\begin{aligned} \hat{z}_1 &\in [\hat{z}_{min}, 1], \\ \hat{z}_2 &\in [\hat{z}_{min}, \hat{z}_1], \text{ and} \\ \hat{z}_3 &\in [\hat{z}_{min}, \hat{z}_2], \end{aligned} \tag{7.25}$$

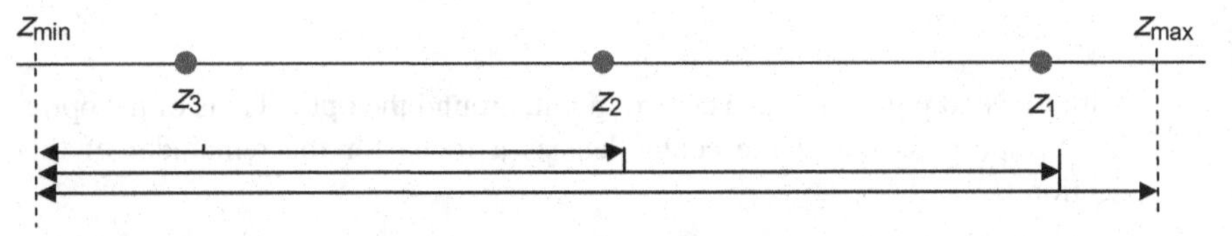

Figure 7.18 Depiction of three points using fractional optimization; thus dynamic constraints ensure that inner points are constrained by outer points in the variable list.

where the hats on the z terms denote a value normalized by the maximum z-value, z_{max}.

Continuing with the development, the actual z-coordinates of z_2 and z_3 are dynamically constrained to the value of z_1, but note the increased utility in that z_2 and z_3 are still variables ranging between z_{min} and z_n, where the n denotes the previous point (i.e., 1 or 2, respectively) that is being queried. However, $\hat{z}_1$ and $\hat{z}_2$ cannot be known *a priori*, thus a slightly altered normalization scheme is required. Rather, while assuming that the minimum z value is 0.0 for each variable, we normalize each dynamic variable by the maximum value of its range to obtain

$$\begin{aligned} &\hat{z}_1 \in [0,1], \\ &\hat{z}_2 \in [0,1], \text{ and} \\ &\hat{z}_3 \in [0,1]. \end{aligned} \tag{7.26}$$

where the hats now indicate that there is normalization by z_{n-1} in each variable, except for z_1, where z_{max} is used for normalization. This normalization means

$$\begin{aligned} &\hat{z}_1 = z_1/z_{max}, \\ &\hat{z}_2 = z_2/z_1, \text{ and} \\ &\hat{z}_3 = z_3/z_2. \end{aligned} \tag{7.27}$$

If the z_{min} terms are not 0.0, then there is added complexity. The constraints from Equation (7.26) are then written as

$$\begin{aligned} &\hat{z}_1 \in [\hat{z}_{1,min}, 1], \\ &\hat{z}_2 \in [\hat{z}_{2,min}, 1], \text{ and} \\ &\hat{z}_3 \in [\hat{z}_{3,min}, 1]. \end{aligned} \tag{7.28}$$

The terminology used in Equation (7.28) can be confusing in the sense that the $\hat{z}_{min}$ functional forms vary like the terms in Equation (7.27),

$$\hat{z}_{1,min} = \frac{z_{min}}{z_{max}} \text{ for } n = 1 \text{ and } \hat{z}_{n,min} = \frac{z_{min}}{z_{n-1}} \text{ for } n \neq 1. \tag{7.29}$$

However, while the upper constraints are normalized to 1, the lower constraints are three different values. We need to further refine our normalization of the three variables in Equation (7.25) by first subtracting out the minimum z-values and then dividing through by the analogous terms as shown in Equation (7.29). The end result is that we obtain constraint ranges as shown in Equation (7.26), but the values of the normalized variables are now

$$\begin{aligned} &\hat{z}_1 = (z_1 - z_{min})/(z_{max} - z_{min}), \\ &\hat{z}_2 = (z_2 - z_{min})/(z_1 - z_{min}), \text{ and} \\ &\hat{z}_3 = (z_3 - z_{min})/(z_2 - z_{min}). \end{aligned} \tag{7.30}$$

The ultimate form is if the minimum and maximum values for each variable are not the same but obey the condition that

$$z_{n-1,\max} \leq z_{n,\max} \text{ and } z_{n-1,\min} \leq z_{n,\min} \text{ for } n \neq 1. \tag{7.31}$$

The normalized variables are then written as

$$\begin{aligned}
\hat{z}_1 &= (z_1 - z_{1,\min})/(z_{1,\max} - z_{1,\min}), \\
\hat{z}_2 &= (z_2 - \delta_2 z_{1,\min})/(\varepsilon_2 z_{1,\max} - \delta_2 z_{1,\min}), \text{ and} \\
\hat{z}_3 &= (z_3 - \delta_2 \delta_3 z_{1,\min})/(\varepsilon_2 \varepsilon_3 z_{1,\max} - \delta_2 \delta_3 z_{1,\min}),
\end{aligned} \tag{7.32}$$

where

$$z_{2,\max} = \varepsilon_2 z_{1,\max} \text{ and } z_{2,\min} = \delta_2 z_{1,\min}, \tag{7.33}$$

and

$$z_{3,\max} = \varepsilon_3 z_{2,\max} \text{ and } z_{3,\min} = \delta_3 z_{2,\min}, \tag{7.34}$$

where the ranges of the ε_n and δ_n terms are [0, 1]. The setup of these functional forms in conjunction with the ranges of the incremental terms ensures that Equation (7.31) is satisfied. In conclusion, you have normalized the lower and upper constraints of the variables pursuant to Equation (7.26), and the variables are described by Equations (7.32)–(7.34).

Additionally, from preliminary optimizations, including the weight factors of control points 4 and 7 the cross sections in almost all cases are linear (see e.g., Fig. 7.13d). This result arises since the weight factors are limited to a range of six orders; therefore, this broad range means that randomly chosen weight values within the optimizer are more likely to be large (i.e., around 500 on average). By taking the logarithm base 10 of the weight factor, it linearlizes the broad range of values such that the average log base 10 value is 0 and the linear value is 1. An offset of 3.1 is also added to this value to ensure that the lower limit is not 0, but rather 0.1.

Using the development from above, we can rewrite the equations that define the constraints for the hybrid optic optimization such that fractional variables are used rather than deterministic coordinates. Note that the maximum y- and z-values for the baseline optic in Figure 7.16 drive the ordering of the fractional variables. This result means that y_5 defines the maximum transverse coordinate, while z_8 defines the maximum axial coordinate. Using these maxima, the equations governing the 10 variables listed above are:

$$\begin{aligned}
y_5 &= 9.88(\hat{y}_5 - 0.1) + 2.801 \\
y_6 &= y_5(\hat{y}_6 - 0.1) \\
y_7 &= y_6(\hat{y}_7 - 0.1) \\
y_4 &= (y_5 - 2.801)(\hat{y}_4 - 0.1) + 2.801 \\
z_8 &= 9.439(\hat{z}_8 - 0.1) + 2.801 \\
z_7 &= (z_8 - 2.801)(\hat{z}_7 - 0.1) + 2.801 \\
z_5 &= (z_7 - 2.801)(\hat{z}_5 - 0.1) + 2.801 \\
z_4 &= (z_5 + 0.46)(\hat{z}_4 - 0.1) - 0.46 \\
w_4 &= 10^{(\hat{w}_4 - 3.1)} \\
w_7 &= 10^{(\hat{w}_7 - 3.1)}
\end{aligned} \tag{7.35}$$

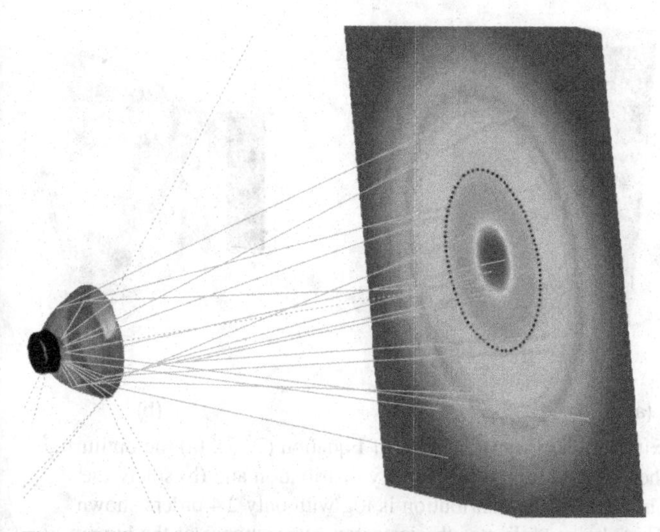

Figure 7.19 Depiction of the initial hybrid optic design and the modeled luminous intensity distribution. The scale of the distribution is log with only 2.4 orders shown from the peak intensity. This choice replicates the instantaneous response of the human visual system. The black, dashed ring indicates the target region of ±10°.

As can be seen in Equation (7.35), each of the coordinate terms relies on a previous point coordinates, except for y_5 and z_8, thus providing the desired fractional optimization that ensures continuity between neighboring points. One can expand out each equation, and, for example, for z_4 we obtain

$$z_4 = [9.439(\hat{z}_8 - 0.1)(\hat{z}_7 - 0.1)(\hat{z}_5 - 0.1) + 3.261](\hat{z}_4 - 0.1) - 0.46 \qquad (7.36)$$

Two separate FOMs are used to optimize the performance of the initial hybrid optic as shown in Figure 7.19, while the optic is shown in Figure 7.16. The dashed black line indicates the desired area of illumination of ±10° from the center of the LED die. There is a significant amount of radiation deposited outside the desired region. The first FOM maximizes the transfer of flux to the target region

$$FOM = \frac{1}{\eta^2}. \qquad (7.37)$$

The resulting intensity distribution is shown in Figure 7.20a and only the optic in Figure 7.20b. The peak intensity is 766 cd, the standard deviation of the intensity distribution is 195 cd, and the transfer efficiency is 66%. The second FOM includes the minimization of the standard deviation of the intensity distribution, σ_{Int}:

$$\text{FOM} = \frac{\sigma_{\text{Int}}}{I_{\text{Peak}}} \frac{1}{\eta^2}, \qquad (7.38)$$

where I_{Peak} is the peak intensity in the distribution and the squaring of the transfer efficiency provides more even weighting between the transfer efficiency and standard deviation parameters. The resulting intensity distribution is shown in Figure

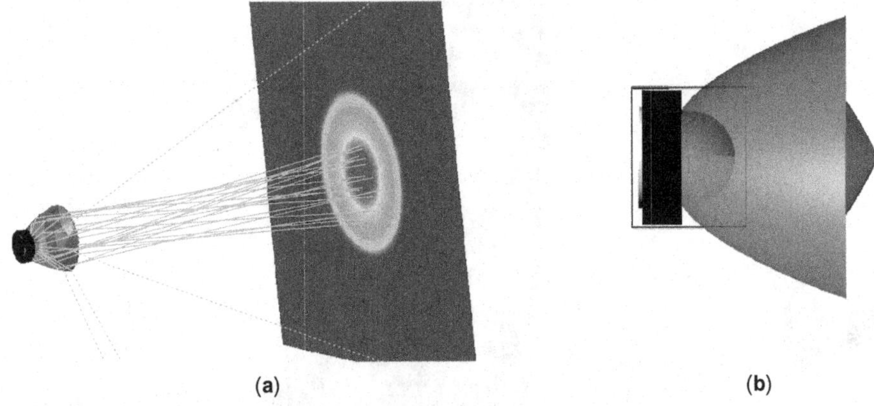

Figure 7.20 For the maximization of transfer FOM of Equation (7.37), (a) the optimized hybrid optic design and the modeled luminous intensity distribution and (b) solely the optimized hybrid optic. The scale of the distribution is log with only 2.4 orders shown from the peak intensity. This choice replicates the instantaneous response of the human visual system. The black, dashed ring indicates the target region of ±10°.

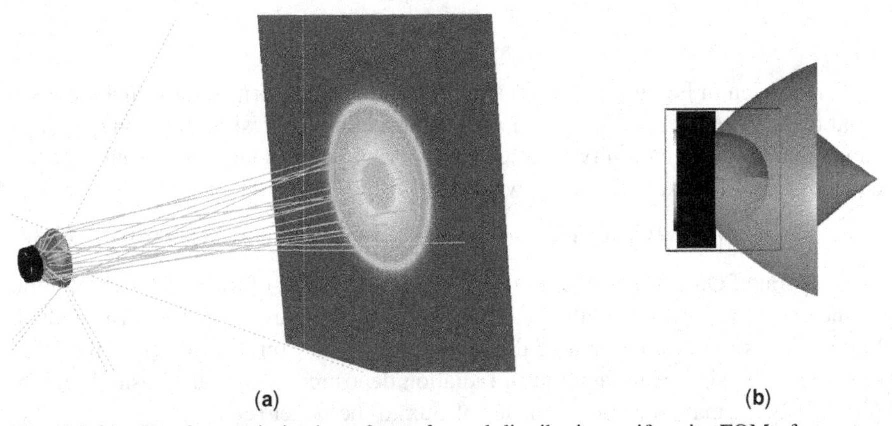

Figure 7.21 For the maximization of transfer and distribution uniformity FOM of Equation (7.38), (a) the optimized hybrid optic design and the modeled luminous intensity distribution and (b) solely the optimized hybrid optic. The scale of the distribution is log with only 2.4 orders shown from the peak intensity. This choice replicates the instantaneous response of the human visual system. The black, dashed ring indicates the target region of ±10°.

7.21a and only the optic in Figure 7.21b. The peak intensity is 451 cd, standard deviation of the intensity distribution is 138 cd, and the transfer efficiency is 46%.

In both cases, the optic is manufacturable with no overlap or other limiting geometry. Interestingly, the central lens takes the form of an axicon, while the side wall is close to parabolic in shape. Additionally, per the conservation of étendue, the case that only addresses the transfer efficiency (Eq. 7.37) provides a larger optic to better control the maximum angle of emission from the optic. The second case that includes the variation of the intensity distribution provides a larger angular extent, losing light from the target region by spilling over the design angle of 10°.

7.5 TOLERANCING

Tolerancing of illumination systems is still in its infancy. There are a number of obstacles, including the parameterization concern expressed in Section 7.4, the lack of standards from national and international bodies, and the lack of software tools to assist in the process. Because of the lack of tools, the designer has to set up the automation scheme to determine the sensitivity of the tolerance parameters, followed by a Monte Carlo analysis using the sensitivity results for all variables. It is expected that illumination tolerancing software tools will increasingly become available over the next few years [22].

Optical designers know that tolerancing of optical systems is based upon the ISO standards, such as ISO 10110-5 and ISO 10110-6 [23, 24]. The -6 standard presents information about optical component centering tolerances, while the -5 standard is in regards to surface form tolerances of optical components. The former is geared to spherical and aspherical component centering. It expects a reference axis and/or positions for each component. Additionally, the text within the formal standard always alludes to image formation. Illumination system optical components can be spherical or described through standard imaging aspheric terms; however, they can also be complex shapes better described through Beziers or NURBS. These complex parameterizations are becoming more common with the requirements of the field for tailored illumination distributions with asymmetric source emission. Also, most illumination components rely on parallel structure in the sense that multiple areas of the component perform different functions (e.g., faceted, freeform, or segmented headlights). Imaging systems tend to use sequential or serial optics. Thus, it is difficult to define a single reference axis or position for illumination systems, but, rather, each facet requires at minimum a single such reference. The -5 standard also expects at all instances imaging type surfaces. Thus, at first blush it is ill suited to explaining surfaces to be used in illumination systems. For example, it describes the tilt or sagittal error introduced by the actual surface with respect to the design surface. It does this by investigating such things as radius of curvature error, which is nonsensical for an illumination surface described by a Bezier or NURBS. In conclusion, the ISO standards as developed are only appropriate for imaging systems. The illumination field requires new standards or extensive modifications to the current standards.

In the next section, the type of errors is presented, including system, gross, process, and roughness. In Section 7.5.2, a short treatment is presented by discussing a source offset system error. In Section 7.5.3, a case study of surface ripple process error arising from injection molding is studied.

TABLE 7.1 Listing of Four Types of Illumination System Errors

Error type	Deviation	Severity	Study	Example
System	Varies	Varies	Parameter	Source offset
Gross	Small	Small	Parameter	Tool error
Process	Medium	Large	Exp. → model	Surface ripple
Roughness	Large	Small	Exp. → model	Rough surface

Included is the level of deviation, severity of the error, how to study the error during tolerancing, and an example of the error.

7.5.1 Types of Errors

Table 7.1 presents a listing of four types of errors that afflict illumination systems: system, gross, process, and roughness. System errors vary from one component to another, such as source-to-optic positions errors and misalignment between components. They are typically explained through a normal distribution. Gross errors are standard across a part due to fabrication blunders. They typically arise as slope and shift errors in the optical components. Process errors occur in the fabrication process, such as injection molding. They are random variations in the parts, and include such things as sinking, warping, or rippling of the surfaces. Roughness errors arise from the inability to perfectly polish a surface. The resulting surface is not completely smooth, so there is microstructure, which leads to scattering. In the table, the deviation from the design path and the severity of the error are given. The deviation indicates the magnitude of the error away from the nominal design specification, while the severity indicates the amount of light that is affected. The severity therefore has a larger impact on overall system performance. The study column indicates how the designer should investigate the error tolerances. Parameter means that the designer should iterate through a series of models to determine the impact on performance. This methodology provides the sensitivity of the system performance on errors of this variable (see the next section). Experiment to model means the designer needs to first make experimental measurements of the error, and then include these measurements in the current and future designs. In Section 7.5.3, process errors are investigated for injection-molded optics. A parameterization protocol for it is proposed.

7.5.2 System Error Sensitivity Analysis: LED Die Position Offset

For an illustration of how to perform a sensitivity analysis, consider the LED shown in Figure 7.22a and then a magnification of the die in the (b) axial and (c) transverse directions. The die can be offset in any direction from its nominal position. In the optical design and analysis software, the position of the die is altered and the effect on system performance is determined. The system can tolerate a prescribed loss in performance (e.g., 10%); therefore, a plot of the system performance versus the variable parameter (e.g., axial die position) gives the tolerance range to this variable.

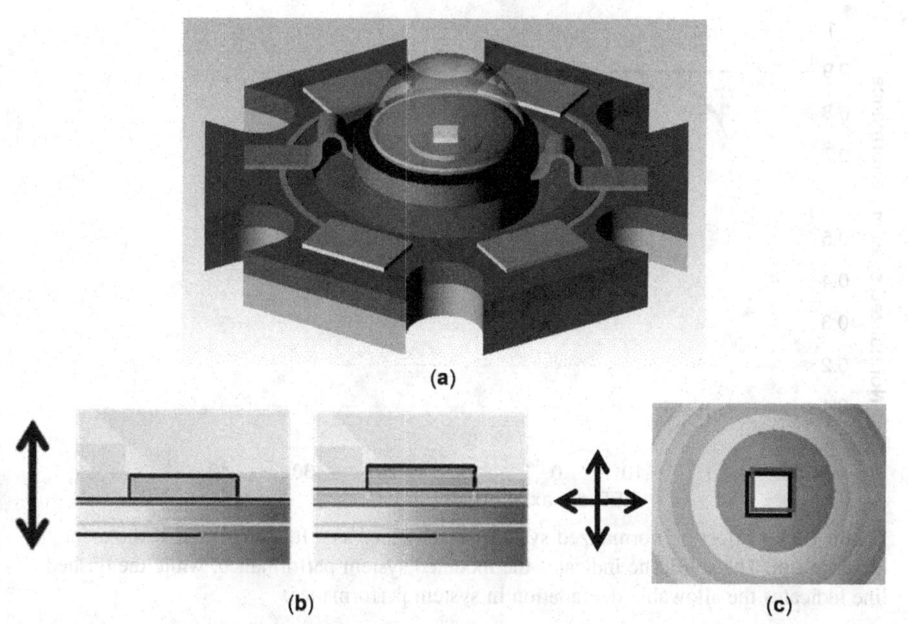

Figure 7.22 (a) Configuration of a high-brightness LED on a star board, (b) magnification of the LED die in its nominal position (left) and axial offset from nominal (right), and transverse offset from nominal (black) to offset (red).

For example, the LED die axial position provides a sensitivity plot as shown in Figure 7.23. The solid blue curve indicates the system performance as the die is moved along the optical axis in both the negative and positive directions. The resulting effect on performance limits the die position to [−8, +13] μm.

This process is repeated for all important parameters. Then, a Monte Carlo tolerance analysis can be performed by randomly selecting within the tolerance ranges for each of the variables. A number of runs are performed to analyze the dependencies of the variables, such that the tolerance ranges can be further restricted, the expected part acceptance can be determined, and/or refinements of the fabrication method need to be explored.

7.5.3 Process Error Case Study: Injection Molding

Process errors do not lend themselves to parameter studies without the investment of experimental measurements. Once a series of experimental measurements are made, a library of data for the method of fabrication is developed. Additionally, this library of data can be used to developed analytic models of the performance effects on incident rays. The fabrication method of injection molding is used to illustrate this technique [22, 25].

Injection molding is a standard fabrication process in the illumination field. It involves injecting, under high pressure, liquid plastic into the void created by the

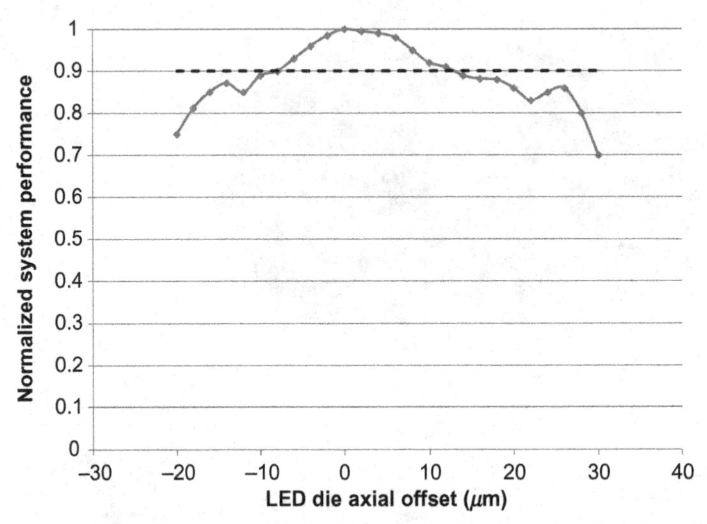

Figure 7.23 Effect on normalized system performance as a function of LED die axial offset in μm. The solid line indicates the modeled system performance, while the dashed line indicates the allowable degradation in system performance.

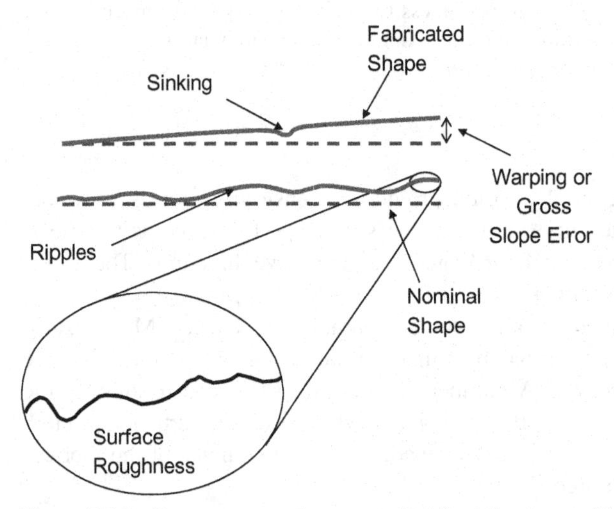

Figure 7.24 Process errors that are typical for injection-molded components including sinking, ripples, warping, and surface roughness. The nominal design shape is shown by the dashed blue line, while the red line is the fabricated shape, and the black inset magnifies the surface roughness.

tooling negative of the part that is to be created. Thus, for a lens, the tooling is two metallic surfaces that define the front, back, and edge surfaces of the lens. The liquid plastic is injected into the void between these two surfaces, and then allowed to cool rapidly. The process time for injection molding is measured in seconds, with the thicker the part the more time that it must spend in the mold annealing. Without sufficient time in the mold, the optical component sees detrimental surface and volume effects, such as shown in Figure 7.24:

- *Sinking (Dimpling).* An indentation in the surface of the part.
- *Ripple.* A near-microstructure waviness to the part caused by localized stresses.
- *Warping.* An altered slope due to stresses causing the part to bend.
- *Roughness.* Micro features due to lack of perfect polishing of the tooling.

Sinking and warping can be alleviated by refinements in the part shape, especially thickness, and process time. Roughness is controlled by improving upon the polishing of the tooling. Ripple is addressed through process time and alteration of the shape of the tooling. Roughness and ripple are always present no matter how refined the component is. Ripple is especially harmful to system performance because it can lead to localized slopes of significant departure from the nominal surface.

To determine the rippling on an injection molded part, a representative flat plastic part is fabricated with prescribed fabrication parameters, including process time, pressure, temperature, and so forth. The surface of the sample is then measured, and a point cloud representing points on the surface is found. These points are processed to determine the minimum, averages and maximum amplitudes, and frequencies of variation across the sample. This process is repeated for samples with different fabrication parameters. This procedure develops the data for the library that can then be assigned as a random perturbation across the surfaces of optical plastic component models. The random perturbation is applied in a deterministic way such that the actual geometry of the optic in software is perturbed. The optic is then analyzed via ray tracing to determine its effect on system performance, such that refined design to alleviate potential issues can be resolved prior to expensive fabrication. In fact, the optical designer can perform a sensitivity analysis with the extent of the library data to determine the tolerance to rippling (i.e., fabrication process time). For example, Figure 7.25 shows a perturbed surface as indicated by the magnification of a small section over the extent of the surface. The maximum amplitude of the ripple is 0.003 mm and the half period is 0.04 mm. A collimated beam is then incident on this seemingly flat surface, but the intensity plot shown in log scale in Figure 7.26 indicates that there is some spread after reflection from this perturbed surface.

Unfortunately, the witness sample method described above is slow to analyze via ray tracing. A potential solution is to replace the deterministic perturbations made on the geometry with scatter models. This step can be done by recognizing that the BSDF is proportional to the intensity distribution scattered from the surface under test

$$\mathrm{BSDF} = \frac{L}{E}, \tag{7.39}$$

where L is the radiance leaving the illuminated surface, E is the irradiance incident on the surface, and BSDF is the bidirectional surface distribution function. The BSDF has units of inverse steradians and represents the scatter as a function of input angle (θ, ϕ) and output angle (θ', ϕ'). The intensity data of results like those of Figure 7.26 are saved to files and then read into scatter fitting programs, as shown in Figure 7.27. In this case, the ABg parameterization method is used to find the A, B, and g

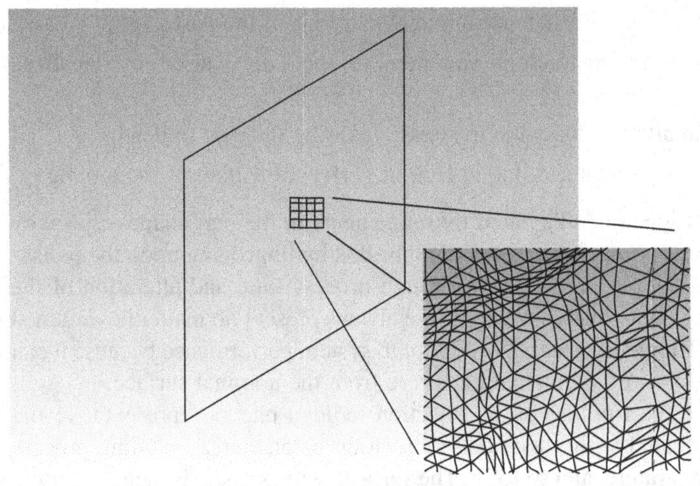

Figure 7.25 A nominally flat surface perturbed with a randomly applied ripple. The inset shows a magnified area of the seemingly flat surface. The maximum amplitude variation for this perturbation is 0.003 mm with a half period frequency of 0.04 mm.

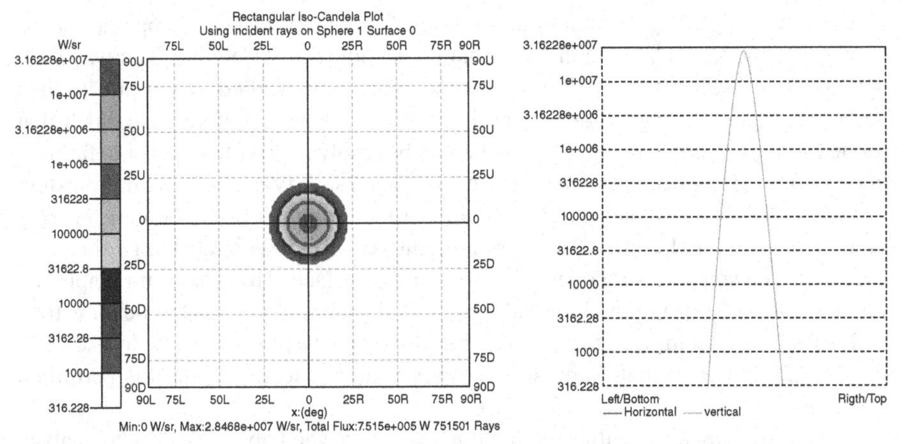

Figure 7.26 The reflected intensity distribution when the surface of Figure 7.25 is illuminated by a collimated beam. Note that the vertical axis is plotted with logarithmically. Note that there is spread of the reflected radiation after reflecting from this nominally flat surface.

parameters (see Fig. 7.27). This process is iterated with all the witness sample data, such that a functional form for these scatter parameters can be found for the period and amplitude values of the surface ripple. The case for a half period of 0.04 mm and varying amplitude of the ripple is shown in Figure 7.28. Note that ABg scatter is parameterized by

$$\mathrm{BSDF} = \frac{A}{B + |\beta - \beta_0|^g}, \tag{7.40}$$

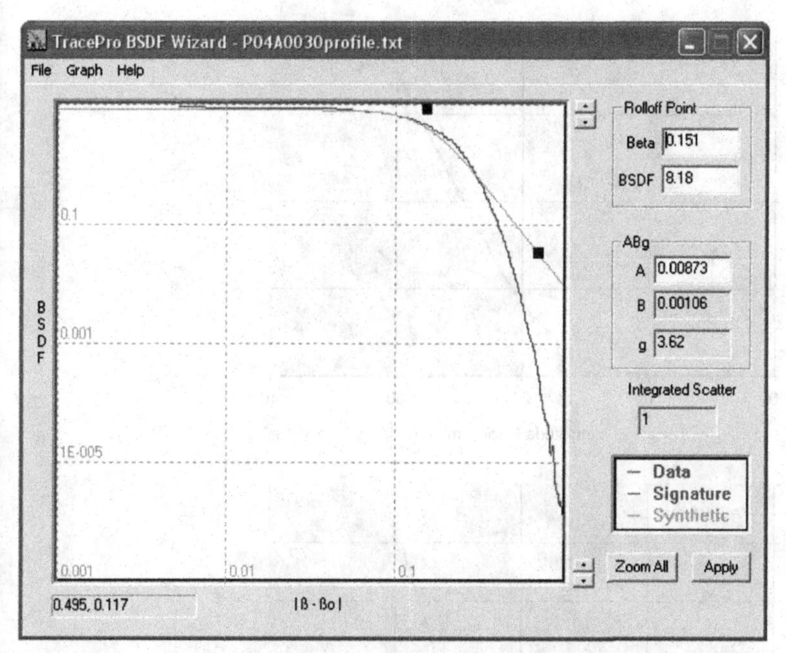

Figure 7.27 ABg scatter fitting of the imported intensity data from Figure 7.26. The imported data is shown in blue, while the fit is shown in green. The determined parameters are shown on the right hand margin.

where $\beta = \sin(\theta)$, $\beta_0 = \sin(\theta_0)$, θ is the scatter angle, θ_0 is the specular angle, g is the slope of the descending curves in Figure 7.27, BSDF(0) = A/B (the vertical axis intercept of the curves), and $B = \beta^g_{\text{rolloff}}$ (the transition point between the horizontal and descending curves in Figure 7.27).

Note that the scatter method cannot be used for software system analysis in the near field with single surface interactions for each ray. Software scatter methods use a Monte Carlo method to perturb the direction of the scattered ray, so for two rays incident at the same point on the surface, they will each see a different (independent) perturbation. Thus, for single interaction or near field results, one must use the witness sample method. For systems where rays have several interactions (e.g., rays trapped in a lightpipe) or those working in the far field, the Monte Carlo aspects are homogenized such that either the witness sample method or scatter method can be used.

This chapter has provided an introduction to illumination system tolerancing, optimization, sampling, and design methods. By no means is it complete, since the field is still in its beginning stages. It provides general principles to guide the optical designer while also providing some advanced examples to highlight the potential that can be included in software models of illumination systems. In future versions of this text, I expect this chapter to be dramatically expanded and broken up into separate chapters as the field progresses.

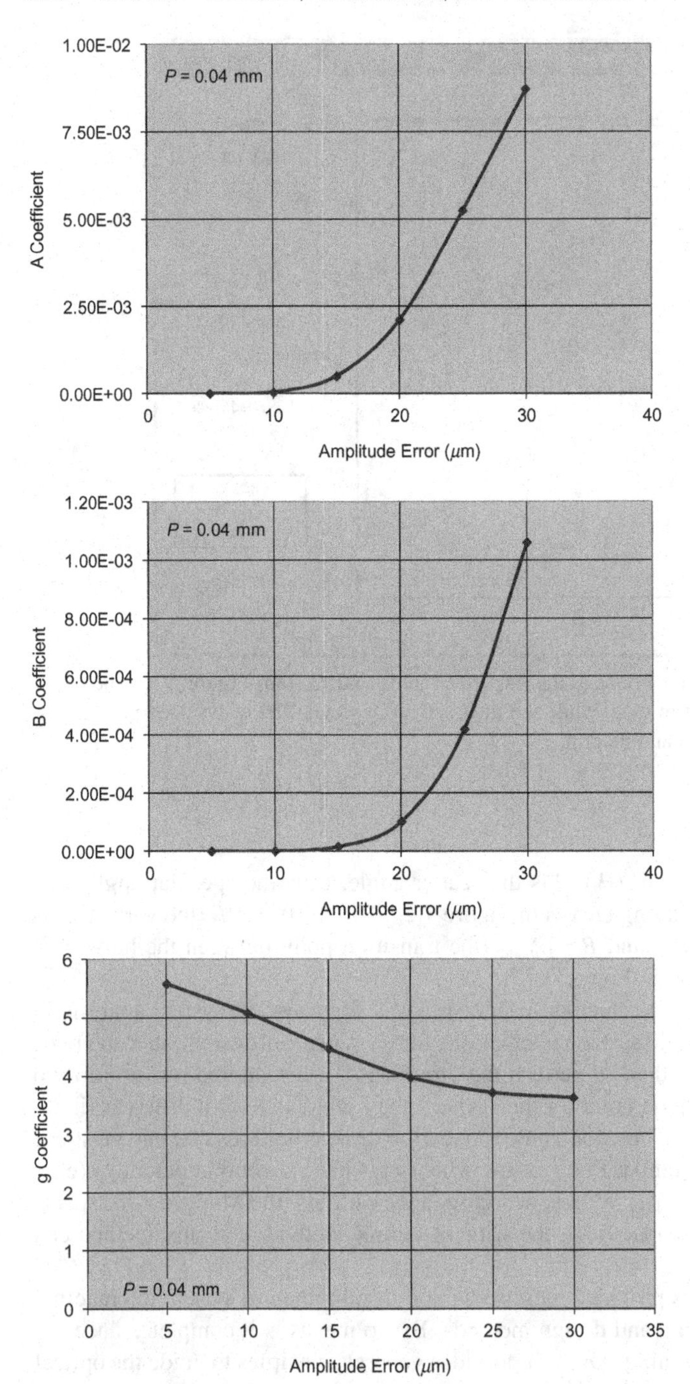

Figure 7.28 The ABg fit data for a ripple half period of 0.04 mm and varying amplitude error (horizontal axes) for (a) A coefficient, (b) B coefficient, and (c) g coefficient.

REFERENCES

1. M.J.J.J.B. Maes and A.J.E.M. Janssen, A note on cylindrical reflector design, *Optik* **88**, 177 (1991).
2. R. Winston and H. Ries, Nonimaging reflectors as functionals of the desired radiance, *J. Opt. Soc. Am. A* **10**, 1902 (1993).
3. P. Benítez, J.C. Miñano, J. Blen, R. Mohedano Arroyo, J. Chaves, O. Dross, M. Hernandez, and W. Falicoff, Simultaneous multiple surface optical design method in three dimensions, *Opt. Eng.* **43**, 1489 (2004).
4. R.J. Koshel and I.A. Walmsley, Non-edge-ray design: improved optical pumping of lasers, *Opt. Eng.* **43**, 1511 (2004).
5. F.R. Fournier, W.J. Cassarly, and J. Rolland, Designing freeform reflectors for extended sources, *SPIE Proc. of Nonimaging Optics: Efficient Design for Illumination and Solar Concentration VI* 7423, 742302 (2009). Society of Photo-Optical Instrumentation Engineers, Bellingham, WA.
6. H. Niederreiter, Low-discrepancy and low-dispersion sequences, *J. Number Theory* **30**, 51 (1988).
7. J. Parent and S. Thibault, Tolerancing panoramic lenses, *SPIE Proc. of Optical System Alignment, Tolerancing and Verification III* **7433**, 74330D (2009). Society of Photo-Optical Instrumentation Engineers, Bellingham, WA.
8. LabSphere, General purpose integrating spheres, labsphere.com/uploads/datasheets/GenPurposeIntegratingSPheres.pdf (accessed 9 September 2012).
9. R.J. Koshel, Aspects of illumination system optimization, *SPIE* Proc. of *Nonimaging Optics and Efficient Illumination Systems I* **5529**, 206 (2004). Society of Photo-Optical Instrumentation Engineers, Bellingham, WA.
10. A. Rose, *Vision: Human and Electronic*, Plenum Press, New York (1973).
11. A. Gupta and R.J. Koshel, Lighting and applications, Chapter 40, in Michael Bass, ed., *Handbook of Optics*, Vol. 2, (McGraw-Hill, New York (2009), pp. 40.1–40.72.
12. L. Piegl and W. Tiller, *The NURBS Book*, 2nd ed., Springer-Verlag, Berlin (1997).
13. A. Gupta, J. Lee, and R.J. Koshel, Design of efficient lightpipes for illumination using an analytical approach, *Appl. Opt.* **40**, 3640 (2001).
14. R.J. Koshel, Optimization of parameterized lightpipes, *SPIE Proc. of the Intl. Opt. Des. Conf. 2006* **6342**, 63420P (2006). Society of Photo-Optical Instrumentation Engineers, Bellingham, WA.
15. R.J. Koshel, Fractional optimization of illumination optics, *SPIE Proc. of Novel Optical Systems Design and Optimization XI* **7061**, 70610F (2008). Society of Photo-Optical Instrumentation Engineers, Bellingham, WA.
16. W.J. Cassarly and M.J. Hayford, Illumination optimization: the revolution has begun, *SPIE Proc. of the Intl. Opt. Des. Conf. 2002* **4832**, 258 (2002). Society of Photo-Optical Instrumentation Engineers, Bellingham, WA.
17. N.E. Shatz and J.C. Bortz, Inverse engineering perspective on nonimaging optical design, *SPIE Proc. of Nonimaging Optics: Maximum Efficiency Light Transfer III* **2538**, 136 (1995). Society of Photo-Optical Instrumentation Engineers, Bellingham, WA.
18. W.J. Cassarly, Illumination merit functions, *SPIE* Proc. of *Nonimaging Optics and Efficient Illumination Systems IV* **6670**, 66700K (2007). Society of Photo-Optical Instrumentation Engineers, Bellingham, WA.
19. J.A. Nelder and R. Mead, A simplex method for function minimization, *Comput. J.* **7**, 308 (1965).
20. W.H. Press, B.P. Flannery, S.A. Teukolsky, and W.T. Vetterling, *Numerical Recipes in C*, Cambridge University Press, Cambridge, UK (1992).
21. R.J. Koshel, Simplex optimization method for illumination system design, *Opt. Lett.* **30**, 649 (2005).
22. R.J. Koshel, Illumination system tolerancing, SPIE *Proc. of Optical System Alignment and Tolerancing I* **6676**, 667604 (2007). Society of Photo-Optical Instrumentation Engineers, Bellingham, WA.
23. International Organization for Standardization, Optics and optical instruments—preparation of drawings for optical elements and systems, Standard ISO 10110-1:1996(E) (International Organization for Standardization, Geneva (1996).
24. R.H. Wilson, R.C. Brost, D.R. Strip, R.J. Sudol, R.N. Youngworth, and P.O. McLaughlin, Considerations for tolerancing aspheric optical components, *Appl. Opt.* **43**, 57 (2004).
25. M.E. Kaminski and R.J. Koshel, Methods of tolerancing injection-molded parts for illumination systems, *SPIE Proc. of Efficient Illumination System Design* **5186**, 61 (2003). Society of Photo-Optical Instrumentation Engineers, Bellingham, WA.

INDEX

Illumination Engineering: Design with Nonimaging Optics, First Edition. R. John Koshel.
© 2013 the Institute of Electrical and Electronics Engineers. Published 2013 by John Wiley & Sons, Inc.

Printed in the United States
By Bookmasters